Fruit Processing

Nutrition, Products, and Quality Management

Second Edition

Edited by

David Arthey
Campden & Chorleywood Food Research Association
Chipping Campden, Gloucestershire
United Kingdom

Philip R. Ashurst
Dr. P.R. Ashurst & Associates
Gooses Foot Estate
Kingstone, Hereford
United Kingdom

AN ASPEN PUBLICATION®
Aspen Publishers, Inc.
Gaithersburg, Maryland
2001

The author has made every effort to ensure the accuracy of the information herein. However, appropriate information sources should be consulted, especially for new or unfamiliar procedures. It is the responsibility of every practitioner to evaluate the appropriateness of a particular opinion in the context of actual clinical situations and with due considerations to new developments. The author, editors, and the publisher cannot be held responsible for any typographical or other errors found in this book.

Aspen Publishers, Inc., is not affiliated with the American Society of Parenteral and Enteral Nutrition.

Library of Congress Cataloging-in-Publication Data

Fruit processing: nutrition, products, and quality management / edited by David Arthey and Philip R. Ashurst.—2nd ed.
p. cm.
Includes index.
ISBN 0-8342-1733-3
1. Fruit. I. Arthey, D. (David) II. Ashurst, P. R.
TP440 .F788 2000
644'.8—dc21
00-059410

Copyright © 2001 by Aspen Publishers, Inc.
A Wolters Kluwer Company
www.aspenpublishers.com
All rights reserved.

Aspen Publishers, Inc., grants permission for photocopying for limited personal or internal use. This consent does not extend to other kinds of copying, such as copying for general distribution, for advertising or promotional purposes, for creating new collective works, or for resale. For information, address Aspen Publishers, Inc., Permissions Department, 200 Orchard Ridge Drive, Suite 200, Gaithersburg, Maryland 20878

Orders: (800) 638-8437
Customer Service: (800) 234-1660

About Aspen Publishers • For more than 40 years, Aspen has been a leading professional publisher in a variety of disciplines. Aspen's vast information resources are available in both print and electronic formats. We are committed to providing the highest quality information available in the most appropriate format for our customers. Visit Aspen's Internet site for more information resources, directories, articles, and a searchable version of Aspen's full catalog, including the most recent publications: **www.aspenpublishers.com**
Aspen Publishers, Inc. • The hallmark of quality in publishing
Member of the worldwide Wolters Kluwer Group.

Editorial Services: Cynthia DeLano
Library of Congress Catalog Card Number: 00-059410
ISBN: 0-8342-1733-3

Printed in the United States of America

Table of Contents

Contributors .... **xi**
Preface .... **xiii**

Chapter 1   Introduction to Fruit Processing 1
R. Barry Taylor

    1.1   Processing on a Global Scale 1
    1.2   Factors Influencing Processing 2
    1.3   Fruit Types for Processing 3
        1.3.1   Pome Fruits 3
        1.3.2   Citrus Fruits 5
        1.3.3   Stone Fruits 5
        1.3.4   Soft Fruits 6
    1.4   Controlling Factors in the Ripening of Fruit 7
        1.4.1   Respiration Climacteric 7
        1.4.2   Ethylene Production 8
    1.5   Biosynthesis of Flavors 9
        1.5.1   Analytical Data 9
        1.5.2   Taste and Aroma 10
        1.5.3   Flavor Formation 10
        1.5.4   Physiological and Biochemical Aspects 14
    1.6   Factors Influencing Fruit Quality and Crop Yield 14
        1.6.1   Fruit Variety 14
        1.6.2   External Factors Affecting Fruit Quality 15
    1.7   Flavor Characteristics 15
    1.8   The Global Market: Threats and Opportunities 16
    1.9   Fruit Processing 17

Chapter 2   Biochemistry of Fruits and Its Implications on Processing 19
Conrad O. Perera and Elizabeth A. Baldwin

    2.1   Introduction 19
    2.2   Minimally or Lightly Processed Fruit Products 19

		2.3	Factors Affecting Shelf Life and Quality of Minimally Processed Fruits	20
		2.4	Physiology and Biochemistry of Fresh-Cut Fruits	20
		2.5	Techniques To Extend Shelf Life of Minimally Processed Fresh Produce	21
		2.6	Enzyme-Catalyzed Reactions during Processing	23
			2.6.1 General Biochemistry of Pectinase Enzymes	23
			2.6.2 The Effects of Pectinase Enzymes on Juice Processing—Cloud Stability	24
			2.6.3 Calcium Pectate Linkages and Texture	25
			2.6.4 Control of Texture in Tomato Processing	25
		2.7	Browning Reactions during Processing	25
			2.7.1 Polyphenol Oxidases and Their Control during Processing	25
			2.7.2 Non-Enzymatic Browning in Fruit Products and Their Control	26
			2.7.3 Browning Reactions Due to Chlorophyll Degradation Reactions	27
		2.8	Development of Bitter Principles in Fruit Products	27
		2.9	Anthocyanins and Their Changes during Processing	28
			2.9.1 Anthocyanin Chemistry	28
			2.9.2 Anthocyanin Color Changes Due to Changes of pH	29
			2.9.3 Anthocyanin Color Changes Due to Temperature	30
			2.9.4 Anthocyanin Color Changes Due to Oxygen	30
			2.9.5 Anthocyanin Color Changes Due to Light	31
			2.9.6 Anthocyanin Color Changes Due to Enzymes	31
			2.9.7 Anthocyanin Color Changes Due to Nucleophilic Agents	31
		2.10	Discoloration during Processing and Storage	32
		2.11	Conclusion	32
Chapter 3	**Fruit and Human Nutrition**			**37**
	Phillip C. Fourie			
		3.1	Introduction	37
		3.2	Composition of Fruits	37
		3.3	The Importance of Fruit in the Human Diet	46
		3.4	Changes in Nutritive Value during Processing	49
Chapter 4	**Storage, Ripening, and Handling of Fruit**			**53**
	Brian Beattie and Neil Wade			
		4.1	Introduction	53
		4.2	Maturity and Ripeness	53
			4.2.1 Climacteric Behavior	54
			4.2.2 Ethylene	56
			4.2.3 Maturity Standards	57

	4.3	Temperature	58
		4.3.1 Field Heat Removal by Precooling	60
		4.3.2 Cool Storage of Fruit	63
		4.3.3 General Requirements of a Coolroom	65
		4.3.4 Mixed Storage	66
	4.4	Storage Atmospheres	66
		4.4.1 Controlled Atmosphere Technology	68
		4.4.2 Atmosphere Generation	68
		4.4.3 Selection and Handling of Fruit for Storage	70
		4.4.4 Modified Atmosphere Technology	71
	4.5	Maintaining Quality	72
		4.5.1 Disease	73
		4.5.2 Disorders	76
		4.5.3 Injury	78
		4.5.4 Managing the Storage, Ripening, and Handling System for Food Safety and Product Quality	79

Chapter 5 Production of Nonfermented Fruit Products **85**
Peter Rutledge

5.1	Introduction	85
5.2	Fruit Quality	85
5.3	Temperate Fruit Juices	87
	5.3.1 Orange Juice	87
	5.3.2 Citrus Juices	91
	5.3.3 Apple Juice	91
	5.3.4 Pear Juice	93
	5.3.5 Stone Fruit Juices	93
	5.3.6 Berry Juices	94
5.4	Tropical Fruit Juices	94
	5.4.1 Pineapple Juice	95
	5.4.2 Papaya Purée	95
	5.4.3 Mango Pulp	95
	5.4.4 Passion Fruit Juice	96
	5.4.5 Guava Pulp	98
	5.4.6 Universal Extraction of Tropical Fruit Juices	99
5.5	Clarification of Fruit Juices	99
5.6	Methods of Preservation	100
	5.6.1 Thermal Treatment of the Juice	100
	5.6.2 Canning	100
	5.6.3 Aseptic Processing	101
	5.6.4 Bottling	102
	5.6.5 Chemical Preservatives	102
	5.6.6 Freezing	102
	5.6.7 Filtration Sterilization	102
	5.6.8 High-Pressure Processing	103

	5.7	Concentration of Fruit Juice	103
		5.7.1 Essence Recovery	103
		5.7.2 Concentration	106
	5.8	Products Derived from Fruit Juice	106
		5.8.1 Fruit Juice Drink	106
		5.8.2 Fruit Nectars	107
		5.8.3 Carbonated Beverages	107
	5.9	Adulteration of Fruit Juice	108

Chapter 6 Cider, Perry, Fruit Wines and Other Fermented Fruit Beverages 111
Basil Jarvis

6.1	Introduction	111
6.2	Cider	112
	6.2.1 A Brief History	113
	6.2.2 Cider and Culinary Apples	113
	6.2.3 Fermentation of Cider	117
	6.2.4 Special Types of Cider	124
	6.2.5 The Microbiology of Apple Juice and Cider	126
	6.2.6 The Chemistry of Cider	133
6.3	Perry (*Poiré*)	135
6.4	Fruit Wines	136
	6.4.1 Fruits Used in Fruit Wine Manufacture	137
	6.4.2 Processing of the Fruit	138
	6.4.3 Fermentation of Fruit Wines	139
	6.4.4 Fruit Pulp Fermentations	141
	6.4.5 Alcohol-Fortified Wines	142
	6.4.6 Sparkling (Carbonated) Fruit Wines	142
6.5	Fruit Spirits and Liqueurs	142
	6.5.1 Fruit Spirits (Fruit Brandies)	142
	6.5.2 Apéritifs and Liqueurs	143
6.6	Miscellany	144

Chapter 7 Production of Thermally Processed and Frozen Fruit 149
Gerry Burrows

7.1	Introduction	149
7.2	Raw Materials	150
7.3	Canning of Fruit	150
	7.3.1 Cannery Hygiene	150
	7.3.2 Factory Reception	151
	7.3.3 Peeling	151
	7.3.4 Blanching	152
	7.3.5 Choice of Cans	152
	7.3.6 Filling	152
	7.3.7 Syrup	152

	7.3.8	Cut Out	153
	7.3.9	Exhausting	153
	7.3.10	Closing	154
	7.3.11	Processing	155
	7.3.12	Can Vacuum	157
	7.3.13	Finished-Pack pH Values	158
7.4	Varieties of Fruit		159
	7.4.1	Apples	159
	7.4.2	Apricots	161
	7.4.3	Bilberries	161
	7.4.4	Blackberries	161
	7.4.5	Black Currants	162
	7.4.6	Cherries	162
	7.4.7	Gooseberries	163
	7.4.8	Grapefruit	163
	7.4.9	Fruit Salad	164
	7.4.10	Fruit Cocktail	164
	7.4.11	Fruit Pie Fillings	165
	7.4.12	Loganberries	165
	7.4.13	Oranges	165
	7.4.14	Peaches	166
	7.4.15	Pears	166
	7.4.16	Pineapple	167
	7.4.17	Plums and Damsons	167
	7.4.18	Prunes	168
	7.4.19	Raspberries	168
	7.4.20	Rhubarb	169
	7.4.21	Strawberries	169
7.5	Bottling		170
7.6	Freezing		170
	7.6.1	Freezing Methods	171
	7.6.2	Storage	172
	7.6.3	Packaging	173
7.7	Aseptic Packaging		174
	7.7.1	Sterilization/Pasteurization	174
	7.7.2	Packaging	175
7.8	Conclusion		175

Chapter 8 The Manufacture of Preserves, Flavorings, and Dried Fruits 177
Roger W. Broomfield

8.1	Preserves		177
	8.1.1	Ingredients	178
	8.1.2	Fruit for Jam Manufacture	178
	8.1.3	Other Ingredients	180
	8.1.4	Product Types and Recipes	182

		8.1.5 Methods of Manufacture	187
	8.2	Fruits Preserved by Sugar: Glacé Fruits	192
	8.3	Fruits Preserved by Drying	196
		8.3.1 Dried Vine Fruit Production	196
		8.3.2 Dried Tree Fruit Production	199
	8.4	Flavorings from Fruits	200
	8.5	Tomato Purée	201

Chapter 9 Packaging for Fruit Products 205
John Bettison

	9.1	Introduction	205
	9.2	Objectives of Packaging	206
	9.3	Heat-Preserved Fruit Products/Beverages	207
		9.3.1 Glass Containers	207
		9.3.2 Metal Cans	213
		9.3.3 Rigid/Semirigid Plastics Containers	216
		9.3.4 Aluminium Foil Trays	218
		9.3.5 Collapsible Tubes	218
		9.3.6 Aseptic Packaging	218
	9.4	Packaging of Frozen Foods	221
		9.4.1 Manufacture of Packaging Materials	221
		9.4.2 Properties of Packaging Materials	221
		9.4.3 Choice of Package	222
	9.5	Packaging of Dehydrated Fruits and Fruit Products	222
	9.6	Packaging of Fresh/Chilled Fruits and Fruit Products	222

Chapter 10 The By-Products of Fruit Processing 225
Ruth Cohn and Leo Cohn

	10.1 Introduction	225
	10.2 By-Products of the Citrus Industry	225
	10.2.1 Citrus Premium Pulp (Juice Cells)	226
	10.2.2 Products Prepared from Peel and Rag	228
	10.2.3 Citrus Oils	232
	10.2.4 Comminuted Juices	234
	10.2.5 Dried Citrus Peel	235
	10.3 Natural Color Extraction from Fruit Waste	236
	10.3.1 Extraction of Color from Citrus Peels	237
	10.3.2 Extraction of Color from Grapes	237
	10.4 Apple Waste Treatment	238
	10.5 Production of Pectin	239
	10.5.1 Characterization of Pectin and Pectolytic Enzymes in Plants	239
	10.5.2 Pectin Enzymes in Plants Used for Production of Pectin	240
	10.5.3 Commercial Pectins and Their Production	240

| | | 10.5.4 | Different Types of Pectin and Their Application . . . . | 244 |
| | | 10.5.5 | Application of Pectin in Medicine and Nutrition . . . . | 245 |

Chapter 11 Quality Management System and Hazard Analysis Critical Control Points 249
Gerry Burrows

11.1	Introduction	249
11.2	Quality System	250
11.3	Role of the Quality Assurance Function	250
11.4	External Accreditation	250
11.5	External Auditing	251
11.6	HACCP	253
11.7	Benefits of an HACCP Approach	254
11.8	HACCP Principles	254
11.9	Conducting an HACCP Study	254

		11.9.1	Determine the Terms of Reference	254
		11.9.2	Select the HACCP Team	255
		11.9.3	Describe the Product	255
		11.9.4	Identify the Intended Use	255
		11.9.5	Construct a Flow Diagram	255
		11.9.6	Establish On-site Verification of the Flow Diagram	256
		11.9.7	List All Hazards Associated with Each Process Step and List the Control Measures for Each Hazard	256
		11.9.8	Apply an HACCP Decision Tree to Each Process Step in Order to Identify CCPs	257
		11.9.9	Establish Target Level(s) and Tolerance for Each CCP . . .	258
		11.9.10	Establish a Monitoring System for Each CCP	258
		11.9 11	Establish a Corrective Action Plan	259
		11.9.12	Establish Verification and Validation Procedures	259
		11.9.13	Establish Documentation and Record Keeping	260
		11.9.14	Review the HACCP Plan	260

| 11.10 | Implementation of the HACCP Plan . . . . | 261 |
| 11.11 | Examples . . . . | 262 |

Appendix 11–A: Canned Rhubarb 264
Appendix 11–B: Cans and Ends Flow Chart 268
Appendix 11–C: Liquor Preparation Flow Chart 271
Appendix 11–D: Retorts Flow Chart 274
Appendix 11–E: Packaging Flow Chart 277

Chapter 12 Water Supplies, Effluent Disposal, and Other Environmental Considerations 281
Michael J.V. Wayman

| 12.1 | Introduction . . . . | 281 |
| 12.2 | Water Sourcing . . . . | 282 |

12.3	Primary Treatment	282
	12.3.1 Screening	282
	12.3.2 Color Removal	283
	12.3.3 Adjustment of pH Value	283
	12.3.4 Filtration	283
	12.3.5 Carbon Adsorption	284
	12.3.6 Primary Disinfection	285
12.4	Secondary Treatment	286
	12.4.1 Boiler Feedwater	286
	12.4.2 Cooling Water	286
	12.4.3 Water for Bottle Washing	286
	12.4.4 Water for Fruit Dressing	286
	12.4.5 Water for Process Use	287
	12.4.6 Water for Special Applications	287
12.5	Effluent Planning	288
	12.5.1 Segregation	289
	12.5.2 Effluent Transfer	289
	12.5.3 Effluent Reception	290
	12.5.4 Treatment Objectives	290
12.6	Effluent Characterization	291
	12.6.1 Suspended Solids	291
	12.6.2 Oxygen Demand	291
	12.6.3 Other Parameters	292
	12.6.4 Effluent Monitoring	292
12.7	Effluent Treatment	292
	12.7.1 Solids Removal	293
	12.7.2 pH Adjustment	293
	12.7.3 Biological Oxygen Demand	294
12.8	Forms of Biological Treatment	295
	12.8.1 The Activated Sludge Process	295
	12.8.2 Percolating Filters	297
	12.8.3 High-Rate Filtration	298
	12.8.4 Mechanical Contacting	298
	12.8.5 Submerged Aerated Media	298
12.9	Tertiary Treatment	299
	12.9.1 Filtration	299
	12.9.2 Solids Removal and Disposal	299
	12.9.3 Recovery and Reuse Options	300
12.10	Environmental Auditing	300
	12.10.1 Baseline Assessment	301
	12.10.2 Period Review	301
	12.10.3 Quality and Environmental Standards	301

Index **303**

Contributors

Elizabeth A. Baldwin, PhD
Research Horticulturist
Citrus and Subtropical Products Laboratory
Agriculture Research Service
U.S. Department of Agriculture
Winter Haven, Florida

Brian Beattie, MSc Agr
New South Wales Agriculture
Horticultural Research and Advisory Station
Gosford, New South Wales
Australia

John Bettison, BSc (Hort), MIFST
Consultant
High Wycombe, Buckinghamshire
United Kingdom

Roger W. Broomfield, ANCFT, FIFST
Malvern, Worcestershire
United Kingdom

Gerry Burrows
Bourne, Lincolnshire
United Kingdom

Leo Cohn
Ramat Gan
Israel

Ruth Cohn
Ramat Gan
Israel

Phillip C. Fourie, PhD
Post-harvest and Processing Technology
ARC Infruitec-Nietvoorbij
Stellenbosch
South Africa

Basil Jarvis, BSc, PhD, DIC, CBiol, FIBiol, FIFST
Department of Food Studies
University of Reading
Reading
United Kingdom

Conrad O. Perera, PhD
Associate Professor
Department of Chemistry
Food Science and Technology Programme
National University of Singapore
Singapore

Peter Rutledge, FAIFST
Food Processing Consultant
North Ryde, New South Wales
Australia

R. Barry Taylor, CChem, FRSC
Danisco Cultor
Denington Estate
Wellingborough, Northamptonshire
United Kingdom

Neil Wade, BSc Agr, Phd
New South Wales Agriculture
Horticultural Research and Advisory Station
Gosford, New South Wales
Australia

Michael J.V. Wayman, BSc, CChem, MRSC, MCIWEM
Technical Director
First Effluent Ltd.
Sutton Coldfield, Birmingham
United Kingdom

Preface

In the years since the first edition of this book was published (1996), interest in fruit and fruit products has shown no sign of diminution. Fruit is of increasing importance in developed countries as both a source of micronutrients and an essential part of a healthy diet. Dietary recommendations in the United States and the United Kingdom suggest, for example, that five to seven portions of fruit and five portions of vegetables are sensible targets for adult daily consumption. The full effects of micronutrients remain unelucidated but appear to rest in their value as sources of fiber and antioxidant substances.

Processed fruit is widely used in many forms and provides both ingredients for further processing as well as a variety of finished products. The trend toward minimal processing gives the consumer easy access to products offering virtually the same benefits as unprocessed source material but with complete convenience.

In developing countries, fruit has other potential. There are many fruits that will grow in most climatic environments, and they often provide an important source of basic nutrients. The further benefit comes from the ability to set up, at relatively modest cost, fruit-processing operations that can provide raw materials for export or be used to substitute imported juice concentrates in local beverage production. The significance of fruit as both basic nutrient and source of raw material for further processing is recognized by many of the aid agencies whose annual reports reflect this in loan funds to relevant projects.

This second edition of the book carries many of the chapters of the original work—a few unmodified and most updated. It also contains some important new contributions, including three new chapters, which add significantly to its worth.

The opening introductory chapter, which sets the scene for the remainder of the book, is followed by a significant new contribution. This deals with the biochemical aspects that have such an important impact on fruit processing. Chapters then follow dealing, respectively, with the significance of fruit on human nutrition and the storage, handling, and ripening of fruit—an essential element in the supply chain for products harvested annually.

Juice production—probably the most common form of fruit processing—is the subject of Chapter 5. In past years, fermentation was the only way of ensuring some form of preservation of fruit juices, and this subject is ably dealt with in the following contribution. As in the first edition, however, the production of both grape juice and wine in the usual sense is excluded from this volume. Wine is a very complex subject and is adequately covered by many other specialist works.

Chapters then follow that deal with thermal processes and freezing for fruit products and the use of fruit in the manufacture of preserves (jams and marmalades) and other more peripheral products.

In many countries, fruit and fruit products are now sold packed (rather than loose), and a significant new contribution to this edition is devoted to the subject of packaging.

The three final chapters deal with the issues of by-products arising from processing; quality systems and the safety of products that can be ensured by use of hazard analysis critical control point (HACCP) evaluation, and last, but certainly not least, water and effluent treatment.

This volume provides a useful general work to enable those with various interests in the processing of fruit to gain a good understanding of the industry. Its readers will be those already involved in the fruit-processing industry and students in the field.

As editors we do, of course, take full responsibility for the book—its contents (or omissions) and the style of presentation. Despite that, the book is in a very real way the work of the contributors, all of whom are acknowledged experts in their own fields. Without their contributions the book would, of course, be nothing, and the editors are most appreciative of all the hard work put in by the contributors to make this work successful.

David Arthey
Philip R. Ashurst

Chapter 1

Introduction to Fruit Processing

R. Barry Taylor

1.1 PROCESSING ON A GLOBAL SCALE

World population continues to rise at a steady rate. United Nations sources estimated a figure of over 6000 million at the millennium. Over half of this figure is represented by the population of Asia, and nearly three-quarters of the world population live in poor and developing countries. The advanced countries rely more and more on imported goods, and this has resulted in a requirement for products vulnerable to deterioration in storage to be harvested, processed, packed, and sent around the globe, often in a matter of hours. This provides trade, employment, and commercial markets on an international scale.

Before we explore the subject of fruit processing, it will be helpful to consider the focus of our labors, the fruit itself and its origins.

The Old Testament book of Genesis starts with its allegorical reference to the tree at the center of the garden of Eden and its forbidden fruit. There is no other description of this given in the text (at least, not in the King James' version of the Bible), yet we all know that it was an "apple" that Eve offered to Adam. In terms of an exact location of this revelation of the weaknesses of man, it has been postulated by theological scholars (Wade, 1901) that the garden of Eden existed as that lush and fertile ground situated between the Tigris and Euphrates rivers. There is evidence that the genus of fruit tree known as malus or apple may have originated in the same geographical region, in the area known today as Iraq. Whether coincidental or not, this rather puzzling piece of information serves to illustrate that man has enjoyed a long-standing relationship with fruit.

During the twentieth century, advances in technology and social-economic factors, including population increases, have dictated the need for rapid communication, more effective transportation, and higher levels of efficiency in the production industries. This is often only answered by automation.

Industrialization, beginning in the Western world during the eighteenth century, has provided the stimulus for both social and technological changes. Although it has brought with it many benefits, particularly for the most advanced nations, there is also a cost, because the purely commercial objectives frequently tend to influence ethical issues and there is a sense of being driven by our own success. It is against this background that the fruit-processing industry finds itself.

Collectively, we are faced with perhaps the most tantalizing of human challenges: how to feed the world! One cannot underestimate the importance of the contribution made by the fruit-growing and -processing industries. Fruit has long been valued as part of the staple diet of many living things; its presence is inexorably linked to the welfare of life on our planet.

The traditional cottage industries and farming practices of the fruit industry have been replaced by highly efficient agricultural techniques providing products of predetermined quality. The term *factory farming*, frequently used in reference to livestock, could be equally well applied to the fruit industry. Plant cultivars are produced and selected to provide maximum yields with the minimum of labor at harvest time. The idyllic scene of a group of country folk picking fruit in an orchard, with ladders and hand-woven baskets, is now part of a bygone era. In many instances, it has been replaced by automated harvesting, where the fruit is mechanically separated from the branches. One such technique utilizes a mechanical clamp that is applied to the tree trunk and literally shakes the fruit into a collecting sheet held below on a circular frame that surrounds the base of the tree. Soft fruits, such as black currants and raspberries, are grown from specially cultured plant stock. Such fruit bushes have been pruned and trained against carefully spaced supports so that mechanical harvesters passing between the rows can collect the crop in a fraction of the time taken by the traditional method of handpicking.

Harvested fruit is sorted, graded, and stored. For soft fruits, long-term storage is preceded by rapid freezing at $-18°C$ to $-26°C$. Block-frozen and individually quick-frozen (IQF) fruit are held for processing. In the case of block-frozen fruit, the structure of the berries will have been greatly altered, and such fruit will be suitable only for juice extraction or comminuting for purée. The IQF fruit will be suitable for whole-fruit products, such as jams, conserves, and yogurts.

The facility to store fruit on a long-term basis enables us to plan and process outside, and in addition to, the normal harvesting periods.

1.2 FACTORS INFLUENCING PROCESSING

In this book the technical and scientific principles upon which the fruit-processing industries depend are discussed.

At all stages of food production, from the harvesting of the raw material to the sale of the finished product, the industry is subject to legislative controls. These are concerned with handling hygiene, nutrition, label declarations, authenticity, additives, export and import tariff controls, packaging, and inspection, to name but a few. Different countries will approach their legal responsibilities in different ways, and this is often a direct result of cultural differences. Anyone who is aware of the problems encountered within the European Union on the subject of harmonization can be under no illusion as to the difficulties involved in reaching agreement on quite minor points of policy, let alone the major concerns. In the United States, the Food and Drug Administration enacts the rules to be passed by Congress, and these will have to become law throughout the 50 states that comprise the union. In spite of the complexities, trade can still flourish on the international scene, through necessity.

The European and US trade blocs with their high levels of importation tend to set rigid rules concerning quality, for obvious reasons, and this has resulted in the producer countries adopting higher manufacturing standards than were previously encountered. International quality standards, such as EN 29000 (ISO 9000), provide a benchmark for the purchaser. Any new enterprise or a change of raw-material sourcing will require careful inspection of the manufacturer's operation before an agreement to proceed can be made.

1.3 FRUIT TYPES FOR PROCESSING

1.3.1 Pome Fruits

The apple, as well as many other fruit types, is no longer restricted to its original indigenous area. Voyages of exploration during the Middle Ages and the opening up of trade routes were instrumental in bringing all manner of botanical specimens both back to Europe and to other continents, where careful husbandry led to the generation of new cultivars and hence to increases in trade. In the United States, there are records as early as 1647 of apples being grafted onto seedling rootstocks in Virginia, and by 1733 apples from America were finding their way into the London markets.

The "pome fruits," of which only the apple and pear are commercially important, are now grown in most temperate regions of the world. Argentina, Australia, Bulgaria, Canada, China, France, Germany, Hungary, Italy, Japan, the Netherlands, New Zealand, Poland, South Africa, Spain, the United Kingdom, and the United States are perhaps the foremost of those countries growing apples on a considerable scale for both home and export use.

Although the apple industry has developed into a major commercial force, it is interesting to note that world markets are dominated by perhaps no more than 20 dessert and culinary varieties, which have been selectively bred to display such characteristics as disease resistance, winter hardiness, appearance (color and shape), and texture, coupled with high average yield (Figure 1–1b). Among these will be found Bramley's Seedling, Brayburn, Cox's Orange Pippin, Delicious, Golden Delicious, Discovery, Granny Smith, Jonathon, and Newtown Pippin.

The main varieties of pears of commercial importance are the Bartlett or Williams Bon Chrétien, the Comice (Doyenné du Comice), and the Conference.

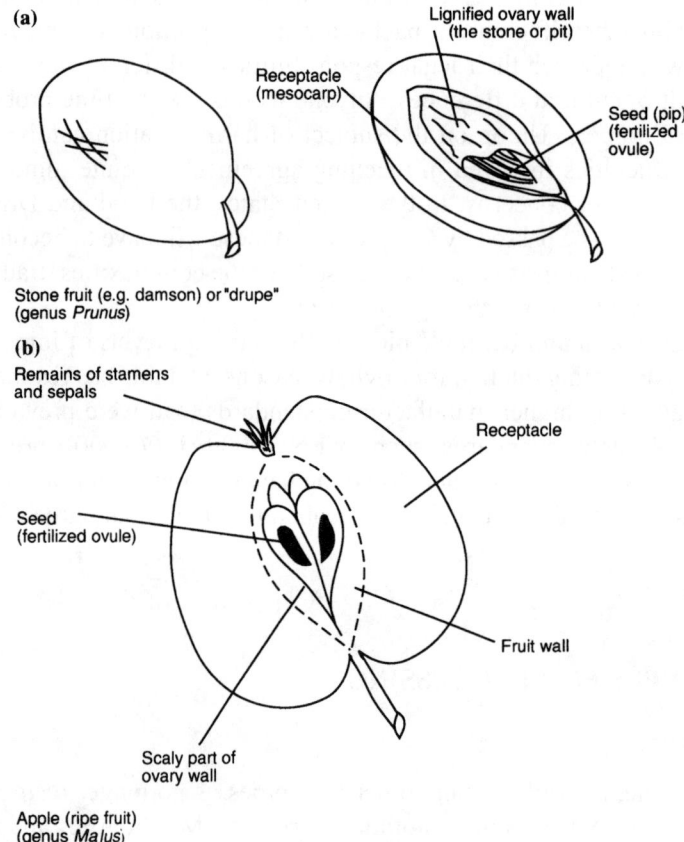

Figure 1–1 (a) A typical stone fruit or drupe, for example the damson (*Prunus*) and **(b)** structure of a ripe apple (*Malus*).

The majority of apples and pears are grown for direct consumption and, therefore, are graded at the point of harvest. Size, shape, color, and freedom from blemishes are of major importance to the retailer. Fruit not meeting such requirements, and yet quite sound, will be suitable for processing, where the selection criteria will relate to ripeness, flavor type (i.e., dessert or culinary), level of acidity, and soluble solids.

Fermentation processes such as those used for cider and perry manufacture may be classed as being traditional. Many of the more highly alcoholic beverages, such as French Calvados or the American apple brandy known as Applejack, are also traditional regional or national products. The ease with which apple juice tends to ferment has undoubtedly promoted its significance among the alcoholic beverages already mentioned, and a vast quantity of fruit is used annually in these products. With modern filtration and evaporation techniques, we are now able to produce aseptically packed concentrates that may be stored for extended periods in readiness for a variety of uses.

1.3.2 Citrus Fruits

As with the pome fruits, citrus varieties are now grown in many parts of the world. Originating in the southern and eastern regions of Asia, China and Chochin China, and the Malayan Archipelago, the citron (*Citrus medica*) is said to have first found its way to Europe during the third century BC, when Alexander the Great conquered western Asia. Later, the orange and lemon were introduced to the Mediterranean regions when the Romans opened up navigable routes from the Red Sea to India. Since then, the cultivation of citrus fruit has spread from Europe to the United States, where notable growing areas are Florida and California, to South America, where Brazil has the largest share of the world's market for oranges and orange juice products, and to South Africa and parts of Australia. Existing markets in the Mediterranean area include Israel, Sicily (lemons), and Spain.

Citrus-processing industries, while primarily centered on juice production, provide many by-products. The citrus varieties have been described as berry fruits in which the hairs inside the ovary walls form juice sacs. These are contained within the characteristically highly colored, oil-bearing peel (flavedo) or epicarp, and further encased in a pithy structure known as the albedo. During removal of the juice, modern processing techniques are able to carry out simultaneous separation of the oil by abrasion of the epicarp to rupture the oil cells. The oil is flushed off with water and recovered by centrifugation. The residual pithy material, peel and albedo, is also used in the production of citrus pectins, as clouding agents for the beverage industry, and in the preparation of comminuted citrus bases.

1.3.3 Stone Fruits

The stone fruits are characterized by a fleshy mesocarp (pulp) surrounding a woodlike endocarp or stone, referred to in agricultural and processing circles as the pit (Figure 1–1a). The epicarp or skin is thin and smooth, except in the case of the peach and apricot, which possess fine hairlike coatings. The stone fruit of commerce all belong to the genus *Prunus* (family Rosaceae) and include the species peach, apricot, plum (and greengage), cherry, and nectarine. Many varieties of plum occur and are found in nearly all the temperature zones.

It is now generally thought that the peach and apricot were first grown and harvested in ancient China, although in relation to Western civilization many fruits originated in the area once known as northern Persia and the Russian provinces south of the Caucasus. Cherries and plums, however, were probably introduced into the Western world sometime in advance of the apricot and peach. Wild cherries and plums would have spread rapidly because of the smaller size of fruit and the comparative ease of transfer by birds.

Unlike the majority of fruit types, the stone fruit is, in effect, a single seed or fertilized ovule surrounded by a fleshy receptacle, which, following the ripening stage, is valued for its texture and flavor. The seed and surrounding ovary are greatly enlarged in comparison with those of other fruits. The growth pattern of the fruit takes place in three stages. Initially, there is an enlargement of the ovary, the outer wall or shell of which remains soft and pliable. When an optimum size is attained, the shell begins to lignify, creating a woodlike protection around the seed, and growth becomes confined to the endosperm and ovary. It is during the third phase that expansion of the mesocarp (edible portion) takes place, and it is at this stage that there is a peak of biochemical activity. Increased sweetness, reduction in acidity, and release of flavor components occur during ripening. This can take place before and after harvest, and with stone fruit care has to be taken to optimize this, as postharvest ripening is generally thought to occur, provided the fruit has reached the correct stage of maturity when picked. If harvested at too early a stage, the fruit will not reach the desired standard of maturity.

The feasibility of using a controlled atmosphere for stone fruit varieties was first mooted during the late 1960s as a means of extending storage time. Controlled atmosphere storage, using low oxygen and/or carbon dioxide, is able to extend the life of the fruit, and this emphasizes the living nature of fruit cells. Different varieties of stone fruit show different responses to controlled atmosphere storage; this was noted by Couey (1960), who suggested an atmosphere of 7% O_2, 7% CO_2, and 86% N_2. This atmosphere delayed ripening in plums and reduced the loss of soluble solids with no impairment of texture during storage at 0–1°C for 6 weeks.

Some plum varieties can be induced to ripen using ethylene treatment. Cherries are less responsive to controlled atmosphere storage, and Porritt and Mason (1965) reported that different combinations of $O_2/CO_2/N_2$ were ineffective for storage of cherries.

The ripening and biochemical activity experienced by many fruits are influenced by natural hormones.

1.3.4 Soft Fruits

The term *soft fruits* includes a number of unrelated fruits that have become grouped together in view of their size and culinary properties rather than for any structural or varietal reason. As might be expected by the term *soft*, these fruits do not store well. They are susceptible to molds and yeasts, and fruit held outside a comparatively short harvest season is usually stored deep-frozen. The soft fruits are also all very prone to bruising at the time of harvest.

Three groups are found under this heading:

1. the berry fruits (genus: *Rubus*): blackberry, raspberry, loganberry, boysenberry, and mulberry (genus: *Morus*)
2. the currants (genus: *Ribes*): gooseberry, currants (black, red, and white), and blueberry (genus: *Vaccinium*)
3. the achenes (false fruit): these are represented by the genus *Fragaria* (the strawberry family)

The Structure of Soft Fruits

All of these fruits fit the description of "soft" (Figure 1–2). The berry fruits consist of aggregates of druplets surrounding a pithy receptacle rather than the situation occurring in the stone fruits, where the whole fruit comprises just one druplet, or rather drupe (originating from the Latin word *druppa* for an overripe olive), which is derived from an enlarged ovary containing a fertilized ovule or seed. The seeds of the currants

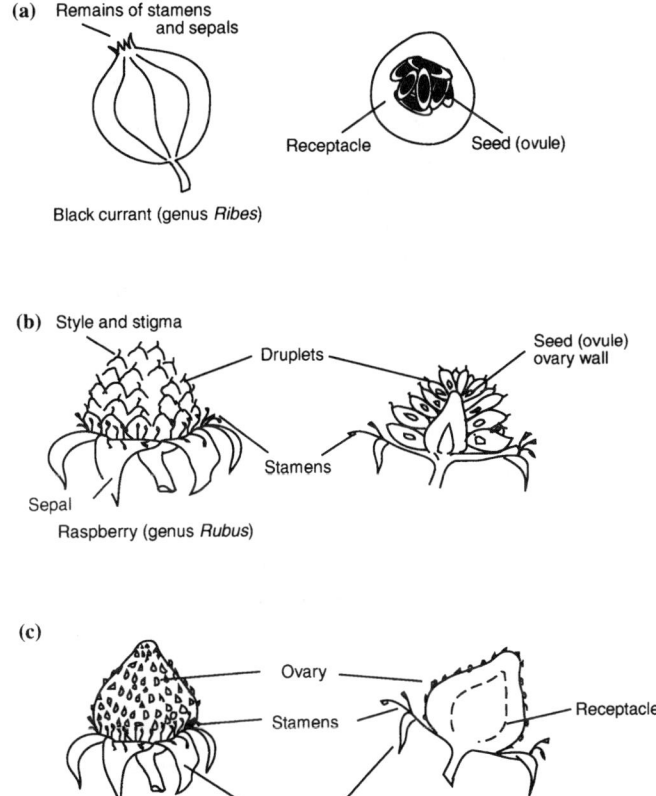

Figure 1–2 The structure of the soft fruits. Berry fruits: **(a)** currants, e.g., black currant (*Ribes*); **(b)** raspberry (*Rubus*); **(c)** achenes, e.g., strawberry (*Fragaria*).

are enclosed in a soft, fleshy pericarp, and the number of seeds is characteristic of the fruit. In the gooseberry, the fruit may appear singly or in small clusters and can be picked at a single stage of maturity as required. The currants, however, produce their fruit on "strigs" or short stems, with the fruit ripening in order along these stems, commencing nearest the main branch and finishing with the terminal fruit. Therefore, even at the most favorable picking time, there will always be unripe and overripe fruit in the harvest. The achenes consist of a large, pulpy, flavorsome receptacle supporting the seeds on the outer surface. There is no fleshy mesocarp surrounding the seeds.

1.4 CONTROLLING FACTORS IN THE RIPENING OF FRUIT

1.4.1 Respiration Climacteric

Respiration climacteric is a well-documented phenomenon characteristic of most but not all fruits during ripening. This stage of the growth phase of the fruit marks a sudden increase in metabolic activity. The term *climacteric rise* was coined by Kidd and West (1922) to describe the sudden upsurge in the evolution of CO_2 that occurred during the ripening of apples (Figure 1–3). Other fruits exhibit the phenomenon to a greater or lesser extent, and considerable variation in pattern has been recorded. For the banana, preclimacteric respiration at 16–24°C can vary from 8 to 50 mg of CO_2 per hour per kilogram of fruit, whereas during the climacteric itself, rates of 60 to about 250 mg/h CO_2 per kilogram of fruit may be achieved (Palmer, 1971). As might be expected, the respiration rate will increase with increasing temperature.

By controlling the atmosphere surrounding fruits during storage, the ripening stage can be advanced or retarded, a factor of extreme importance in the marketing of fruit on a commercial scale.

1.4.2 Ethylene Production

The gas ethylene is used to induce the climacteric phase in many fruits prepared for market. Kidd and West (1933) showed that the vapors produced by ripe apples when passed across unripe apples would stimulate respiration into the climacteric phase and cause the unripe fruit to ripen. Later it was shown that the active principle in the vapors was ethylene (Gane, 1935).

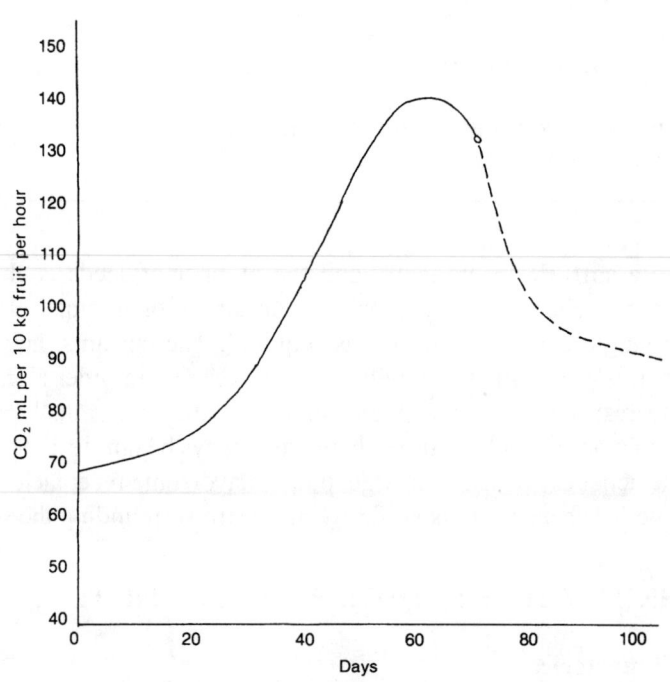

Figure 1–3 Respiration climacteric, characterized by the evolution of CO_2 during the ripening of apples.

While the harvesting and marketing of any fruit crop is a critical operation, it is perhaps more so in the banana industry, where it is essential to control maturation of the fruit at all stages. Bananas are extremely susceptible to the effect of endogenous ethylene, and the picking of green banana fruit hastens ripening, apparently by lowering the threshold of sensitivity to ethylene (Burg and Burg, 1965). To prevent this ripening effect during transportation and storage, the temperature is reduced to a level where ripening is not initiated by exposure to ethylene. The optimum storage temperature for bananas is about 13°C and, in practice, it is necessary to protect the fruit from the effects of ethylene following harvest for only 2–3 days until cooled to the optimum transportation temperature.

In addition to temperature control, the ripening effect can be delayed for extended periods by storing the green bananas in an atmosphere of 1–10% O_2, 5–10% CO_2, or a combination of low O_2 and high CO_2 (Young et al., 1962; Mapson and Robinson, 1966). In this way, bananas can be stored for weeks or months while the onset of ripening is delayed.

Other fruits for which controlled atmosphere storage is of major importance are pineapples, tomatoes, and melons, and here the object is to meet the demands of the retail trade. Where used for processing purposes, the fruit is frequently sold in the form of pulp.

1.5 BIOSYNTHESIS OF FLAVORS

1.5.1 Analytical Data

The flavor identity of each fruit variety plays an important part in its success on the commercial market, and while every effort is made during the farming of fruit to maintain and standardize flavor quality, the mechanisms and biological pathways by which specific flavoring components are formed are still largely unexplored.

Modern analytical methods enable the examination and identification of the chemical nature of each flavor, and for most fruit types there are extensive libraries, or lists, of chemicals that have been detected by mass spectral analysis. Perhaps the most widely used reference lists for *Volatile Compounds in Food* are those produced by the TNO-CIVO Food Analysis Institute of Holland, while many analytical laboratories and scientific organizations produce and maintain their own lists. TNO-CIVO has issued regular updates on its listings since 1963. Since then improvements in analytical techniques have resulted in the discovery of many compounds that occur at levels previously below the limit of detection. For example, the survey carried out in 1967 by Nursten and Williams (1967) reported on 137 compounds that had been identified in strawberry fruit aroma; the TNO sixth edition report of 1989 lists no fewer than 351 volatile compounds, including 129 esters (Maarse and Visscher, 1989).

Currently there are about 5000–6000 known flavoring components (not all of these isolated from fruit), whereas the flavorist may work from a list of no more than 500–600 in creating flavor matches of an acceptable standard. However, during the 1980s there has been a general requirement in the Western world for "naturalness" in flavorings, and this has renewed interest in biological syntheses as well as in more effective methods of isolation of flavor components from natural sources.

1.5.2 Taste and Aroma

The human organoleptic and gustatory senses will respond to remarkably low levels of stimuli. The flavor components responsible for characteristics in fruits may be present in extremely low concentrations, of the order of parts per million (ppm, 10^{-6}) or parts per billion (ppb, 10^{-9}), where the threshold of sensory detection may be achieved for many flavor chemicals. Most artificial flavorings are formulated to be used at a dose rate of about 0.1%. In the isolation of natural flavors from fruit sources by purely physical means, it is far more difficult to reach these levels of concentration, as interaction of components can occur during processing, resulting in degradation of flavor. Undoubtedly future development will take place to solve this problem within the constraints of manufacturing costs, but meanwhile for the majority of fruit aroma concentrates typical dosage rates are about 0.5%.

While much is known about the identity of chemicals responsible for flavor, the pathways by which these chemicals have been formed have aroused less interest, and there is little widespread understanding as to how they are produced.

1.5.3 Flavor Formation

In 1975, a systematic approach to the biosynthesis of flavors was made (Tressl et al., 1975). Five basic metabolic pools, covering the metabolism of carbohydrates, lipids, amino acids, terpenes, and cinnamic acid, were cited, and examples were given of pathways leading to the formulation of mono- and sesquiterpenes, branched aliphatic esters, alcohols, acids, phenolic acids and ethers, and carbonyl compounds.

The flavor development period occurs during the climacteric rise in respiration; this is effectively the ripening stage (see Figure 1–4). Minute quantities of carbohydrates, lipids and protein, and amino acids are converted to volatile flavors. The rate of flavor formation increases after the respiration climacteric and continues following the harvesting of the fruit up to the time that senescence sets in. Identification of the optimum stage of flavor development is often a highly subjective matter; for example, the European palate will often balk at the sight of anything other than a yellow banana, whereas those originating from countries in which the banana is indigenous will require the skin of this fruit to be almost black before committing it to their taste buds! Judging from the analytical flavor profile of "overripe" bananas, the latter opinion must be correct (see Figure 1–5).

The chemicals responsible for characteristic aroma and flavor profiles of the different fruits may be grouped as follows.

Acids

Citric and malic acids are the most plentiful and widely dispersed of the nonvolatile carboxylic acids, and they are followed by tartaric; malonic; fumaric and gluconic (plums); ascorbic (vitamin C); and traces of benzoic, salicylic (bilberries), shikimic, and

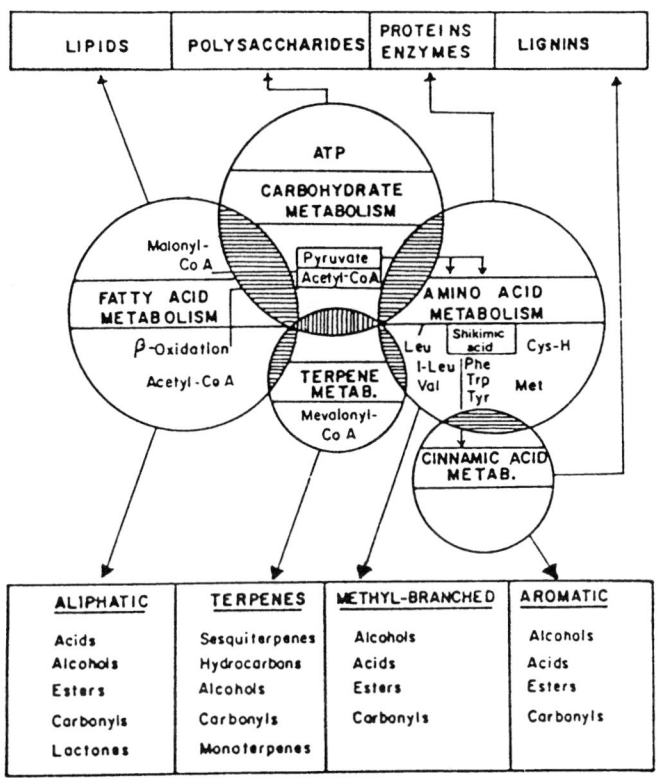

Figure 1–4 Biosynthesis of fruit volatiles. Taken from Tressl et al., 1975.

---

quinic acids. Acids present in the volatile fractions of fruit are limited to two, namely, formic and acetic acids.

Carbonyls

Carbonyls make a significant contribution to the aroma and flavor of most fruits and are of major importance in certain instances, for example, benzaldehyde in stone fruits, 5-hydroxy-2-methyl furfural in pineapples, acetaldehyde in oranges (i.e., orange essence oil), and furfural in strawberries.

Esters

These are the most important of the natural components that provide the flavor of fruits; they are highly variable and numerous and are responsible for the characterization of fruit type.

Lactones

Lactones possess a functional carbonyl group and are often described as cyclic esters. They provide characteristic aromatic notes in many fruits, in particular peach and apricot (Tang and Jennings, 1971; Romani and Jennings, 1971).

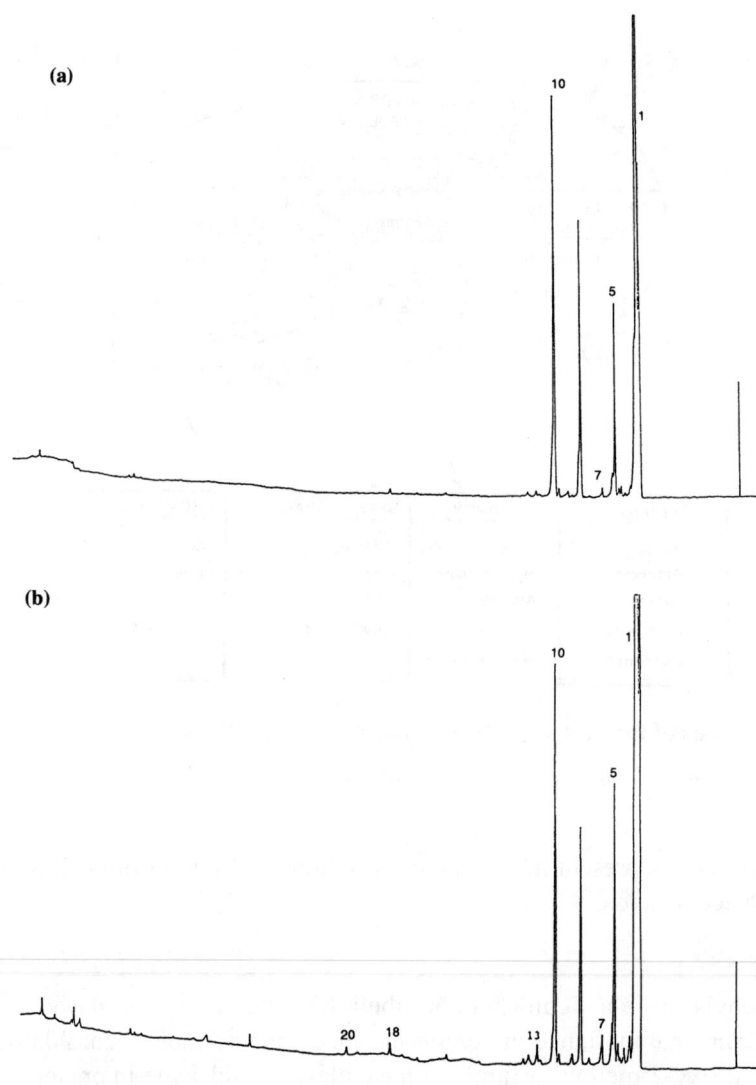

Figure 1–5 Peak identification for gas chromatograms showing profile changes in banana volatiles isolated from the fruit at various stages of maturity. **(a)** Underripe flesh; **(b)** ripe flesh;

Banana volatiles are labeled as follows: 1, ethanol; 2, ethyl acetate; 3, isobutanol; 4, *n*-butanol; 5, 3-methylbutan-1-ol; 6, 2-pentanol; 7, 3-methylbutanol; 8, isobutyl acetate; 9, butyl acetate; 10, *trans*-2-hexanol; 11, ethyl 2-methyl butyrate; 12, hexanol; 13, 3-methyl butylacetate; 14, 2-heptanol; 15, isobutyl-*n*-butyrate; 16, *n*-butyl-*n*-butyrate; 17, butyl-2-methylbutyrate; 18, 3-methylbutylbutyrate; 19, 3-methylbutyl-2-methylbutyrate; 20, 3-methylbutyl-3-methylbutyrate.

Introduction to Fruit Processing 13

Figure 1–5 continued

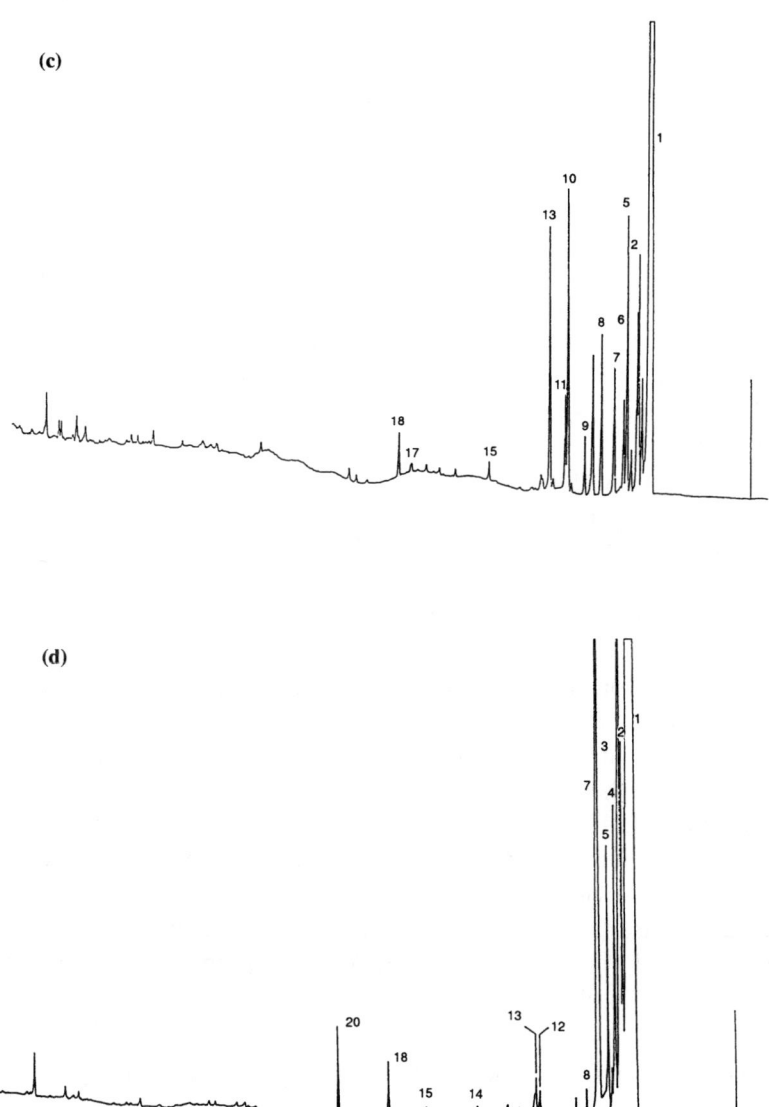

(c) well-ripened flesh; and (d) overripe flesh.

Banana volatiles are labeled as follows: 1, ethanol; 2, ethyl acetate; 3, isobutanol; 4, *n*-butanol; 5, 3-methylbutan-1-ol; 6, 2-pentanol; 7, 3-methylbutanol; 8, isobutyl acetate; 9, butyl acetate; 10, *trans*-2-hexanol; 11, ethyl 2-methyl butyrate; 12, hexanol; 13, 3-methyl butylacetate; 14, 2-heptanol; 15, isobutyl-*n*-butyrate; 16, *n*-butyl-*n*-butyrate; 17, butyl-2-methylbutyrate; 18, 3-methylbutylbutyrate; 19, 3-methylbutyl-2-methylbutyrate; 20, 3-methylbutyl-3-methylbutyrate.
Source: Research Department, Borthwicks Flavours Ltd, Wellingborough.

Phenols

Phenols are present in many fruits in the form of tannins; these polymeric phenolic substances are associated with the sensation of astringency. They are largely undefined in terms of structure and are usually detected and measured by various color reactions. Active tannins decrease during ripening and in the ripe pulp have reduced to below 20% of their preclimacteric values.

1.5.4 Physiological and Biochemical Aspects

As the following chapters will indicate, fruit processing is very much a science-based operation, and knowledge of the physiological and biochemical aspects of each fruit variety is advisable before subjecting it to any form of mechanical or enzymatic change. Industrial processing, therefore, has become highly efficient and competitive in improving quality and presentation of the highly varied and extensive fruit product range.

The normal use of the term *fruit* refers to the receptacle, that is, the casing around the true fruit or seed. The seed occurs as the result of sexual reproduction of the plant species concerned, and different functional components of the plant itself provide the protective environment or receptacle within which it can fully develop and eventually reproduce. For example, the edible part of an apple is the receptacle, which not only protects its fruit, the pips or seeds, but also provides an attractive packaging to animals to encourage ingestion and subsequent transportation of the seed to other locations.

Most small fruits (the word fruit is now being used in the more general sense) are readily transported and their seeds distributed via animal feces. The eventual function of the receptacle in the larger fruits is to provide a bed of rotting humus in which their seeds can germinate.

The mechanism of attraction is a vital part of the reproductive cycle of plants and, just as the flower provides a perfume to lure certain insects into effecting pollination, so does the fruit itself, when mature, produce both aroma and flavor as an attractant to many animal and bird species. These organoleptic properties are also responsible for the commercial value placed on fruit by humans.

1.6 FACTORS INFLUENCING FRUIT QUALITY AND CROP YIELD

1.6.1 Fruit Variety

While each fruit type possesses its own characteristic flavor, there are varietal differences that have to be carefully assessed by the growers and processors to ensure success in the marketplace. The range of texture, of color tones, of depth or freshness of flavor, and the stability of these characteristics in any specific processing method to be employed are points to be considered by the processor when selecting raw materials. The grower of fruit for the fresh market is primarily concerned by visual factors such as size, shape, color, and absence from blemishes and disease. The fruit processor, however, while showing preference toward high croppers on economic grounds, often has to apply higher and more restrictive standards of composition. Although the appearance of the fruit is still desirable for most processing needs, factors such as quality of flavor, texture, color, and nutritional value are going to be of major importance and, therefore, it is essential that the grower and processor maintain close liaison during the lead-up to harvest so that optimization of yields can be effected.

For many years in the United Kingdom, strawberry cultivation was dominated by the variety Cambridge Favourite, a reasonably heavy cropper with good flavor. Also, the fruit was relatively easy to harvest, being readily detached from the plant. This particular variety is noted for its distinct lack of color, the flesh of the fruit being white at the center as the color components are located on the inside of the fruit surface. Initially the poor color had little or no effect on its use for jam production, as strawberry is notorious for browning during the cooking process, and brown tones in the end-product were generally accepted as status quo. Where appropriate and as the product demanded, the processor could always add color either as natural anthocyanin extract or as one or more of the permitted artificial colorings (e.g., carmoisine and amaranth). Today, the use of these colors may be prohibited for legal reasons.

In recent years, new varieties of strawberry have appeared where, by careful selection and plant breeding, the color has been enhanced while maintaining an acceptable flavor character. Varieties such as Totem and the more popular Senga Sengana provide an uninterrupted zone of color across the entire fruit, and this, combined with improvements in processing, ensures a good color in the finished product. This is particularly valuable in fruit juice concentrates, where a high-quality standard is required. In addition, the use of artificial colorings is no longer an option when the product is to be declared 100% natural.

1.6.2 External Factors Affecting Fruit Quality

Other factors that affect fruit quality are geographical location of the growing area, climatic changes, and the selection of the degree of maturity of the fruit at the point of harvest or, pertinently, at the point of processing. Storage characteristics of fruit vary with type; soft fruits, for instance, are extremely vulnerable to damage during handling, and if not used within a few days of harvesting they need to be stored in deep-frozen conditions (–18°C to –26°C). The pome fruits (e.g., apples and pears), however, are stored under controlled atmosphere, whereby maturation is effectively programmed to meet marketing requirements.

Storage conditions are understandably most critical as the maturation of the fruit is accompanied by changes in both the composition of flavor characteristics and the structural consistency or texture.

1.7 FLAVOR CHARACTERISTICS

There are four main groups contributing to the flavor characteristics in fruits. These are organic acids, sugars, bitter or astringent principles, and volatile flavor constituents. The organic acids are usually citric, malic, or lactic acids, and these provide a tartness in flavor and give the thirst-quenching properties to soft drinks. Sugars contribute sweetness and body, and astringency relates to certain phenolic compounds (e.g., tannins, saponins, naringin, and hesperidin). These components are all present to a greater or lesser degree and constitute the background for perhaps the main characterization factor, the aroma volatiles.

At the point of optimum flavor quality, occurring at the peak maturity, the fruit is highly vulnerable to deterioration. Processing is required to ensure preservation and/or presentation in a form suitable for marketing.

Processing will of necessity involve major structural changes in the fruit (e.g., slicing, pulping, pressing, heating, and freezing), and this will be accompanied by a change in organoleptic characteristics as flavor compounds are affected in situ. Such changes may be beneficial, for instance, in the manufacture of certain conserves where the jammy flavor obtained by the extended cooking of some varieties of fruit may be preferred to the flavor achieved by a more natural approach where aroma volatiles normally lost during the initial cooking in the jam tun are replaced at a later stage in production to capture the whole fruit flavor.

1.8 THE GLOBAL MARKET: THREATS AND OPPORTUNITIES

Now that major fruit types (citrus, pome, and berry fruits) can be grown in many parts of the world, it is perhaps not remarkable that a glut in production in a particular region can have far-reaching effects on world prices, often sending them plummeting in an effort to remain competitive.

Under the agriculture policy of the European Union, it is the practice in such situations to adopt an interventionist approach by guaranteeing the producer a minimum price for his harvest, which is then effectively "dumped" out of the market. In Europe there have been many instances of "apple mountains" and "wine lakes" as a result of overproduction.

By comparison, a dearth in production can lead to equally dire results, inviting underhanded practices, whereby fruit pulps and juice concentrates are "stretched" by the addition of sugars, acids, and flavor top-notes, thus artificially increasing yields. Such practices became fairly commonplace in the early and mid-1980s and eventually resulted in the development of new analytical techniques to identify and combat production of highly sophisticated copies of the real thing. Perhaps the most effective of the new techniques is a method known as site-specific nuclear isotopic fraction by nuclear magnetic resonance (SNIF NMR), which has been developed into a powerful analytical tool.

SNIF NMR represents the state of the art in the analysis of many naturally sourced products, including wines, spirits, fruit juices, and flavors. It has rapidly become an important arbiter in the area of authenticity testing of products derived from, or said to be derived from, fruit sources. A typical scenario could involve the high-value red fruit concentrates (e.g., strawberry, black currant, raspberry, and blackberry). Concentrates are currently available on the market that have been produced by adulteration of the named source material with concentrates prepared from other berries, inferior-quality fruit, or even from the lower-value deodorized and decolorized apple or pear juice concentrates. Adjustments are made by addition of citric (sometimes malic) acid and elderberry concentrate, or red grape extract is used to intensify color. Depending on the extent of the intended sophistication, aroma top-note chemicals are added to boost the flavor as deemed necessary. While the finished blend may be natural, it is nevertheless nonauthentic as regards the named fruit source. In the early 1980s, it would have been difficult to challenge "natural copies" in terms of their authenticity, but nowadays it is beginning to be too costly to produce a "copy" that can get past the analyst.

In conjunction with SNIF NMR techniques, high-performance liquid chromatography (HPLC) is used for anthocyanin determinations. Anthocyanin colors are widely distributed in nature, and their distinctive profiles in many fruits allow their use as markets for the identification and classification of juices.

1.9 FRUIT PROCESSING

This chapter has drawn attention to some of the factors to consider when working with fruit, and the following pages will frequently return to such parameters in regard to specific processing requirements. The processing operation, correctly designed, will need to succeed on three counts: (1) the plant should be adequately sized for the task and in good working order; (2) the operator should be properly trained and follow the correct instructions; and (3) the raw-material fruit should be to the right specification. While all three are essential, it is the last aspect that will always be of paramount importance.

REFERENCES

Burg, S.P., and Burg, E.A. (1965). Relationship between ethylene production and ripening in bananas. *Botanical Gazette*, **126**, 200.
Couey, P.M. (1960). Effects of temperature and modified atmosphere on the storage life, ripening behaviour and dessert quality of Eldorado plums. *Proceedings of the American Society for Horticultural Sciences*, **75**, 207.
Gane, R. (1935). Formation of ethylene by plant tissues and its significance in the ripening of fruit. *Journal of Pomology*, **13**, 351.
Kidd, F., and West, C. (1922). *Report of the Food Investment Board for 1921*, p. 17.
Kidd, F., and West, C. (1933). *Report of the Food Investment Board for 1932*, p. 55.
Maarse, H., and Visscher, C.A., eds. (1989). *TNO-CIVO Volatile Compounds in Food*, 1(16), 189.
Mapson, L.W., and Robinson, J.E. (1966). Relation between oxygen tension biosynthesis of ethylene, respiration and ripening changes in banana fruit. *Journal of Food Technology*, **1**, 215–225.
Nursten, H.E., and Williams, A.A. (1967). Fruit Aromas: a survey of components. *Chemistry and Industry*, 486–497.
Palmer, J.K. (1971). In *The Biochemistry of Fruits and Their Products*, ed. A.C. Hulme, Vol. 2, Academic Press, New York, p. 77.
Porritt, S.W., and Mason, J.L. (1965). Controlled atmosphere storage of sweet cherries. *Proceedings of the American Society for Horticultural Sciences*, **87**, 128.
Romani, R.J., and Jennings, W.G. (1971). In *The Biochemistry of Fruits and Their Products*, ed. A.C. Hulme, Vol. 2, Academic Press, New York, pp. 427–428.
Tang, C.S., and Jennings, W.G. (1971). Lactone compounds of apricots. *Journal of Agricultural and Food Chemistry*, **16**, 252–254.
Tressl, R., Holzer, M., and Apetz, M. (1975). Biogenesis of volatiles in fruit and vegetables. In *Proceedings of the International Symposium on Aroma Research, Zeist, The Netherlands, May 1975*, eds. H. Maarse and P.J. Groenen, pp. 41–62.

Wade, E.W. (1901). *Old Testament History*, Vol. 1, 40. Methuen, London.

Young, R.E., Romai, R.J., and Biale, J.B. (1962). Carbon dioxide effects on fruit respiration. Response of avocados, bananas and lemons. *Plant Physiology (Lancaster)*, **37**, 416.

Chapter 2

Biochemistry of Fruits and Its Implications on Processing

Conrad O. Perera and Elizabeth A. Baldwin

2.1 INTRODUCTION

Fruits are living organs and undergo biological and biochemical activity even after they are separated from their plants. This process is known as respiration. The quality of fresh fruits or the products processed from fresh fruits is governed by a number of preharvest and postharvest factors (Kader and Barrett, 1996). The important preharvest factors that influence fruit quality are the following: the genetics—selection of the right cultivars and rootstocks; climate—temperature, light, and wind factors; and cultural practices—soil type, soil nutrient and water supply, pruning, thinning, and pest control. Postharvest factors that influence fruit quality are environmental—temperature, relative humidity, and atmospheric composition of storage; handling methods—postharvest handling systems involving the channels through which the harvested fruit reach the processor or consumer; and time period between harvesting and consumption—delays between harvesting and cooling or processing may cause losses in fruit quality (Kader and Barrett, 1996). The biological and biochemical processes involved in preharvest and postharvest fruits have a profound influence on the quality attributes of fresh and processed fruit. This chapter will focus on the biochemical processes involved mainly after harvest and their influences on the quality attributes of minimally and traditionally processed products.

2.2 MINIMALLY OR LIGHTLY PROCESSED FRUIT PRODUCTS

Minimal or light processing refers to trimming, peeling, sectioning, slicing, and coring of fruits (Baldwin et al., 1995; Shewfelt, 1987). Initially, lightly or minimally processed fruits were used in the food service industry, then by restaurants, supermarkets, and warehouse stores (Watada et al., 1996). Recently, the demand for minimally processed fruits (also known as "fresh cuts") by individual consumers has grown due to busy lifestyles, increased purchasing power, and health-conscious trends. Currently, the fresh-cut industry is the most rapidly growing portion of fresh produce sales (Sloan, 1995). These specialty products from fruits are more problematic than vegetables but are gaining in popularity, for example, mango, melon, peeled and cut or cored pineapple, and sliced apple. Cut pineapple and cut apple are usually available as packaged products.

Minimally processed fruits are more perishable than the unprocessed raw materials from which they are made (Huxsoll and Bolin, 1989). Processing renders the products less, rather than more stable as is the case with most processing systems (Rolle and Chism, 1987; Shewfelt, 1987). Therefore, there is the question of where to do the processing, and hence the debate of store-produced versus regionally produced versus nationally shipped fresh-cut products. Processors, for example, are often located in regions of consumption (i.e., large metropolitan areas). National processors have developed regional processing facilities, such as the Fresh Express' string of plants in the United States; the latest trend is to prepare products close to the source of production. Nevertheless, some supermarkets still prefer to use in-store precutting as a disposal method for damaged or spoiled fruit (Prevor, 1996).

2.3 FACTORS AFFECTING SHELF LIFE AND QUALITY OF MINIMALLY PROCESSED FRUITS

Quality of fresh-cut product is dependent on variety, maturity and quality at harvest, harvesting and processing technique, storage temperature, and microbial control. Sometimes intrinsic quality is sacrificed for better shipping and shelf-life quality in terms of variety selection and harvest maturity decisions (Shewfelt, 1987). Ideally, the product should be of the best quality to start with, harvested with optimal quality, and subjected to minimal damage or bruising. Quality is best maintained if the product is chilled to optimum temperature levels, which are maintained until purchased by the consumer ("cold chain"). Stringent sanitary conditions should be observed at the processing plant for good microbial control. The processing should incur minimal damage to the product. For example, clean slices are better than tearing or chopping, and cutting equipment should be kept sharp. The cut product should be washed to remove cellular contents from the surface that favor microbial growth and then as much moisture as possible should be removed prior to packaging.

2.4 PHYSIOLOGY AND BIOCHEMISTRY OF FRESH-CUT FRUITS

As living tissue, fresh-cut products have a limited energy supply. Respiration converts the stored energy into usable energy to sustain life. Basically, the higher the respiration, the shorter the shelf life (Rolle and Chism, 1987). Respiration increases with storage temperature. The temperature coefficient (Q^{10}) of respiration rates was found to range from 2.0 to 8.6 among various fresh-cut fruits and vegetables. This means that a difference in 5°C storage temperature increased respiration dramatically for some fresh-cut products (Watada et al., 1996). Therefore, suppression of respiration is important for all fresh produce, but especially for fresh cuts. Respiration can be reduced by low temperature and/or by controlled atmosphere (CA) or modified atmosphere (MA) storage conditions of relatively low O_2 and high CO_2. In addition, many fruits produce elevated levels of ethylene as they ripen and/or upon wounding. Ethylene promotes ripening, softening, color changes, and senescence. Synthesis of ethylene by fresh produce can also be reduced by low temperatures, CA and MA storage. Use of these storage techniques is limited by the products' susceptibility to chilling and CO_2 injury, and tolerance for low O_2 (Huxsoll and Bolin, 1989; Rolle and Chism, 1987).

In addition, minimally processed produce are essentially wounded tissues, and thus undergo reactions designed to repair the damage. Unfortunately, many of these reactions are deleterious to quality (Rolle and Chism, 1987). The wounding of living plant tissue results in an increase in respiration and ethylene production (Philosoph-Hadas et al., 1991), both of which will result in shortened shelf life. The ethylene will also accelerate ripening, softening, and senescence (Philosoph-Hadas et al., 1991), which leads to membrane damage, while the respiration will use up energy reserves. Senescence involves oxidative breakdown of membrane lipids and depletion of linoleic, linolenic, and polyunsaturated fatty acids that are substrates for lipoxygenase. Wounding also induces other metabolic pathways that result in secondary metabolites that can cause discoloration, off-flavor, and texture changes. Cell disruption at the cut surface leads to decompartmentation of enzymes and substrates. For example, phenylalanine, ammonia-lyase and polyphenol oxidase contribute to browning reactions when combined with oxygen and monophenolic substrates (Sapers, 1993). Cell wall enzymes such as *exo-* and *endo*-polygalacturonase, β-galactosidase, and pectin methylesterase (some induced by ethylene) can digest cell walls, resulting in texture changes (Wong et al., 1994; King and Bolin, 1989; Rolle and Chism, 1987), and lipoxygenase can degrade membrane lipids (Siedow, 1991).

Finally, almost all minimally processed foods have all or part of the peel or outer protective coating removed. This allows entrance of spoilage organisms and dehydration of the fruit tissue. Dehydration may be partially responsible for some of the softening that is observed in fresh-cut produce. Use of edible coating or plastic packaging is necessary to retard moisture loss by providing a barrier to water vapor, resulting in a high relative humidity environment, as well as to minimize microbial contamination (Watada et al., 1996; Baldwin et al., 1995, 1996).

It was noted that green tomato fruit explants underwent similar processes as did the intact fruit, such as softening, color change, and respiratory activity (Parkin, 1987). These fresh tissues contain living cells like their whole fruit/vegetable counterpart, but there are some important differences due to wounding. However, this points to the possibility of using unripe fruit pieces that are allowed to ripen after processing while in the marketing chain.

2.5 TECHNIQUES TO EXTEND SHELF LIFE OF MINIMALLY PROCESSED FRESH PRODUCE

Washing fresh cuts is important to control microbial loads that include mesophilic microflora, lactic acid bacteria, coliforms, fecal coliforms, yeasts and molds, and pectinolytic microflora (Nguyen-the and Carline, 1994). Fresh cuts are generally rinsed in 50–200 ppm chlorine or chlorine dioxide, which may also aid in reduction of browning reactions (Brecht et al., 1993), but some organisms survive in broken cells or tissue, stomata, or under trichomes not penetrated by the chemical (Breidt and Fleming, 1997; Watada et al., 1996). It has been proposed that lactic acid bacteria could be useful as a biological control agent to inhibit competitively pathogenic bacteria on minimally processed produce (Breidt and Fleming, 1997). Other methods of sanitation such as use of ozone and ultraviolet radiation are currently being explored.

Methods used to stabilize and extend shelf life of minimally processed fruits include low-temperature storage, chemical preservatives, mild heat treatments, modification of pH, reduction of water activity (a_w), irradiation, and CA or MA storage (Huxsoll and Bolin, 1989). The problem with low temperature is the tolerance level of the produce. Some produce is susceptible to chilling injury, which often appears as a pitting of the peel. In the case of minimally processed produce where the peel may have been removed, sometimes a lower temperature can then be used than would be the case for the intact counterpart. Low-temperature injury can also occur as a reduction in aroma, as is the case with tomatoes (Kader et al., 1978; McDonald et al., 1996). Nevertheless, the coldest temperature tolerated by the minimally processed produce is the optimal temperature to be maintained in the "cold chain" from harvest to consumer. This will best maintain the quality and safety of these specialty products.

Chemical preservatives can help extend the shelf life of minimally processed products, as was shown for cut apple (Baldwin et al., 1996). The problem with chemical preservatives is the impact on the consumers' impression of product wholesomeness (Huxsoll and Bolin, 1989). The exceptions would be nutritional additives such as ascorbic acid (vitamin C) or tocopherol (vitamin E) to reduce browning; citric acid, as a chelator or acidulant; or various forms of calcium as a firming agent. Calcium can also stabilize membrane lipid structural components and thus play a role in delaying senescence. Use of these compounds was successful in extending the shelf life of strawberries, pears (Rosen and Kader, 1989), apples (Baldwin et al., 1996; Sapers, 1993), and carambola slices (Weller et al., 1997). Polyamine, a naturally occurring peptide in plant tissue that has a positive charge, may act similarly to calcium. These compounds were shown to increase firmness of strawberry slices (Ponappa et al., 1993).

Mild heat treatments can inactivate some enzymes that promote deterioration or discoloration, and can reduce microbial populations, but these treatments can also alter the color, texture, and flavor of some fresh produce.

Modification of pH is usually toward the acid side to reduce microbial populations using acidulants such as citric acid or ascorbic acid. However, this can result in a sour flavor. Reduction of water activity can be accomplished using sugars, salts, or polyols with relatively high osmotic pressure that are infused into the fruit tissue. Limited osmotic dehydration was used to extend shelf life of fruit pieces in conjunction with polysaccharide coatings (Wong et al., 1994). However, as with pH modification, flavor can be affected.

Ionizing radiation has been explored to extend the shelf life of minimally processed fruits. However, the dose required to affect microbial organisms may also induce undesirable texture changes or other damage to produce tissue (Huxsoll and Bolin, 1989).

Use of CA, MA (relatively low O_2 and high CO_2), or MA packaging (MAP) can extend shelf life of whole produce by reducing respiration; ethylene production; ethylene-induced processes, such as color changes, softening, and senescence; as well as preventing dehydration and microbial contamination (Kader, 1986; Zagory and Kader, 1988). Low O_2 can favor growth of human pathogens due to lack of competition from spoilage organisms that have been reduced by low temperature and low O_2 conditions. Use of temperature and gas mixtures alters the microecology of a food (Vankerschaver et al., 1996; Brackett, 1987). The oxygen demand of fresh produce is dependent on respiration rates, which in turn, are dependent on temperature. Therefore, MAP designed for cold-temperature storage, under conditions of temperature abuse, can result in anaerobic conditions and possible growth of dangerous anaerobes. Levels of CO_2, however, can be microbially static above 10%, and atmospheres with 30% CO_2 were shown to extend shelf life by visual appearance, but allowed growth of *Escherichia coli* to higher levels than air controls (Diaz and Hotchkiss, 1996). Nevertheless, products are often packaged in semipermeable plastic films that can be manufactured with different gas transmission rates. Polyvinylchloride is used mostly for overwrap packaging. Polypropylene, polyethylene, and ethylene vinyl acetate can be combined in laminated or coextruded films for sealed packages of varying permeability characteristics (Zagory, 1997). Vacuum packaging and gas flushing with certain gas mixes such as O_2, CO_2, and N_2 are generally used.

Some novel gas mixtures, such as high O_2, argon, and nitrous oxide, are being explored for the MAP of fresh-cut products (Day, 1994). A new technology in packaging is the incorporation of a patch containing side-chain crystallizable polymers. These acrylic polymers with fatty acid-based side chains of various lengths alter in gas permeability characteristics in response to temperature. This Inteliner technology was developed by Landec Corporation of Menlo Park, CA (Anonymous, 1996).

Use of coatings can help stabilize fresh-cut products. A polysaccharide/protein coating carrying antioxidants, acidulants, and preservatives was shown to reduce water loss and browning in apples (Baldwin et al., 1996), which was further enhanced when combined with vacuum packaging. In enzyme-peeled citrus, a carnauba wax emulsion coating retarded fluid leakage from the peeled segments (Baker and Hagenmaier, 1997). Finally, addition of ethylene absorbents (such as potassium permanganate) to packages to reduce ethylene levels reduced the rate of softening in kiwifruits and bananas (Abe and Watada, 1991).

2.6 ENZYME-CATALYZED REACTIONS DURING PROCESSING

2.6.1 General Biochemistry of Pectinase Enzymes

Three types of pectinase enzymes are well documented (Pilnik and Voragen, 1989, 1993). Of these, only pectin methylesterase and polygalacturonase are found in fruits. The other type of pectinase enzyme, pectate lyase, is found in microorganisms. These enzymes are classified according to their mode of attack on the galacturonan part of the pectic molecule as follows.

Polygalacturonase (poly-α-1,4-galacturonide glycano-hydrolase, EC 3.2.1.15)

Polygalacturonase hydrolyzes the α-1,4-glycosidic bond between the anhydrogalacturonic acid units. There are *exo-* and *endo-* polygalacturonases. The *exo* type hydrolyzes bonds at the ends of the pectic polymer, while the *endo* type acts on the interior of the polymer. This hydrolysis of the pectic polymers in fruits results in a softening in texture of these products.

Pectin Methylesterase (pectin pectylhydrolase, EC 3.1.1.11)

Pectin methylesterase hydrolyzes the methyl ester bonds of pectin to give pectic acid and methanol. The enzyme is sometimes also referred to as pectin esterase. Hydrolysis of pectin to pectic acid in the presence of Ca^{++} results in a firming of texture due to the formation of cross-bridges between Ca^{++} and carboxyl groups of the pectic acid.

Pectic Acid Lyase (poly 1,4-α-D-galacturonide lyase, EC 4.2.2.2) and Pectin Lyase

Pectic acid lyases split the glycosidic linkages next to free carboxyl groups by β-elimination. Similarly, pectin lyase splits the glycosidic linkages in pectin molecules by β-elimination. Pectic acid lyase and pectin lyase both exist in the *endo* and *exo* forms as in the case of polygalacturonase.

2.6.2 The Effects of Pectinase Enzymes on Juice Processing—Cloud Stability

Cloud particles in juices, essentially composed of protein nuclei carrying outward positive charges, are coated with negatively charged pectin molecules (Yamasaki et al., 1964). Partial degradation of this pectin coating by polygalacturonase enzyme results in the aggregation of oppositely charged particles, thus destabilizing the cloud. If pectin methylesterase is not inhibited directly after juice extraction by heat inactivation or by freezing, the native pectin will be de-esterified and will be coagulated by Ca^{++} in the juice (Joslyn and Pilnik, 1961; Versteeg et al., 1978). On standing, the juice then separates into a clear supernatant and sediment. In the case of juice concentrates, calcium pectate gels may be formed, resulting in poor reconstitution. Addition of exogenous enzymes such as polygalacturonase and pectin lyase can be effective in overcoming this problem. Baker (1980) demonstrated that the soluble pectin is either directly degraded into low-molecular-weight fractions by the added pectin lyase, or the low-methoxy pectin formed by endogenous pectin methylesterase is degraded by polygalacturonase to yield low-molecular-weight products.

Exogenous pectic enzymes are useful as processing aids in juice manufacture. The expression of juice from pulp is facilitated by the addition of exogenous enzymes. They reduce the viscosity by breaking down the soluble pectins, thus improving the ease of pressing and juice yield. No endogenous enzymes are added if cloudy juice is required.

Clarification of juices by pectin degradation is also important in the manufacture of high Brix concentrates to avoid gelling and the development of haze.

2.6.3 Calcium Pectate Linkages and Texture

Thermal processing, freezing, minimal processing, and other processing operations generally cause the plant tissues to soften, because of the modification of the cellular structure. Stability and integrity of these tissues depend on the maintenance of cellular structures. Pectins are important components of the cell wall structure, and pectic substances are involved in the stabilization of these cell walls by cross-linking the free carboxyl groups via polyvalent cations. Fruits generally have considerable amounts of Ca^{++} present in their tissues. However, in processing, calcium salts at concentrations of 0.1% to 0.25% as calcium are frequently added to maintain structural integrity. The most commonly used calcium salts in the food industry are calcium chloride, calcium citrate, calcium lactate, and calcium sulfate. Calcium chloride and calcium sulfate contribute a bitter flavor at concentrations in excess of 1%.

Firmness of fruits can also be manipulated with the use of pectin methylesterase. As discussed earlier, pectin methylesterase demethylates pectin molecules to pectic acid, which has high affinity to form calcium bridges with other such molecules. Fruit preparations that are added mainly to dairy products and other food preparations must have special characteristics such as fruit identity, fruit firmness, and stability of the dispersing phase of the preparation. This can be only partially obtained by the addition of calcium salts alone, as discussed earlier. However, application of pure pectin methylesterase on whole fruit or fruit pieces allows the demethylation of fruit pectin in situ. The demethylated pectin gels at acid pH with calcium present to give the right fruit firmness (Fauquembergue and Grassin, European patent, 1997).

2.6.4 Control of Texture in Tomato Processing

Pectin methylesterase and polygalacturonase are two enzymes that are abundantly present in many fruits, especially in tomato. Immediate heat inactivation of these enzymes is necessary in order to obtain highly viscous juices or pulps. For instance, in the manufacture of tomato juice/paste, crushing fresh tomatoes into circulating hot tomato pulp is known as "hot break," which produces pulps having high viscosity. Such high-viscosity pulps are used in the manufacture of sauces, soups, ketchup, and similar products. In the case where tomato ingredients are used for color and flavor only, and consistency is provided by other ingredients, "cold break" juice is the starting material. In the cold-break process, the crushed tomatoes are held for a period of time for the pectin to be broken down by polygalacturonase and pectin methylesterase, before application of a heat treatment.

2.7 BROWNING REACTIONS DURING PROCESSING

2.7.1 Polyphenol Oxidases and Their Control during Processing

Polyphenol oxidase is widely distributed in fruits. It has been given two entries in the International Union of Biochemistry classification, namely as EC 1.14.18.1, mono-phenol mono-oxygenase, and EC 1.10.3.1, catechol oxidase. Common trivial names for mono-phenol mono-oxygenase are tyrosinase, phenolase, and cresolase. Common trivial names for catechol oxidase are diphenol oxidase, *o*-diphenolase, phenolase, polyphenol oxidase, and tyrosinase (Enzyme Nomenclature, 1984). Polyphenol oxidases can catalyze many reactions involving phenolic compounds found naturally in many fruits. When the mono-phenol *p*-cresol is the substrate, it is oxidized to 4-methyl catechol, a diphenol. The oxygen for the hydroxylation comes from the atmosphere. Thus the enzyme acts as a monooxygenase. When the diphenol cresol is the substrate, it is dehydrogenated to *o*-benzoquinone. Quinones are highly reactive compounds and undergo further oxidative polymerization to form brown-colored pigments (Whitaker, 1996).

Polyphenol oxidase is a copper-containing enzyme. Sodium bisulfite and thiol compounds are known to remove the essential Cu^{++} from the active site of the enzyme, thus inactivating it. Ascorbic acid is known to inactivate the enzyme by destroying the histidine molecules at the active site (Osuga and Whitaker, 1994). The enzyme can also be inhibited by chelation of the copper with the addition of citric acid. Removal of one or more of the required substrates or the cofactor will eliminate the enzyme activity. Thus it is well known that polyphenol oxidase browning in fruits can be prevented by elimination of oxygen. This can be done by processing fruit products under vacuum or under an inert gas such as nitrogen.

Another way of preventing browning caused by polyphenol oxidase is to reduce the initial product *o*-benzoquinone back to the original substrate before it can undergo non-enzymatic oxidative polymerization to brown pigments. The reducing agents usually employed in the fruit-processing industry are sodium bisulfite and ascorbic acid. These compounds are also shown to indirectly inactivate the enzyme by free radical degradation of the histidine molecule at the active site, and by reducing the cofactor Cu^{++} to Cu^{+}, thereby causing the cuprous ion to dissociate more readily from the enzyme (Osuga and Whitaker, 1994).

2.7.2 Non-Enzymatic Browning in Fruit Products and Their Control

Fruit juices usually undergo a number of non-enzymatic browning reactions depending on their composition, concentration, and storage conditions (Eskin et al., 1971). Maillard browning reactions occur between reducing sugars and α-amino groups of amino acids, peptides, and proteins. The reaction between amino acids and α-dicarbonyls, known as the Strecker degradation in the Maillard reaction, also leads to brown pigment formation. Lipid oxidation can give rise to reducing substances that can react with amino acids to form brown pigments and off-flavors.

Ascorbic acid degradation in juice concentrates such as kiwifruit can take place readily at room temperature, leading to brown discoloration during storage (Wong and Stanton, 1989). Ascorbic acid destruction in fruit juices can take place under aerobic or anaerobic conditions under normal processing temperatures (Rojas and Gerschenson, 1997). They found that under anaerobic conditions, the destruction of ascorbic acid, and thus the browning reactions, increase with increase in pH from 3.5 to 5.0. Addition of tin or lysine increased ascorbic acid loss and browning. Under aerobic conditions, the degradation of ascorbic acid was retarded in the presence of glucose.

Caramelization of sugars can take place at relatively high temperatures under acid or alkaline conditions in the absence of amino acids, giving rise to non-enzymatic browning. Caramelization is associated with unpleasant, burned, and bitter flavors. In fruit juice concentrates and dried fruits, these reactions can take place under processing, packaging, and storage conditions, giving rise to undesirable color, flavor, and textural changes.

In general, non-enzymatic browning reactions can be inhibited or controlled in a number of ways.

Temperature—Lowering the temperature will lower the reaction rates of the above reactions. Thus the shelf life can be prolonged by low-temperature storage.
Moisture content—The non-enzymatic browning reactions are moisture dependent, thus controlling the moisture content of the finished product will prolong shelf life. However, this may not be suitable for all fruit products.
Gas packaging—Exclusion of oxygen will generally reduce the possibility of an oxidative type of reaction, thus giving rise to prolonged shelf life.
Chemical inhibitors—Sulfites, thiols, and calcium salts are used extensively in fruit processing to control non-enzymatic browning.

2.7.3 Browning Reactions Due to Chlorophyll Degradation Reactions

During fruit ripening the chlorophylls, which are present in all unripe fruit, break down, following the chloroplast disintegration. Kiwifruit and avocado are two common fruits in which the chlorophyll color still remains after the fruit is ripened. Chlorophylls are porphyrins containing the basic tetrapyrrol ring, of which one is reduced. The four rings are coordinated with Mg^{++}. The Mg^{++} is easily replaced in acid solutions found in the fruit by H^+, giving rise to dull olive green color, and further degradation leads to brown pigment formation (Gross, 1991).

Chlorophylls in kiwifruit are all lost during canning at 100°C for 5 minutes, and in frozen purée stored at −18°C, over two-thirds was lost within 36 days (Robertson, 1985). However, a later study showed that below −18°C, there is no significant loss of chlorophyll color during storage for up to 12 months (Venning et al., 1989). Perera and Venning (1988) developed a patented process for drying that maintained the vitamin C and green color in dried kiwifruit slices. The process involved vacuum infusion of sugar syrup with potassium citrate at pH of 4.2–5.0, and drying the resultant product below 45°C. Drying was conducted in two stages, first in air and then under vacuum to optimize the drying rate. In order to maintain the green color, the final water activity of the product should be below 0.23.

2.8 DEVELOPMENT OF BITTER PRINCIPLES IN FRUIT PRODUCTS

A number of stone fruit juices develop a bitter taste during processing (Tien and Fang, 1991). The bitter taste of Japanese apricot (Mei) juice is due to a compound known as amygdalin (D-mandelonitrile-β-D-glucosido-6-β-D-glucoside). It is a cyanogenic compound that occurs naturally in the members of the *Prunus* genus of the Rosaceae family (Wu et al., 1993). Debittering can be effected by hydrolysis of amygdalin by β-glucosidase. β-glucosidase occurs naturally in the kernel of these stone fruits, and

Chang and Wu (1985) suggested a simple way of preparing a crude extract that can be added to the juice before pasteurization, allowing sufficient time for the enzymatic reaction to take place.

Bitter-tasting peptides are produced when fruits containing protease enzymes such as kiwifruit are mixed with milk products (Bachmann and Farah, 1982). Therefore, fresh pulps such as kiwifruit, pineapple, and fig, which contain the powerful protease enzymes actinidin, bromalin, and ficin, respectively, should not be mixed with milk products. These fresh fruit pulps containing protease enzymes also will not form gelatine gels due to their activity on proteins. Heat treatment to inactivate these protease enzymes will ensure setting of gels.

Bitter compounds occur naturally in some fruits. The bitter compound limonin is well known in certain citrus cultivars (Hasegawa and Maier, 1990; Puri, 1990). It is also known that rhamnosyltranferase catalyzes the production of the bitter flavanone-glucosides, naringin and neohesperidin, in citrus (Bar-Peled et al., 1991).

Limonoids are a group of chemically related triterpene derivatives that occur commonly in citrus. Limonin is a highly oxygenated triterpene derivative that is intensely bitter, which adversely affects citrus juice quality. There is usually no limonin bitterness in intact fruit. However, in the juice-extraction process, a nonbitter precursor, limonic acid, α-ring lactone, is converted to limonin under the acid conditions found in the juice (Puri, 1990). This conversion is accelerated by heat such as that found during pasteurization and concentration. Limonin is found in all citrus fruits. The detection limit of limonin in water is about 1 ppm, but in orange juice it is about 6 ppm.

Naringin is a neohesperidoside and is most commonly found in grapefruit. It is generally found in the flavedo, albedo, segment membrane, and juice sacs and is relatively water-soluble (Puri, 1990). In grapefruit juice, a naringin concentration of 300–500 ppm gives the characteristic desirable bitter flavor. Various adsorption agents such as cyclodextrin, polystyrene divinylbenzine, and amberlite have been used to debitter citrus juices (Manlan et al., 1990; Shaw and Buslig, 1986; Puri, 1990).

2.9 ANTHOCYANINS AND THEIR CHANGES DURING PROCESSING

2.9.1 Anthocyanin Chemistry

Anthocyanins belong to the subgroup known as flavonoids, which are part of the large group of compounds generally referred to as phenolics. Anthocyanins are responsible for a wide range of colors in plants, ranging from blue to orange. They are considered flavonoids because they possess the characteristic C_6-C_3-C_6 carbon skeleton. They are glycosides of polyhydroxy and polymethoxy derivatives of 2-phenylbenzopyrylium or flavylium cation (Figure 2–1).

The differences between the individual anthocyanins are based on the number of hydroxyl groups in the molecule; the degree of methylation of these hydroxyl groups; the nature, number, and position of glycosylation; and the nature and number of aromatic or aliphatic acids attached to the glucosyl residue (Wilska-Jeszka, 1997). An increase in the hydroxyl groups tends to deepen the color to a more bluish shade. An increase in the methoxyl groups tends to increase the redness. This deepening of the hue is the

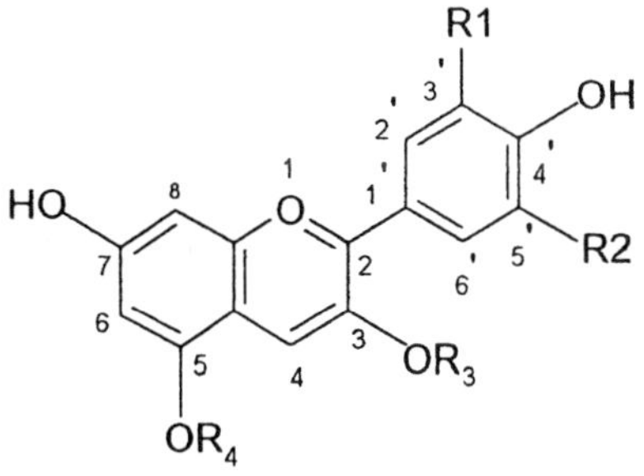

Figure 2–1 The structure of the flavylium cation.

result of a bathochromic shift, which means that the light absorption shifts from violet through red to blue. Bathochromic effects are produced by auxochrome groups, which by themselves have no chromophoric properties but cause deepening of the hue when attached to the flavylium cation. The auxochrome groups in anthocyanins are electron donors such as the hydroxyl and methoxyl groups. Changes in pH, metal complex formation, and copigmentation can also cause changes in hue of the anthocyanin pigments.

The structure of anthocyanin molecules has a profound effect on color intensity and stability during processing, and the rate of degradation varies with different anthocyanins because of the differences in their structures (von Elbe and Schwartz, 1996; Jackman and Smith, 1992). In general, increased hydroxylation decreases stability and increased methylation increases stability. Increase in the number of hydroxyl groups in the B-ring shifts the absorption maximum to longer wavelengths and the color changes from orange to bluish red. The methoxyl groups replacing hydroxyl groups reverse this trend. The hydroxyl group at C-3 is particularly significant because it shifts the color from yellow-orange to red. Anthocyanins containing pelargonidin, cyanidin, or delphinidin aglycones are less stable than those containing petunidin or malvidine aglycones. The stability of the latter two is due to the blocking of the reactive hydroxyl groups. Increased glycosylation as with monoglucosides and diglucosides increases stability of anthocyanins.

2.9.2 Anthocyanin Color Changes Due to Changes of pH

In an aqueous medium, anthocyanins can exist in four possible structural forms depending on the pH. They are the blue quinonoidal base (A), the red flavylium cation (AH$^+$), the colorless carbinol pseudobase (B), and the colorless chalcone (C) (von Elbe and Schwartz, 1996). In a solution of malvidin-3-glucoside at low pH, the flavylium structure dominates, while at pH 4–6 the colorless carbinol dominates. Anthocyanins show their greatest tinctorial strength at approximately pH 1.0, when the pigment molecules are in the un-ionized state. At pH 4.5, anthocyanins in fruit juices are slightly bluish. If yellow flavonoids are present, as is common in fruits, the juice will be green in color.

2.9.3 Anthocyanin Color Changes Due to Temperature

Anthocyanins are degraded by temperature. The rate of degradation is influenced by the presence of oxygen, pH, and structural conformation. The structural conformations that lead to pH stability discussed earlier also increase thermal stability. Thus highly hydroxylated anthocyanins are less stable to heat than highly methylated glycosylated or acylated anthocyanidins.

$$\text{(A) quinonoid (blue)} \leftrightarrow \text{(AH}^+\text{) flavylium (red)} \leftrightarrow \text{(B) carbinol base (colorless)} \leftrightarrow \text{(C) chalcone (colorless)}$$

The four known structural forms of anthocyanins in aqueous solutions are shown above. Heating shifts the equilibrium toward the colorless chalcone and the reverse reaction is slower than the forward reaction. The exact mechanism of thermal degradation of anthocyanin has not been fully elucidated. Hrazdina (1971) identified coumarin diglycoside as a common degradation product of anthocyanidin 3,5-diglycosides and proposed the pathway shown by Jackman and Smith (1992). The mechanism involves the transformation of the flavylium cation to a quinonoidal base, and then to several intermediate products before finally breaking down to a coumarin derivative and a compound corresponding to the B-ring of the anthocyanidin. Anthocyanidin 3-glycosides do not form coumarin derivatives (Hrazdina, 1971). Markakis et al. (1957) showed that the first step in their thermal degradation involves the formation of a colorless carbinol pseudobase and subsequent opening of the pyrylium ring to form the chalcone, before hydrolysis of the glycidic bond. Adams (1973) postulated another mechanism for heat deterioration of anthocyanidin 3-glycoside. When this compound is heated, at pH 2–4, it undergoes hydrolysis of the glycosidic bonds, followed by conversion of the aglycone to a chalcone and subsequent formation of an α-diketone. The thermal degradation products of cyanidin 3-glycoside include chalcones and α-diketone, protochuic acid, quercetin, and phloroglucinaldehyde (Jackman and Smith, 1992). These primary breakdown products are presumed to lead to the formation of brown-colored products.

2.9.4 Anthocyanin Color Changes Due to Oxygen

The oxidative effects of molecular oxygen on anthocyanins have been well documented (Jackman and Smith, 1992; Adams, 1973; Clydesdale et al., 1978). Oxygen and temperature seem to accelerate the destruction of anthocyanins (Nebesky et al., 1949). The deleterious effects of oxygen on anthocyanin color stability during processing of fruit juices were reported by Daravingas and Cain (1968) and Starr and Francis (1968). It has been known that when grape juice is hot filled, complete filling will delay degradation of color from purple to dull brown. Similar observations have been made with other anthocyanin-containing juices. The stability of anthocyanins with respect to water activity is less well documented. However, it seems from the work of Khachik et al. (1986) that the greatest stability of anthocyanins was found in the water activity range of 0.63–0.79. Reports of ascorbic acid and anthocyanin undergoing degradation simultaneously in fruit juices suggested direct interaction between ascorbic acid and anthocyanin (Poei-Langston and Wrolstad, 1981). However, this was discounted by Jackman and Smith (1992), who suggested that the ascorbic acid–induced degradation of anthocyanins results from the indirect oxidation by the hydrogen peroxide formed during aerobic oxidation of ascorbic acid. Ascorbic acid and oxygen have been shown to act synergistically in anthocyanin degradation (Keith and Powers, 1965). Maximum pigment losses in anthocyanin-containing juices occur at high concentrations of oxygen and ascorbic acid. A nucleophilic attack at the C-2 position of anthocyanins by hydrogen peroxide cleaves the pyrylium ring and leads to the formation of various colorless esters and coumarin derivatives (Hrazdina and Franzese, 1974). These oxidation products may take part in further degradation reactions or are polymerized to form brown precipitates.

2.9.5 Anthocyanin Color Changes Due to Light

Anthocyanins are generally unstable when exposed to ultraviolet or visible light and other sources of ionizing radiation. Their decomposition appears to be mainly photooxidative because *p*-hydroxybenzoic acid has been identified as a minor degradation product (Sweeny et al., 1981). The ability of light to yield an anthocyanin-excited state via electron transfer would appear to predispose these pigments to photochemical decomposition.

2.9.6 Anthocyanin Color Change Due to Enzymes

The enzymes that are implicated in the oxidative discoloration of anthocyanins are termed *anthocyanases*. Depending on their activity, two distinct groups, found in the oxidative discoloration of anthocyanins in plant tissues, are glycosidases and polyphenol oxidases. Glycosidases hydrolyze the glycosidic bonds of the anthocyanins to yield free sugar and aglycone. The aglycone is unstable and spontaneously transforms to colorless derivatives (Forsyth and Quesnel, 1957). Polyphenol oxidase acts on anthocyanins in the presence of *o*-diphenols to produce oxidized anthocyanins (Peng and Markakis, 1963), which may subsequently react with each other, or with amino acids or proteins to yield brown-colored polymers.

Steam blanching prior to subsequent processing has been proven to be effective in destroying the endogenous anthocyanase activity (Siegel et al., 1971). Wrolstad et al. (1990) showed that storage and packing in high-sugar syrups (concentration greater than 20%) resulted in inhibition of the anthocyanase activity in fruits.

Polyphenol oxidase activity in various anthocyanin-containing fruit extracts has been effectively inhibited by sulfur dioxide, bisulfite, dithiothreitol, phenylhydrazine, and cysteine (Goodman and Markakis, 1965; Cash et al., 1976). Ascorbic acid also has a protective effect on anthocyanin degradation by polyphenol oxidase.

2.9.7 Anthocyanin Color Changes Due to Nucleophilic Agents

The decolorizing effect of sulfur dioxide, which is used extensively in the processing of fruit and vegetables, results from the formation of colorless C-4 adjuncts of anthocyanins (Jurd, 1964). This reaction is reversible and acidification to pH 1.0 will restore the red color. Anthocyanin-bisulfite complexes are relatively stable. The bisulfite moiety presumably deactivates the C-3 glycosidic bond, thus preventing its hydrolysis and subsequent formation of brown degradation products. The red anthocyanin color in fruit pulps will be decolorized by sulfur dioxide, but during drying, due to evaporative loss of sulfur dioxide and shifts in equilibrium, the red colors will be regenerated.

2.10 DISCOLORATION DURING PROCESSING AND STORAGE

The pink discoloration often found in canned juices is believed to be due to the conversion of leucoanthocyanin (proanthocyanadins) to anthocyanin when heated in acidic conditions (Luh et al., 1960). Tin and polyphenols have also been implicated in the development of pink discoloration in canned fruit and fruit products (Mathew and Parpia, 1970; Chandler and Clegg, 1970). Hwang and Cheng (1986) studied the pink discoloration in canned lychee juice and proposed a possible pathway as follows:

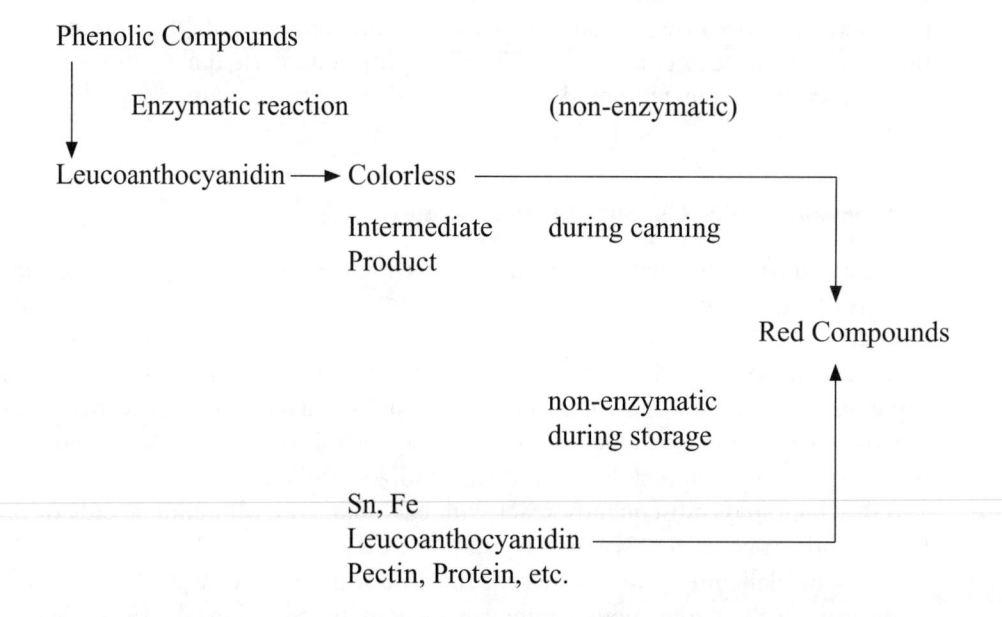

Phenolic Compounds
 │   Enzymatic reaction   (non-enzymatic)
 ▼
Leucoanthocyanidin ⟶ Colorless ─────────────────────────┐
                     Intermediate   during canning   │
                     Product   │
                                                        ▼
                                               Red Compounds
                                                        ▲
                      non-enzymatic   │
                      during storage   │
                                                        │
              Sn, Fe   │
              Leucoanthocyanidin ───────────────────────┘
              Pectin, Protein, etc.

The same authors recommended that shortening the duration between peeling, depitting, and heating, and immersion of lychee flesh in sodium bisulfite solution prior to thermal processing will decrease the degree of pink discoloration. Shi and Luh (1999) suggested the addition of tartaric acid or 30% sugar syrup containing 0.1–0.15% citric acid to achieve a pH of 4.5 to control this undesirable pink discoloration in canned lychees.

2.11 CONCLUSION

In this chapter, we have considered the biochemistry of fruits in general and its implications on some aspects of processing. The study of physiology and biochemistry of fruits as they mature and ripen is important from the point of view of extending the shelf life of fresh and minimally processed fruit products.

The biochemistry of pectinase enzymes is well known, and its impact on ripening and senescence of a number of fruits has been extensively studied. An in-depth knowledge of the action of these enzymes and their control has led to the production and processing of high-quality, stable orange juice concentrates, cloudy and clear apple juice concentrates, and many other juices and beverage products.

Generally, fruits are good sources of anthocyanin color, and the biochemistry and stability of anthocyanins have been dealt with in greater detail than carotenoids or chlorophylls, which are found to a much lesser degree in ripe fruits.

The enzymatic and non-enzymatic browning reactions and their control during processing and storage, including minimal processing, have been well documented.

Fruits are gaining universal acceptance and play an important rule in providing essential minerals, vitamins, fiber, and other phytochemicals for healthy living. It is well established that the consumption of these products tends to reduce the risks of major health hazards facing modern-day society.

Although much work has been done on temperate fruits such as pip fruits, berry fruits, and stone fruits, very little is known about the tropical fruits in terms of their physiology, composition, and biochemistry. Such studies may lead to greater availability, acceptance, and commercialization of tropical and subtropical fruits.

REFERENCES

Abe, K., and Watada, A.E. (1991). Ethylene absorbent to maintain quality of lightly processed fruits and vegetables. *Journal of Food Science*, **6**,1589–1592.

Adams, J.B. (1973). Thermal degradation of anthocyanins with particular reference to the 3-glycosides of cyanidin in acidified aqueous solutions at 100°C. *Journal of the Science of Food and Agriculture*, **24**, 747–762.

Anonymous. (1996). These packages can change permeability. *Fresh Cut*, July, 14–16.

Bachmann, M.R., and Farah, Z. (1982). Occurrence of bitter taste in mixtures of milk proteins and kiwi fruit (*Actinidia chinensis*). *Lebensmittel-Wissenschaft Technologie*, **15**, 157–158.

Baker, R.A. (1980). The role of pectin in citrus quality. In *Citrus Nutrition and Quality*, eds. S.N. Nagy and J.A. Attaway. ACS Symposium Series 143. American Chemical Society, Washington, DC, pp. 109–128.

Baker, R.A., and Hagenmaier, R.D. (1997). Reduction of fluid loss from grapefruit segments with wax microemulsion coatings. *Journal of Food Science*, **62**, 789–792.

Baldwin, E.A., Nisperos-Carriedo, M.O., and Baker, R.A. (1995). Use of edible coatings to preserve quality of lightly (and slightly) processed products. *Critical Reviews in Food Science and Nutrition*, **35**, 509–524.

Baldwin, E.A., Nisperos, M.O., Chen, X., and Hagenmaier, R.D. (1996). Improving storage life of cut apple and potato with edible coating. *Postharvest Biology and Technology*, **9**, 151–163.

Bar-Peled, M., Lewinshon, E., Fluhr, R., and Gressel, J. (1991). DP-rhamnose:flavanone-7-o-glucoside-2-o-rhamnosyltransferase: purification and characterization of an enzyme catalyzing the production of bitter compounds in citrus. *Journal of Biological Chemistry*, **266**, 20953–20959.

Brackett, R.E. (1987). Microbiological consequences of minimally processed fruits and vegetables. *Journal of Food Quality*, **10**, 195–206.

Brecht, J.K., Sabaa-Srur, A.U.O., Sargent, S.A., and Bender, R.J. (1993). Hypochlorite inhibition of enzymic browning of cut vegetables and fruit. *Acta Horticulturae*, **343**, 341–344.

Breidt, F., and Fleming, H.P. (1997). Using lactic acid bacteria to improve the safety of minimally processed fruits and vegetables. *Food Technology*, **51**(9), 44–51.

Cash, J.N., Sistrunk, W.A., Stotte, C.A. (1976). Characteristics of concord grape polyphenoloxidase involved in juice colour loss. *Journal of Food Science*, **41**, 1398–1402.

Chandler, B.V., and Clegg, K.M. (1970). Pink discolouration in canned pears: role of tin in pigment formation. *Journal of the Science of Food and Agriculture*, **21**, 315–319.

Chang, C.M., and Wu, L.S. (1985). Debittering of Japanese apricot (*Prunus mume*) juice with kernel extract. *Journal of the Chinese Agricultural and Chemical Society (Taiwan)*, **23**(3,4), 282–287.

Clydesdale, F.M., Main, J.H., Francis, F.J., and Damson, R.A. Jr. (1978). Concord grape pigments as colorant for beverages and gelatin desserts. *Journal of Food Science*, **43**, 1687–1692, 1697.

Daravingas, G., and Cain, R.F. (1968). Thermal degradation of black raspberry anthocyanin pigments in model systems. *Journal of Food Science*, **33**, 138–142.

Day, B.P.E. (1994). High oxygen modified atmosphere packaging for fresh prepared produce. *Postharvest New Information*, **7**(3), 31–34.

Diaz, C., and Hotchkiss, J.H. (1996). Comparative growth of *Escherichia coli* 0157:H7, spoilage organisms and shelf-life of shredded iceberg lettuce stored under modified atmospheres. *Journal of the Science of Food and Agriculture*, **70**, 433–438.

Enzyme Nomenclature. (1984). Nomenclature Committee of the International Union of Biochemistry. Academic Press, Orlando, FL.

Eskin, N.A.M., Henderson, H.M., and Townsend, R.J. (1971). Browning reactions in foods. In *Biochemistry of Foods*, Ch. 3. Academic Press, New York, pp. 69–108.

Fauquembergue, P.C., and Grassin, C.M. (1997). Use of pectinesterase in the treatment of fruit and vegetables. European patent # EP0624 062 granted on 23/07/1997, publication date 17 November 1994.

Forsyth, W.G.C., and Quesnel, V.C. (1957). Cacao polyphenolic substances, 4: the anthocyanin pigments. *Biochemical Journal*, **65**, 177–179.

Goodman, L.P., and Markakis, P. (1965). Sulphur dioxide inhibition of anthocyanin degradation by phenolase. *Journal of Food Science*, **30**, 135–137.

Gross, J. (1991). *Pigments in Vegetables: Chlorophylls and Carotenoids*. Van Nostrand Reinhold, New York, p. 351.

Hasegawa, S., and Maier, V.P. (1990). Biochemistry of limonoid citrus juice bitter principles and biochemical debittering processes. In *Bitterness in Food and Beverages: Developments in Food Science 25*, ed. R.L. Rouseff. Elsevier Science Publishing, Amsterdam, pp. 293–308.

Hrazdina, G. (1971). Reactions of the anthocyanidin-3,5-diglucosides: formation of 3,5-di-(o-beta-D-glucosyl)-7-hydroxy coumarin. *Phytochemistry*, **10**, 1125–1130.

Hrazdina, G., and Franzese, A.J. (1974). Structure and properties of the acylated anthocyanins from *Vitis* species. *Phytochemistry*, **13**, 225–229.

Huxsoll, C.C., and Bolin, H.R. (1989). Processing and distribution alternatives for minimally processed fruits and vegetables. *Food Technology*, **43** (February), 124–128.

Hwang, L.S., and Cheng, Y.C. (1986). Pink discoloration in canned lychees. In *Role of Chemistry in the Quality of Processed Food*, eds. O.R. Fennema, W.H. Chang, and C.Y. Li. Food & Nutrition Press Inc., Westport, CT.

Jackman, R.L., and Smith, J.L. (1992). Anthocyanins and betalains. In *Natural Food Colorants*, eds. G.A.F. Hendry and J.D. Houghton. Blackie and Sons Ltd., Glasgow, Scotland, pp. 183–241.

Joslyn, M.A., and Pilnik, W. (1961). Enzymes and enzyme activity. In *The Orange: Its Biochemistry and Physiology*, ed. W.B. Sinclair. University of California, Davis, CA, pp. 373–435.

Jurd, L. (1964). Reactions involved in sulphite bleaching on anthocyanins. *Journal of Food Science*, **29**, 16–19.

Kader, A.A., Morris, L.L., Stevens, M.A., and Albright-Holton, M. (1978). Composition and flavor quality of fresh market tomatoes as influenced by some postharvest handling procedures. *Journal of the American Society for Horticultural Science*, **103**, 6–13.

Kader, A.A. (1986). Biochemical and physiological basis for effects of controlled and modified atmospheres on fruits and vegetables. *Food Technology*, **40**, 99–104.

Kader, A.A., and Barrett, D.M. (1996). Classification, composition of fruits and postharvest maintenance of quality. In *Processing Fruits: Science and Technology*. Part 1. *Biology, Principles, and Applications*. eds. L.P. Somogyi, L.P. Ramaswamy, and Y.H. Hui. Technomic Publishing Co. Inc., Lancaster, PA, pp. 1–24.

Keith, E.S., and Powers, J.J. (1965). Polarographic measurement and thermal decomposition of anthocyanin compounds. *Journal of Agricultural and Food Chemistry*, **13**, 577–579.

Khachik, F., Beecher, G.R., and Whittaker, N.F. (1986). Separation identification and quantitation of the major carotenoid and chlorophyll constitution in extracts of several green vegetables by liquid chromatography. *Journal of Agricultural and Food Chemistry*, **34**, 603–616.

King, A.D., and Bolin, H.R. (1989). Physiological and microbiological storage stability of minimally processed fruits and vegetables. *Food Technology*, **43**, 132–139.

Luh, B.S., Leonard, S.J., and Patel, D.S. (1960). Pink discolouration in Bartlett pears. *Food Technology*, **14**, 53–56.

Manlan, M., Matthews, R.F., Rouseff, R.L., Littell, R.C., Marshall, M.R., Moye, H.A., and Teixeira, A.A. (1990). Evaluation of the properties of polystyrene divinlbenzene adsorbants for debittering grapefruit juice. *Journal of Food Science*, **55**, 440–445, 449.

Markakis, P., Livingston, G.E., and Fellers, C.R. (1957). Quantitative aspects of strawberry-pigment degradation. *Food Research*, **22**, 117–129.

Mathew, A.G., and Parpia, H.A.B. (1970). Polyphenols of cashew kernel testa. *Journal of Food Science*, **35**, 140–142.

McDonald, R.E., McCollum, T.G., and Baldwin, E.A. (1996). Prestorage heat treatments influence free sterols and flavor volatiles of tomatoes stored at chilling temperature. *Journal of the American Society for Horticultural Science*, **121**, 531–536.

Nebesky, E.A., Esselsen, W.B., Jr., McConnell, J.E.W., and Fellers, C.R. (1949). *Food Research*, **14**, 261–274.

Nguyen-the, C., and Carline, F. (1994). The microbiology of minimally-processed fresh fruits and vegetables. *CRC Critical Reviews in Food Science and Nutrition*, **34**, 371–401.

Osuga, D., and Whitaker, J. R. (1994). Mechanism of some reducing compounds that inactivate polyphenol oxydase. In *Enzymatic Browning and Its Prevention*, eds. C.Y. Lee and J.R. Whitaker. ACS Symposium Series 600, American Chemical Society, Washington, DC, pp. 210–222.

Parkin, K.L. (1987). A new technique for the long-term study of the physiology of plant fruit tissue slices. *Physiologia Plantarum*, **69**, 472–476.

Peng, C.Y., and Markakis, P. (1963). Effect of phenolase on anthocyanins. *Nature*, **199**, 597–598.

Perera, C.O., and Venning, J. (1988). New Zealand patent #223971.

Philosoph-Hadas, S., Meir, S., and Aharoni, N. (1991). Effect of wounding on ethylene biosynthesis and senescence of detached spinach leaves. *Physiologia Plantarum*, **83**, 341–246.

Pilnik, W., and Voragen, A.G.J. (1989). Effect of enzyme treatment on the quality of processed fruit and vegetables. In *Quality Factors in Fruit and Vegetables: Chemistry and Technology*, ed. J.J. Jen. ACS Symposium Series 405, American Chemical Society, Washington, DC, pp. 250–269.

Pilnik, W., and Voragen, A.G.J. (1993). Pectic enzymes in fruit juice manufacture. In *Enzymes in Food Processing*, 3rd ed., eds. T. Nagodawithana and G. Reed. Academic Press Inc., San Diego, CA, pp. 363–399.

Poei-Langston, M.S., and Wrolstad, R.E. (1981). Colour degradation in an acid-anthocyanin-flavanol model system strawberry. *Journal of Food Science*, **46**, 1218–1236.

Ponappa, T., Scheerens, J.C., and Miller, R. (1993). Vacuum infiltration of polyamines increases firmness of strawberry slices under various storage conditions. *Journal of Food Science*, **58**, 361–364.

Prevor, J. (1996). Overcoming fresh-cut obstacles. *Produce Business*, January, 6.

Puri, A. (1990). Removal of bitter compounds from citrus products by adsorption techniques. In *Bitterness in Food and Beverages, Developments in Food Science 25*. ed. R.L. Rouseff, Elsevier Science Publishing, Amsterdam, pp. 324–336.

Robertson, G.L. (1985). Changes in chlorophyll and pheophytin concentrations of kiwifruit during processing and storage. *Food Chemistry*, **17**, 25–32.

Rojas, A.M., and Gerschenson, L.N. (1997). Influence of system composition on ascorbic acid destruction at processing temperatures. *Journal of the Science of Food and Agriculture*, **74**(3), 369–378.

Rolle, R.S., and Chism, W. (1987). Physiological consequences of minimally processed fruits and vegetables. *Journal of Food Quality*, **10**, 157–177.

Rosen, J.C., and Kader, A.A. (1989). Postharvest physiology and quality maintenance of sliced pear and strawberry fruits. *Journal of Food Science*, **54**, 656–659.

Sapers, G.M. (1993). Browning of foods: control by sulfites, antioxidants and other means. *Food Technology*, **47**, 75–84.

Shaw, P.E., and Buslig, B.S. (1986). Selective removal of bitter compounds from grapefruit juice and from aqueous solution with cyclodextrin polymers and with amberlite XAD-4. *Journal of Agricultural and Food Chemistry*, **34**, 837–840.

Shewfelt, R.L. (1987). Quality of minimally processed fruits and vegetables. *Journal of Food Quality*, **10**, 143–156.
Shi, J.X., and Luh, B.S. (1999). Fruit products. In *Asian Foods: Science & Technology*, eds. C.Y.W. Ang, K. Liu, and Y.W. Huang. Technomic Publishing Co. Inc., Lancaster, PA, pp. 275–316.
Siedow, J.N. (1991). Plant lipoxygenase: structure and function. *Annual Review of Plant Physiology*, **42**, 145–188.
Siegel, A., Markakis, P., and Bedford, C.L. (1971). Stabilization of anthocyanins in frozen tart cherries by blanching. *Journal of Food Science*, **36**, 962–963.
Sloan, A.E. (1995). Fresh cut gets fresh. *Food Technology*, **49**(5), 38–40.
Starr, M.S., and Francis, F.J. (1968). Oxygen and ascorbic acid affect on the relative stability of four anthocyanin pigments in cranberry juice. *Food Technology*, **22**, 1293–1295.
Sweeny, J.G., Wilkinson, M.M., and Iacobucci, G.A. (1981). Effect of flavonoid sulfonates on the photobleaching of anthocyanins in acid solution food colorants. *Journal of Agricultural and Food Chemistry*, **29**(3), 563–567.
Tien, Y.Y., and Fang, T.T. (1991). Extraction, fractionation and identification of bitter substances in Mei fruit (*Prunus mume* Sieb. Et Zucc.). *Journal of the Chinese Agricultural and Chemical Society (Taiwan)*, **37**(2), 38–49.
Vankerschaver, K., Willocx, F., Smout, C., Hendrickx, M., and Tobback, P. (1996). The influence of temperature and gas mixtures on the growth of the intrinsic micro-organisms on cut endive: predictive versus actual growth. *Food Microbiology*, **13**, 427–440.
Venning, J.A., Burns, D.J.W., Hoskin, K.M., Nguyen, T., and Stec, M.G.H. (1989). Factors influencing the stability of frozen kiwifruit pulp. *Journal of Food Science*, **54**, 396–400, 404.
Versteeg, C., Rombouts, F.M., and Pilnik, W. (1978). Purification and some characteristics of two pectinesterase isozymes from orange. *Lebensmittel-Wissenschaft Technologie*, **11**, 264–274.
Von Elbe, J.H., and Schwartz, S.J. (1996). Colorants. In *Food Chemistry*, 3rd ed., ed. O.R. Fennema. Marcel Dekker Inc., New York, pp. 651–722.
Weller, A., Sims, C.A., Matthews, R.F., Bates, R.P., and Brecht, J.K. (1997). Browning susceptibility and changes in composition during storage of carambola slices. *Journal of Food Science*, **62**, 256–260.
Watada, A.E., Ko, N.P., and Minott, D.A. (1996). Factors affecting quality of fresh-cut horticultural products. *Postharvest Biology and Technology*, **9**, 115–125.
Whitaker, J.R. (1996). Enzymes. In *Food Chemistry*, 3rd ed., ed. O.R. Fennema. Marcel Dekker Inc., New York, pp. 431–530.
Wilska-Jeszka, J. (1997). Food colorants. In *Chemical and Functional Properties of Food Components*, ed. Z.E. Sikorski. Technomic Publishing Co., Lancaster, PA, pp. 191–210.
Wong, D.W.S., Camirand, W.M., and Pavlath, A.E. (1994). Development of edible coatings for minimally processed fruits and vegetables. In *Edible Coatings and Films to Improve Food Quality*, eds. J.M. Krochta, E.A. Baldwin, and M.O. Nisperos-Carriedo. Technomic Publishing Co., Lancaster, PA.
Wong, M., and Stanton, D.W. (1989). Nonenzymatic browning in kiwifruit juice concentrate systems during storage. *Journal of Food Science*, **54**(3), 669–673.
Wrolstad, R.E., Skrede, G., Lea, P., and Enersen, G. (1990). Influence of sugar on anthocyanin pigment stability in frozen strawberries. *Journal of Food Science*, **55**, 1064–1065.
Wu, J.S.-B., Sheu, M.J, and Fang, T.T. (1993). Oriental fruit juices: carambola, Japanese apricot (mei), lychee. In *Fruit Juice Processing Technology*, eds. S. Nagy, C.S. Chen, and P.E. Shaw. Agscience Inc., Auburndale, FL, pp. 595–619.
Yamasaki, M., Yasuri, T., and Arima, K. (1964). Pectic enzymes in the clarification of apple juice. Part 1. Study on the clarification reaction in a simplified mode. *Agricultural and Biological Chemistry*, **28**, 779–787.
Zagory, D. (1997). Modified atmosphere packaging. In *Packaging Technology*, 2d ed., eds. A.L. Brody and K.S. Marsh. John Wiley & Sons, New York, NY, pp. 650–656.
Zagory, D., and Kader, A.A. (1988). Modified atmosphere packaging of fresh produce. *Food Technology*, **42**(9), 70–77.

Chapter 3

Fruit and Human Nutrition

Phillip C. Fourie

3.1 INTRODUCTION

Fruits contain a wide range of different compounds and, therefore, show considerable variations in composition and structure. Each individual fruit is composed of living tissues that are metabolically active, and it is constantly changing in composition. The rate and extent of such changes depend on the physiological role and stage of maturity of the fruit concerned (Salunkhe, Bolin, and Reddy, 1991). The nutritive value of fruit also depends on its composition. However, although fruit contributes significantly to the daily nutrient needs of the individual, the composition of fruit is such that fruit per se is not recommended as a sole source of nutrition but forms a significant part of a balanced diet. Also, it can be used advantageously to supplement deficiencies of other foods. In this chapter, the composition and importance of fruit in the diet, and the influence of processing on nutritional value of temperate fruits, citrus fruits, and tomatoes, are discussed. Tomatoes are included because they are botanically classified as a fruit, although they are marketed and consumed as a vegetable (Wills, Lim, and Greenfield, 1984).

3.2 COMPOSITION OF FRUITS

Fruits are composed of both macronutrients (such as water, protein, carbohydrate, and fat) and micronutrients (such as vitamins and minerals). These components are essential nutrients that are needed by the human body for growth, development, and maintenance of living tissue. The amount of each of these nutrients required by the body depends on factors such as age, mass, gender, health, and physical activity. The requirements listed in nutritional tables merely indicate approximate daily requirements for healthy individuals.

Water is the most abundant nutrient in fruit (more than 80%), ranging from 82% in grapes to 90% in strawberries (Table 3–1) and 93% in tomatoes (Table 3–2). The low value reported for figs (24%) is because in this specific case the product was semidried (Table 3–1). In normal circumstances, values up to 86% are reported for figs (Wills, Lim, and Greenfield, 1987). However, the maximum water content varies between individual fruit of the same kind because of structural differences. It may also be affected by cultural conditions, which influence structural differentiation (Salunkhe et al., 1991).

Proteins usually contribute less than 1% to the fresh mass of fruit. Proteins are composed of amino acids, nine of which are classified as essential for the human diet, namely, valine, threonine, tryptophan, isoleucine, methionine, leucine, lysine, phenylalanine, and histidine. These essential amino acids are required by humans because they cannot be synthesized by the human body in sufficient quantities and therefore must be consumed regularly. A protein containing all nine essential amino acids is known as a complete protein (Potter, 1986). However, it is not sufficient for a complete protein merely to contain the essential amino acids; they must also be fully available to the body in the correct ratios. If the total amount of nitrogen supplied by protein in the diet is adequate, the body can synthesize the non-essential amino acids in sufficient quantities. The protein content of fresh fruit is calculated by multiplying the total nitrogen content by a factor of 6.25. This figure is based on the fact that protein contains about 16% nitrogen and that all nitrogen, not considering other simple nitrogenous substances that may be present in the uncombined form, is present as protein. Simple nitrogenous substances such as asparagine and glutamine and their related acids, aspartic and glutamic acids, are abundant in citrus fruit, tomatoes, and strawberries. The presence of asparagine is also very high in apples and pears; oranges are rich in proline (Salunkhe et al., 1991).

Carbohydrate consists of polysaccharides such as starch, cellulose, hemicellulose, and pectic material and also of disaccharides and monosaccharides, such as the sugars sucrose, fructose, and glucose. The amount of each of these constituents can change drastically during ripening of fruit. Sugars are usually abundant when the fruit reaches its full maturity. In fruit containing starch, all of the starch is then fully hydrolyzed. These sugars are mostly glucose and fructose, but in fruit such as peaches, nectarines, and apricots the main sugar is sucrose (Wills, Scriven, and Greenfield, 1983). Apples and pears are rich in fructose. Traces of other mono- and disaccharide sugars, such as xylose, arabinose, mannose, galactose, and maltose, may also be present in fruit in small amounts. Sorbitol, a polyol related in structure to the sugars, which is well known for its laxative effect, is present in relatively high concentrations in pears and plums, while no sorbitol has been reported in strawberries (Wrolstad and Shallenberger, 1981). Cellulose, hemicellulose, and pectic material are the cell wall components of fruit. Pectin can be used commercially for the manufacture of jams and jellies and can be extracted from the white, spongy layer of citrus fruit skins, especially grapefruit and lemons, as well as from apples (see Chapter 10). The total carbohydrate value reported in the tables of this chapter includes starch and total sugars. These values for

Table 3-1 Description and Some Compositional Data on Temperate Fruits per 100 Grams

Fruit	Description	Edible Portion	Water (g)	Protein (g)	Carbo-hydrate (g)	Energy Value (kcal)	Energy Value (kJ)	Total Nitrogen (g)	Starch (g)	Total Sugars (g)	Dietary Fiber Southgate Method (g)	Dietary Fiber Englyst Method (g)
Apples	Flesh and skin	1.00	84.5	0.4	11.8	47	199	0.06	Tr*	11.8	(2.0)	1.8
Apricots	Flesh and skin	1.00	87.2	0.9	7.2	31	134	0.14	0	7.2	1.9	1.7
Cherries	Flesh and skin	1.00	82.8	0.9	11.5	48	203	0.14	0	11.5	1.5	0.9
Figs	Semidried	1.00	23.6	3.3	48.6	209	889	0.52	0	48.6	11.4	6.9
Grapes	White, black, and seedless	1.00	81.8	0.4	15.4	60	257	0.06	0	15.4	0.8	0.7
Nectarines	Flesh and skin	1.00	88.9	1.4	9.0	40	171	0.20	0	8.0	2.2	1.2
Peaches	Flesh and skin	1.00	88.9	1.0	7.6	33	142	0.16	0	7.6	2.3	1.5
Pears	Flesh and skin	1.00	83.8	0.3	10.0	40	169	0.05	0	10.0	N*	2.2
Plums	Flesh and skin	1.00	83.9	0.6	8.8	36	155	0.09	0	8.8	2.3	1.6
Strawberries	Flesh and pips	0.95	89.5	0.8	6.0	27	113	0.13	0	6.0	2.0	1.1

*Tr, trace; N, not determined.

Source: Reprinted from Holland et al., *The Composition of Foods*, 5th Edition, pp. 280–311, © 1992, The Royal Society of Chemistry. Crown copyright material is reproduced with the permission of the Controller of Her Majesty's Stationery Office.

Table 3–2 Description and Some Compositional Data on Citrus Fruit and Tomatoes per 100 Grams

Fruit	Description	Edible Portion	Water (g)	Protein (g)	Carbohydrate (g)	Energy Value (kcal)	Energy Value (kJ)	Total Nitrogen (g)	Starch (g)	Total Sugars (g)	Dietary Fiber Southgate Method (g)	Dietary Fiber Englyst Method (g)
Grapefruit	Flesh only	1.00	89.0	0.8	6.8	30	126	0.13	0	6.8	(1.6)	1.3
Lemons	Whole, without pips	0.99	86.3	1.0	3.2	19	79	0.16	0	3.2	4.7	N*
Oranges	Flesh only	1.00	86.1	1.1	8.5	37	158	0.18	0	8.5	1.8	1.7
Tangerines	Flesh only	1.00	86.7	0.9	8.0	35	147	0.14	0	8.0	1.7	1.3
Tomatoes	Flesh, skin, and seeds	1.00	93.1	0.7	3.1	17	73	0.11	Tr*	3.1	1.3	1.0

*Tr, trace; N, not determined.

Source: Reprinted from Holland et al., The Composition of Foods, 5th Edition, pp. 280–311, © 1992, The Royal Society of Chemistry. Crown copyright material is reproduced with the permission of the Controller of Her Majesty's Stationery Office.

total carbohydrate vary from 3% in lemons and tomatoes (Table 3–2) to about 15% in grapes (Table 3–1).

Dietary fiber, by definition, comprises the structural materials of plant cells that are resistant to the digestive enzymes of the stomach. It includes the structural polysaccharides of the cell wall and lignin (Ross, English, and Perlmutter, 1985). The dietary fiber content of fresh fruit generally falls within the range 0.7–4.7% (Tables 3–1 and 3–2). Values quoted for dietary fiber in this chapter have been determined by the methods of Southgate and Englyst and also that of Cummings. The Southgate values are generally higher than the noncellulosic polysaccharide values of Englyst and Cummings because they include lignin. High dietary fiber values are usually associated with fruit of a lower water content or with fruit containing seeds in the edible portion. The fiber content can be reduced by removing the peel of the fruit. In apples, the reduction is approximately 11% and in pears it is approximately 34% (Jones et al., 1990).

The total fat content of fruit is generally below 1% and varies with the commodity (Tables 3–3 and 3–4). Apart from the fact that fats (oils) serve as sources of energy, the body also requires small quantities of unsaturated fatty acids. At least one of these, linoleic acid, is an essential fatty acid. However, fruit is not a good source of fat, and of the temperate fruits, lemons and tomatoes have the highest fat content (both 0.3%) (Tables 3–3 and 3–4). The most important sources of energy in food are carbohydrates and fats (oils). In fruit, the most significant contribution to energy value is made by

Table 3–3 The Total Fat, Fatty Acids, and Cholesterol Composition of Temperate Fruits per 100 Grams

Fruit	Description	Fat (g)	Saturated (g)	Monounsaturated (g)	Polyunsaturated (g)	Cholesterol (mg)
Apples	Flesh and skin	0.1	Tr*	Tr	0.1	0
Apricots	Flesh and skin	0.1	Tr	Tr	Tr	0
Cherries	Flesh and skin	0.1	Tr	Tr	Tr	0
Figs	Semidried	1.5	N*	N	N	0
Grapes	White, black, and seedless	0.1	Tr	Tr	Tr	0
Nectarines	Flesh and skin	0.1	Tr	Tr	Tr	0
Peaches	Flesh and skin	0.1	Tr	Tr	Tr	0
Pears	Flesh and skin	0.1	Tr	Tr	Tr	0
Plums	Flesh and skin	0.1	Tr	Tr	Tr	0
Strawberries	Flesh and pips	0.1	Tr	Tr	Tr	0

*Tr, trace; N, not determined.

Source: Reprinted from Holland et al., *The Composition of Foods,* 5th Edition, pp. 280–311, © 1992, The Royal Society of Chemistry. Crown copyright material is reproduced with the permission of the Controller of Her Majesty's Stationery Office.

Table 3–4 The Total Fat, Fatty Acids, and Cholesterol Composition of Citrus Fruits and Tomatoes per 100 Grams

Fruit	Description	Fat (g)	Saturated (g)	Monounsaturated (g)	Polyunsaturated (g)	Cholesterol (mg)
Grapefruit	Flesh only	0.1	Tr*	Tr	Tr	0
Lemons	Whole, without pips	0.3	0.1	Tr	0.1	0
Oranges	Flesh only	0.1	Tr	Tr	Tr	0
Tangerines	Flesh only	0.1	Tr	Tr	Tr	0
Tomatoes	Flesh, skin, and seeds	0.3	0.1	0.1	0.2	0

*Tr, trace.

Source: Reprinted from Holland et al., *The Composition of Foods*, 5th Edition, pp. 280–311, © 1992, The Royal Society of Chemistry. Crown copyright material is reproduced with the permission of the Controller of Her Majesty's Stationery Office.

---

carbohydrates. Proteins and organic acids can also serve as a source of energy, but the body will preferably use carbohydrates and fats. The energy value is measured in heat units called calories and is expressed as kilocalories (kcal) or kilojoules (kJ) (kilocalories × 4.2 = kilojoules). In the temperate fruits, the energy values for tomatoes (73 kJ) and lemons (79 kJ) are the lowest (Table 3–2), while cherries (203 kJ) and grapes (257 kJ) are the best sources of energy (Table 3–1).

Minerals and vitamins cannot be synthesized by the human body and must be provided by the diet, although only small amounts are required daily; therefore, they are called micronutrients. Fruit contains a variety of mineral elements, and about 14 are considered to be essential nutritional constituents: calcium, sodium, zinc, iodine, copper, phosphorus, potassium, sulfur, fluoride, manganese, iron, magnesium, cobalt, and chloride. Several other elements occur in low concentrations in human tissues and fluids, but it has not yet been established that they are essential nutrients. Although fruits are not rich in minerals, potassium is the most abundant mineral available in fruit (Tables 3–5 and 3–6) and occurs mainly in combination with the various organic acids (Hugo, 1969). In fruit the pH of the tissue is controlled by the potassium/organic acid balance, and high concentrations of potassium can contribute to high blood pressure of humans. In strawberries and tomatoes, calcium, magnesium, phosphorus, and chloride are the only minerals occurring in amounts exceeding 10 mg/100 g of fresh fruit tissue, but the mineral content usually varies considerably within fruit from a specific producing area. Calcium is always present in the pectic material in the cell walls of the fruit, magnesium is present in the chlorophyll molecules, and phosphorus can play an important part in carbohydrate metabolism. In general, minerals contribute widely

Table 3-5 The Inorganic Constituents of Temperate Fruits per 100 Grams

Fruit	Description	Na (mg)	K (mg)	Ca (mg)	Mg (mg)	P (mg)	Fe (mg)	Cu (mg)	Zn (mg)	Cl (mg)	Mn (mg)	Se (µg)	I (µg)
Apples	Flesh and skin	3	120	4	5	11	0.1	0.02	0.1	Tr*	0.1	Tr	Tr
Apricots	Flesh and skin	2	270	15	11	20	0.5	0.06	0.1	3	0.1	(1)	N*
Cherries	Flesh and skin	1	210	13	10	21	0.2	0.07	0.1	Tr	0.1	(1)	Tr
Figs	Semidried	62	970	250	80	89	4.2	0.30	0.7	170	0.5	Tr	N
Grapes	White, black, and seedless	2	210	13	7	18	0.3	0.12	0.1	Tr	0.1	(1)	1
Nectarines	Flesh and skin	1	170	7	10	22	0.4	0.06	0.1	5	0.1	(1)	3
Peaches	Flesh and skin	1	160	7	9	22	0.4	0.06	0.1	Tr	0.1	(1)	3
Pears	Flesh and skin	3	150	11	7	13	0.2	0.06	0.1	1	Tr	Tr	1
Plums	Flesh and skin	2	240	13	8	23	0.4	0.10	0.1	Tr	0.1	Tr	Tr
Strawberries	Flesh and pips	6	160	16	10	24	0.4	0.07	0.1	18	0.3	Tr	9

*Tr, trace; N, not determined.

Source: Reprinted from Holland et al., *The Composition of Foods*, 5th Edition, pp. 280–311, © 1992, The Royal Society of Chemistry. Crown copyright material is reproduced with the permission of the Controller of Her Majesty's Stationery Office.

Table 3-6 The Inorganic Constituents of Citrus Fruits and Tomatoes per 100 Grams

Fruit	Description	Na (mg)	K (mg)	Ca (mg)	Mg (mg)	P (mg)	Fe (mg)	Cu (mg)	Zn (mg)	Cl (mg)	Mn (mg)	Se (µg)	I (µg)
Grapefruits	Flesh only	3	200	23	9	20	0.1	0.02	Tr*	3	Tr	(1)	N
Lemons	Whole, without pips	5	150	85	12	18	0.5	0.26	0.1	5	N*	(1)	N
Oranges	Flesh only	5	150	47	10	21	0.1	0.05	0.1	3	Tr	(1)	2
Tangerines	Flesh only	1	120	31	8	12	0.2	0.01	0.1	1	Tr	N	N
Tomatoes	Flesh, skin and seeds	9	250	7	7	24	0.5	0.01	0.1	55	0.1	Tr	2

*Tr, trace; N, not determined.

Source: Reprinted from Holland et al., *The Composition of Foods*, 5th Edition, pp. 280–311, © 1992, The Royal Society of Chemistry. Crown copyright material is reproduced with the permission of the Controller of Her Majesty's Stationery Office.

to the quality of fruit products. Calcium, for example, can influence the texture and storage life of fruit.

Vitamins are nutrients required for specific functions in the body. If the vitamins are not consumed in sufficient quantities, deficiency diseases develop. Considerable differences in vitamin content are reported among fruit species and varieties as well as within the same variety grown under different environmental conditions. Climate, soil, and fertilizer practices also affect the levels of vitamins in fruits. Vitamins are broadly classified as either fat soluble (A, D, E, and K) or water soluble. The water-soluble vitamins are thiamine (B_1), riboflavin (B_2), niacin, pyridoxine (B_6), folic acid, biotin, pantothenic acid, cobalamin (B_{12}), and ascorbic acid (C). Fruit is especially known as a source of ascorbic acid, although tropical fruits are a better source than the similar products grown in temperate regions. Strawberries and black currants are good sources of ascorbic acid; citrus fruits contain moderate amounts; while apples, pears, cherries, and plums contain very little (Tables 3–7 and 3–8). An important environmental factor controlling the level of ascorbic acid is sunlight. Generally, the greater the amount of sunlight during growth, the greater the ascorbic acid content (Salunkhe et al., 1991). Tomatoes with heavier shading have been shown to have lower ascorbic acid levels than those with more light exposure (Bradley, 1972). Fruit possesses an advantage over many vegetables because ascorbic acid is present in the acidic environment (where it is more stable) provided by fruit juices, compared with the more neutral environment in vegetables. A further advantage is that fruit is eaten raw and the possible loss of ascorbic acid through cooking is prevented (Bradley, 1972).

Vitamin D, which is also a fat-soluble vitamin, is absent in fruit, while vitamin E occurs only in small quantities in some fruits (Tables 3–7 and 3–8). Fruit is a moderate to poor source of the members of the vitamin B group. Plums (Table 3–7) and tomatoes (Table 3–8) are a good source of niacin, and strawberries (Table 3–7), oranges, and grapefruit (Table 3–8) have significant amounts of folate. Vitamin B_{12} is the only vitamin in the B group that is not present in fruit. The remainder of the nutrients are minor compositional components, but they also play an important role in the appearance, color, taste, and aroma of fruit.

Vitamin A is fat soluble and does not occur as such in fruit, although certain fruit carotenoids can be converted to vitamin A in the body. These pigments are referred to as provitamin A. Most of the provitamin A in fruit is in beta-carotene, with less amounts in alpha-carotene, gamma-carotene, and other carotenoid pigments (Bradley, 1972). Fruits, in general, are not good sources of carotene, but in temperate regions apricots and plums are moderate sources, while peaches, nectarines, and tangerines contain only small amounts. Of the more than 50 dietary carotenoids, lycopene, found primarily in tomatoes and tomato products, is the most prevalent in the diet in concentrations of 3.1 to 7.7 mg/100g of ripe fruit. It is found predominantly in the chromoplast of plant tissues (Nguyen and Schwartz, 1999). Lycopene represents 90% of the total carotenoid fraction in tomatoes, but does not yield vitamin A potential (Tables 3–7 and 3–8).

Organic acids play an important role in fruit taste through the sugar/acid ratio. Sugar provides sweetness and the organic acids sourness. The main organic acids present in fruit are citric and malic acids. Citrus fruit, strawberries, pears, and tomatoes are dominated by citric acid, while apples, plums, cherries, and apricots produce primarily

Table 3-7 The Vitamin Content of Temperate Fruits per 100 Grams

Fruit	Description	Retinol (µg)	Carotene (µg)	Vitamin D (µg)	Vitamin E (mg)	Thiamine (mg)	Riboflavin (mg)	Niacin (mg)	Tryptophan 60 (mg)	Vitamin B_6 (mg)	Vitamin B_{12} (µg)	Folate (µg)	Pantothenate (mg)	Biotin (µg)	Vitamin C (mg)
Apples	Flesh and skin	0	18	0	0.59	0.03	0.02	0.1	0.1	0.06	0	1	Tr*	1.2	6
Apricots	Flesh and skin	0	405	0	N*	0.04	0.05	0.5	0.1	0.08	0	5	0.24	N	6
Cherries	Flesh and skin	0	25	0	0.13	0.03	0.03	0.2	0.1	0.05	0	5	0.26	0.4	11
Figs	Semidried	0	(59)	0	N	0.07	0.09	0.7	0.4	0.24	0	8	0.47	N	1
Grapes	White, black, and seedless	0	17	0	Tr	0.05	0.01	0.2	Tr	0.10	0	2	0.50	0.3	3
Nectarines	Flesh and skin	0	58	0	N	0.02	0.04	0.6	0.3	0.03	0	Tr	0.16	(0.2)	37
Peaches	Flesh and skin	0	58	0	N	0.02	0.04	0.6	0.2	0.02	0	3	0.17	(0.2)	31
Pears	Flesh and skin	0	18	0	0.50	0.02	0.03	0.2	Tr	0.02	0	2	0.07	0.2	6
Plums	Flesh and skin	0	295	0	0.61	0.05	0.03	1.1	0.1	0.05	0	3	0.15	Tr	4
Strawberries	Flesh and pips	0	8	0	0.20	0.03	0.03	0.6	0.1	0.06	0	20	0.34	1.1	77

*Tr, trace; N, not determined.

Source: Reprinted from Holland et al., *The Composition of Foods*, 5th Edition, pp. 280–311, © 1992, The Royal Society of Chemistry. Crown copyright material is reproduced with the permission of the Controller of Her Majesty's Stationery Office.

Table 3-8 The Vitamin Content of Citrus Fruits and Tomatoes per 100 Grams

Fruit	Description	Retinol (µg)	Carotene (µg)	Vitamin D (µg)	Vitamin E (mg)	Thiamine (mg)	Riboflavin (mg)	Niacin (mg)	Tryptophan 60 (mg)	Vitamin B_6 (mg)	Vitamin B_{12} (µg)	Folate (µg)	Pantothenate (mg)	Biotin (µg)	Vitamin C (mg)
Grapefruit	Flesh only	0	17	0	(0.19)	0.05	0.02	0.3	0.1	0.03	0	26	0.28	(1.0)	36
Lemons	Whole, without pips	0	18	0	N*	0.05	0.04	0.2	0.1	0.11	0	N	0.23	0.5	58
Oranges	Flesh only	0	28	0	0.24	0.11	0.04	0.4	0.1	0.10	0	31	0.37	1.0	54
Tangerines	Flesh only	0	71	0	N	0.05	0.01	0.1	0.1	0.05	0	15	0.15	N	22
Tomatoes	Flesh, skin, and seeds	0	640	0	1.22	0.09	0.01	1.0	0.1	0.14	0	17	0.25	1.5	17

*N, not determined.

Source: Reprinted from Holland et al., *The Composition of Foods*, 5th Edition, pp. 280–311, © 1992, The Royal Society of Chemistry. Crown copyright material is reproduced with the permission of the Controller of Her Majesty's Stationery Office.

malic acid. In peaches, the two acids are in equal amounts. Grapes differ from other fruit in that tartaric acid is the main acid present. In all of these fruits, there is a considerable decrease in acidity during ripening of the fruit. Bitterness is not common in fruit, but certain flavonoids, such as naringin in grapefruit and limonene in citrus fruit, are intensely bitter. The aroma of fruit is a key factor in assessing quality as well as identity (Salunkhe et al., 1991). The major chemical compounds associated with aroma are esters of aliphatic alcohols and short-chain fatty acids. However, except for some fruits in which the major volatile compound is isoamyl acetate, there is still uncertainty as to the chemical substance responsible for the specific odor of most fruit. Chlorophylls, carotenoids, and anthocyanins are pigments that are responsible for the color of fruit. Chlorophyll provides the green color of, for example, apples; carotenoids, mainly β-carotene; and lycopene, the yellow to orange color of citrus fruit, apricots, and peaches and the red color of tomatoes. Anthocyanins provide the naturally red, blue, or purple color of apples, plums, grapes, and strawberries.

Enzymes are important in fruit because of the chemical changes that they initiate. Ficin is a proteolytic enzyme occurring only in figs. This enzyme is present in the latex of the fig and reacts with proteins of the human skin when the figs are handled. This reaction causes dermatitis, and some individuals are so adversely affected that they cannot ingest fresh figs (Macrae, Robinson, and Sadler, 1993). Phenoloxidases in apples, pears, grapes, strawberries, and figs are responsible for the discoloration of cut surfaces when exposed to air. Citrus fruit and tomatoes are rich in pectin esterase, and pears and tomatoes in polygalacturonase, both being pectolytic enzymes responsible for softening during ripening.

3.3 THE IMPORTANCE OF FRUIT IN THE HUMAN DIET

Nutrients frequently consumed in suboptimal concentrations by humans are protein, calcium, iron, vitamin A, thiamine (vitamin B_1), riboflavin (vitamin B_2), and ascorbic acid. Members of this group are often known as the critical nutrients. Fruit contains most of these critical nutrients, and sometimes some of these nutrients occur in higher concentrations than in other foods. Therefore, fruit plays an important role in balancing the human diet, mainly because the composition of fruit differs markedly from other foods of plant and animal origin (Hugo, 1969). The contribution of fruit to the protein requirement is slight, and it is not a good source of calcium, thiamine, and riboflavin. However, some fruit is particularly rich in iron, carotene, and ascorbic acid, and lack of fruit in the diet will lead to a deficit in the vitamin C and β-carotene content (Walter, 1994). Therefore, fruit with a low kilojoule content is frequently recommended by dietitians to be included in a kilojoule-restricted diet.

Research studies have shown that ascorbic acid of natural origin is superior to the synthetic product, due to the presence of certain flavonoid compounds in fruit that influence the blood circulation, increasing the permeability and the elasticity of the capillary vessels. This action is known as vitamin P activity. These flavonoid substances are not classified as vitamins because several compounds have this property and no severe deficiency diseases develop if these substances are not present in food. There are indications that these flavonoids have a protective action against infections of the respiratory system. Unfortunately, they are readily decomposed in the body so that it is impossible to maintain high enough concentrations in the blood.

There are considerable differences of opinion as to the minimum daily requirement of ascorbic acid. In some countries, a daily allowance of 20 mg is considered sufficient, but in other countries quantities of up to 70 mg/d are recommended. Portions of 250 g of strawberries, oranges, and grapefruit contain considerably more than the minimum daily requirements of ascorbic acid, while most of the other fruit can provide more than half the daily requirement. Vitamin C is also quite often added to fruit juices in addition to the natural level of this vitamin (Walter, 1994). Deficiencies in vitamin C tend to occur in temperate regions during the winter months and in tropical regions during periods of drought (Salunkhe et al., 1991).

Portions of 250 g of apricots should provide sufficient vitamin A to satisfy the daily requirements. Peaches contain approximately 50% of the daily vitamin A requirements, and 250-g portions of plums, strawberries, grapes, and oranges contain more than 10% of the daily requirements. The bioavailability of certain other vitamins can differ greatly according to the foodstuff. The fat-soluble vitamins such as vitamin A are absorbed in significant amounts only if the food consumed also contains lipids (Walter, 1994).

In the mature fruit stage, the organic acids and sugar occur in fruit in characteristic ratios. In most cases, the pH is lower than 4.2. Ripe fruit is higher in sugar content, but processors prefer unripe or partially ripened fruit because the juice is extracted more readily (Akhavan and Wrolstad, 1980). However, when fresh, matured fruit is eaten, a mixture of natural sugars and not a single sugar is consumed (Koeppen, 1974). As a result, fruit and fruit beverages are particularly appetizing. Although fruits contain fructose, glucose, and sucrose in various combinations, those containing high concentrations of free sugars should be restricted in the diet of diabetic patients (Nahar, Rahman, and Mosihuzzaman, 1990). The response of blood glucose to fructose is low, while the response to sucrose is intermediate between that to fructose and that to glucose (Miller, Colagiuri, and Brand, 1986). However, oral fructose can be insulinogenic in humans when blood glucose levels are elevated (Reiser et al., 1987). Therefore, dietitians need more detailed information regarding the content of individual sugars in fruits when planning special diets for diabetics.

The minerals present in fruit are basic by nature, with the result that the residues of metabolism are alkaline. In comparison with other foods, fruit has a relatively high potassium to sodium ratio. Although some fruits, such as apricots, pears, peaches, plums, and strawberries, contain oxalic acid, the concentration of this acid is not sufficient to reduce the absorption of calcium from the fruit or from other foods eaten together with the fruit. Acids occurring in fruit are almost completely metabolized in the body. From experiments carried out with animals, it has been established that acids from fruit favor the development of dental caries. In rats fed different types of fruit juices, a relationship between the acidity of the juice and the development of caries was established. In humans this phenomenon has not yet been studied thoroughly. It has, however, been established in experiments carried out in different countries that children eating apples regularly were much less prone to dental caries and their gums were healthier than those children who did not eat apples regularly. The texture of apples is such that it has a cleansing action on the teeth.

The dietary fiber hypothesis has tended to ascribe many health conditions characteristic of modern industrialized societies to too little dietary fiber and has made claims for the therapeutic value of fiber-enriched diets in curing these conditions. However, in many cases the scientific evidence to support a claim is inconclusive, and research is continuing. Failure to distinguish between the different physiological effects to different fiber components may have obscured real relationships between specific disease conditions (Gurr and Asp, 1996). The presence of about 10% (on a dry mass basis) of indigestible material or crude fiber in food is required for normal digestion. In human nutrition, this fact is often overlooked. Fruit contains 0.3–5.7% (2.6–33% on a dry mass basis) indigestible material, consisting of cellulose and related substances. Inclusion of fruit in the diet is therefore essential for supplying dietary fiber. It provides an indigestible matrix that stimulates the activity of the intestines and helps to keep the intestinal muscles in working order. Chronic constipation is probably largely alleviated by an increased intake of fresh fruits. However, much of the fibrous materials of stone and pome fruits are lost when these fruits are peeled. Whatever the mechanism, dietary fiber does play a part in a reduction of cholesterol levels in blood. However, all sources of dietary fiber do not have the same effects on blood serum cholesterol levels (Salunkhe et al., 1991). In recent years, lack of fiber in the diet has been shown to cause colon cancer in humans (Ross et al., 1985). The influence of dietary fiber on energy balance is complex. In some people, high-fiber diets may reduce appetite, increase satiety, and reduce overall energy intake. This is not a universal finding, however, and generalizations about the effectiveness of fiber in weight control are unwise (Gurr and Asp, 1996).

Fruit and fruit beverages can also be used advantageously as snacks (especially with children) instead of sweets and synthetic beverages. Fruit eaten as snacks should not adversely affect food intake during regular meals, because fruit is digested relatively quickly. Because of their ease of digestion, high vitamin content, and appetizing effects, fruit juices are particularly useful to the indisposed. Obviously, the nature of the illness should be taken into account. In the treatment of obesity, fruit and fruit juice can be included advantageously in the diet in a regimen where only fruit and fruit juice are consumed on certain days of the week. On such days, the kilojoule intake is low, while the high potassium-to-sodium ratio assists the excretion of excessive salt and water. The consumption of fruit stimulates the excretion of water, with the result that the functions of the heart are eased. For this reason, fruit and fruit juices are indispensable in the treatment of dropsy, chronic heart diseases, and poor blood circulation. Fruit and fruit juices are useful in the diet of patients suffering from kidney complaints, gastritis, gall bladder and liver complaints, and high fever.

Many therapeutic drugs in use in modern medicine originated as plant extracts. Therefore, it is not surprising that certain fruit components exert pharmacological or therapeutic effects. Limonin and nomilin (and other limonoids) are present in citrus fruit such as oranges, lemons, and grapefruit. These compounds are believed to have a role in inhibiting the development of certain forms of cancer. Prunes contain hydroxyphenylisatin derivatives, which stimulate colonic smooth muscles, explaining their traditional use as a laxative (Macrae et al., 1993). Sorbitol in pears and plums also serves as a laxative.

Several components with antioxidant activity naturally occur in fruit. These components include ascorbic acid, tocopherols (vitamin E), carotenoids, and other flavonoid components. Antioxidants such as bioflavonoids and chalcones occur in citrus fruits and phenolic acid in red grape varieties. Although these substances have no known nutritional function, they may be important to human health because of their antioxidant potency. Ascorbic acid in strawberries, tomatoes, and citrus fruits; carotenoids such as β-carotene in apricots, peaches, and plums; lycopene in tomatoes; and β-cryptoxanthin in citrus fruits are all good sources of antioxidant vitamins (Langseth, 1995). The antioxidant activity of tocopherols is believed to be the main source of their vitamin activity (Macrae et al., 1993). It is the major antioxidant in all cellular membranes, and it protects polyunsaturated fatty acids against oxidation (Langseth, 1995). Vitamin C has specific physiological functions that are not dependent on its antioxidant properties (Macrae et al., 1993). However, it is believed to be the most important antioxidant in extracellular fluids, and it has many known intracellular activities as well. Some carotenoids can act as precursors of vitamin A; others cannot. However, this property is unrelated to their antioxidant activity (Langseth, 1995). β-carotene has antioxidant properties that help neutralize free radicals, which are reactive and highly energized molecules formed through normal biochemical reactions or through external factors such as air pollution or cigarette smoke. β-carotene is mainly active as a quencher of singlet oxygen, which can induce precancerous changes in cells. In addition to their individual effects, antioxidants interact in synergistic ways and have sparing effects in which one antioxidant protects another against oxidative destruction. For example, ascorbic acid reinforces the antioxidant effect of vitamin E by regenerating the active form of the vitamin after it has reacted with a free radical. This beneficial interaction has been demonstrated in biological fluids as well as in model systems. Vitamin E can also protect the β-carotene molecule from oxidation and thus may have a sparing effect on this antioxidant. Vitamin E and the mineral selenium appear to act synergistically. High intakes of fruit are associated with a protective effect against cancer, whereas low intakes are associated with increased risks of cancers at most body sites other than the prostate. Fruit is the principal source of ascorbic acid and carotenoids, and it is believed that these antioxidants are major contributions to the apparent cancer-protective effects of these foods (Langseth, 1995).

3.4 CHANGES IN NUTRITIVE VALUE DURING PROCESSING

Probably the single most important factor in the quality of fruit is the maturity of the fruit at harvest. It is important to distinguish between the stage that represents optimal eating quality and the stage of full biological maturation. In certain fruits, such as oranges, the two stages coincide. In fruit such as peaches, ascorbic acid decreases with advanced maturity. Ascorbic acid synthesis also continues during ripening of detached tomatoes, although the ascorbic acid content in a ripe tomato tends to be higher in fruit that is harvested at a later stage of maturity.

Carotenoids tend to increase during ripening of fruits such as peaches and pears, although much of the green to yellow color change is the result of unmasking of carotenoids through a loss of chlorophyll. Major changes have also been observed in the mineral content of grapes during maturation on the vine (Shewfelt, 1990). Contents of total sugar increase with ripening, especially in cherries and plums (Hulme, 1971). Any type of harvesting technique that results in bruising generally leads to an increase in degradative reactions and a loss of nutritional quality, although the extent of nutrient loss varies widely among fruit species (Shewfelt, 1990). Preservation is a convenient method of storing fruit for use in periods when the fresh products are not available. The characteristics of fruit are usually altered to such an extent during processing that the processed products do not necessarily resemble the fresh products. Some fruits, such as clingstone peaches and sultana (grapes), are grown almost entirely for canning and drying, respectively. If processed and stored properly, the nutritive value of these processed fruits is comparable to that of the fresh products.

Apart from the effect of the addition of sugar, water, and some other ingredients, usually only slight changes occur in the nutritive value of fruit during processing. In the case of dried fruit and fruit juice concentrates, the evaporation of water results in a concentration of the fruit constituents. Dietary fiber content will be affected by processing methods such as peeling, juice extraction, and drying, as well as by cultivar and variation in horticultural practices (Jones et al., 1990). Heating does not generally change the total dietary fiber content, but wet processing, such as cooking and blanching, may change some fiber properties; for example, the amount of soluble fiber in fruit may be increased by partial breakdown of pectins (Gurr and Asp, 1996).

Application of poor processing techniques to fruit may result in considerable losses of ascorbic acid. This vitamin is readily destroyed by oxidation, especially in the presence of certain enzyme systems and metal ions. In sound fruit, ascorbic acid is not subjected to enzymatic oxidation. When fruit is peeled or macerated, however, this vitamin is subjected to oxidation because of the presence of gaseous oxygen and oxidative enzymes. In deciduous fruits, the enzymatic change is characterized by browning of the fruit tissue. The appearance of a brown color is usually an indication that practically all of the ascorbic acid has been destroyed in that part of the fruit. Ascorbic acid is also rapidly destroyed when exposed to heat, especially in the presence of light or air and at neutral pH (6–7). It is relatively stable in more acidic foods. This is an important advantage because acidic products such as tomato and citrus juices are important sources of this vitamin (Institute of Food Technologists, 1986). This means that during canning of fruit a relatively small portion of the ascorbic acid is destroyed, with the result that most canned fruits are usually good sources of this vitamin. Canned fruit often contains more ascorbic acid than the corresponding fresh fruit stored under unfavorable conditions. Nutrient losses caused by canning depend on factors such as type of fruit, type of container, and the severity of the heat process. Minimum nutrient losses will occur if proper processing and cooking techniques are used. The processing methods for tomato sauce and tomato juice can cause an ascorbic acid loss of up to 34% (Farrow, Kemper, and Chin, 1979). Loss of vitamins also occurs in some canned fruit products after processing. The loss of ascorbic acid depends mainly on storage conditions. Canned fruit can be stored at room temperature for long periods without appreciable changes. This, of course, is one of the advantages of such products.

Most fruit cans are exhausted prior to canning. The fruit is filled in a container, and it usually passes through steam before sealing to reduce oxygen in the head space and to prevent oxidative changes and off-flavors. Blanching may also be applied before freezing, juicing, or, in some cases, dehydration of fruit. The fruit may be blanched by exposure to near-boiling water, steam, or hot air for 1 to 10 minutes. Blanching inactivates enzyme systems that degrade flavor and color and that cause vitamin losses during subsequent processing and storage. Blanching also removes air from the fruit tissues, destroys some of the contaminating microorganisms present, and causes softer texture. Nutrient losses caused by blanching result directly from leaching of water-soluble vitamins into the water used during processing. Blanching with steam, hot air, or microwaves does not require immersion in water and reduces leaching of vitamins (Institute of Food Technologists, 1986).

Fruit may be dried by sun drying or various evaporation processes. During dehydration, losses of vitamin C may vary from 10% to 50% and that of vitamin A from 10% to 20% (Institute of Food Technologists, 1986). The addition of sulfur dioxide during the drying of fruit has been claimed to have a beneficial effect on the retention of ascorbic acid and carotene because it inhibits oxidation and prevents enzymatic browning. Thiamine, however, is destroyed by the presence of sulfur dioxide. Because the thiamine content of fruit is usually low, more value is attached to the retention of ascorbic acid and carotene. Despite the long history of utilization of sulfur dioxide, there is today an awareness of potential hazards associated with exposure to sulfur dioxide gas, and several cases of respiratory diseases and atmospheric pollution have been reported (Wedzicha, 1984). The amount of sulfur dioxide permitted in dried fruit is controlled through regulations and varies from country to country. Sulfur dioxide levels in dried fruit also decrease during storage.

The nutritive value of fruit beverages depends mainly on the type of fruit used, methods of processing, and the degree of dilution. As in canned fruit, the vitamin content of fruit beverages is lower than that of the original fruit. However, vitamin C loss is greater in orange juice than in grapefruit juice at the same storage temperature as a result of non-enzymatic aerobic and anaerobic reactions. In the preparation of fruit nectars, only part of the fibrous material is removed. Clear beverages, however, contain no fibers. During clarification, the pectic substances are decomposed with the aid of pectolytic enzymes (Hugo, 1969). Methyl alcohol is produced, but in all experiments carried out so far it has been found that the concentration of this alcohol is so low that no health hazard exists. Fruit juices are also packed aseptically with high-temperature/short-time techniques, and this provides increased nutrient retention and fruit juice quality (Institute of Food Technologists, 1986).

In the freezing of strawberries and other berries, no pretreatments are applied and practically no changes in nutritive value occur during processing and storage if proper packaging and processing conditions are used. During thawing, however, losses may occur. Freeze drying of fruit, which is carried out in the absence of oxygen, does not detrimentally affect vitamin C retention (Institute of Food Technologists, 1986).

Irradiation of fruits is approved in some countries. Irradiation is carried out to control maturation and insect infestation. Experimental evidence indicates that nutrient retention of irradiated fruit is comparable to that of heat-processed fruits.

ACKNOWLEDGMENT

The writer thanks Arrie van der Schyf of Langeberg Foods (Pty) Ltd. for information on tomatoes.

REFERENCES

Akhavan, I., and Wrolstad, R.E. (1980). Variation of sugars and acids during ripening of pears and in the production and storage of pear concentrate. *Journal of Food Science,* **45**, 499–501.
Bradley, G.A. (1972). Fruits and vegetables as world sources of vitamin A and C. *Hortscience,* **7**(2), 141–143.
Farrow, R.P., Kemper, K., and Chin, H.B. (1979). Natural variability in nutrient content of California fruits and vegetables. *Food Technology,* February, 52–54.
Gurr, M.I., and Asp, N.G. (1996). *Dietary Fibre*. ILSI Press, Washington, DC.
Holland, B., Welch, A.A., Unwin, I.D., Buss, D.H., Paul, A.A., and Southgate, D.A.T. (1992). In *McCance and Widdowson's The Composition of Foods,* 5th ed. The Royal Society of Chemistry, Cambridge, England.
Hugo, J.F. Du T. (1969). Review of literature on the health value of fruit and fruit juices. *Deciduous Fruit Grower,* **19**(3), 62, 73–75.
Hulme, A.C. (1971). *Biochemistry of Fruits and Their Products,* Vol. II. Academic Press, London.
Institute of Food Technologists. (1986). Effects of food processing on nutritive values. *Food Technology*, December, 109–116.
Jones, G.P., Briggs, D.R., Wahlqvist, M.L., Flentje, L.M., and Shiell, B.J. (1990). Dietary fibre content of Australian foods. 3. Fruits and fruit products. *Food Australia,* **42**(3), 143–145.
Koeppen, B.H. (1974). Why we must get to know fruit sugars. *Food Industries in South Africa,* **26**(12), 23–25.
Langseth, L. (1995). *Oxidants, Antioxidants and Disease Prevention*. ILSI Europe, Brussels, Belgium.
Macrae, R., Robinson, R.K., and Sadler, M.J. (1993). *Encyclopaedia of Food Science, Food Technology and Nutrition,* Vol. 3. Academic Press, Harcourt Brace Jovanovich Publishers, London, pp. 2083–2091.
Miller, J.J., Colagiuri, S., and Brand, J.C. (1986). The diabetic diet: information and implications for the food industry. *Food Technology in Australia,* **38**(4), 155–157.
Nahar, N., Rahman, S., and Mosihuzzaman, M. (1990). Analysis of carbohydrates in seven edible fruits of Bangladesh. *Journal of the Science of Food and Agriculture,* **51**, 185–192.
Nguyen, M.L., and Schwartz, S.J. (1999). Lycopene: chemical and biological properties. *Food Technology,* **53**(2), 38–43.
Potter, N.N. (1986). *Food Science,* 4th ed., Ch. 4. Van Nostrand Reinhold, New York.
Reiser, S., Powell, A.S., Yang, C., and Canary, J.J. (1987). An insulinogenic effect of oral fructose in humans during postprandial hyperglymia. *American Journal of Clinical Nutrition,* **45**, 580.
Ross, J.K., English, C., and Perlmutter, C.A. (1985). Dietary fiber of selected fruits and vegetables. *Journal of the American Dietetic Association,* **85**, 1111–1116.
Salunkhe, D.K., Bolin, H.R., and Reddy, N.R. (1991). *Storage, Processing and Nutritional Quality of Fruits and Vegetables,* Vol. I, 2nd ed., Ch. 6. CRC Press, Boca Raton, FL.
Shewfelt, R.L. (1990). Sources of variation in the nutrient content of agricultural commodities from the farm to the consumer. *Journal of Food Quality,* **13**, 37–54.
Walter, P. (1994). Vitamin requirements and vitamin enrichment of foods. *Food Chemistry,* **49**,113–117.
Wedzicha, B.L. (1984). *Chemistry of Sulfur Dioxide in Foods*, Ch. 7. Elsevier Applied Science, New York.
Wills, R.B.H., Lim, J.S.K., and Greenfield, H. (1984). Composition of Australian foods. 22. Tomato. *Food Technology in Australia,* **36**(2), 78–80.
Wills, R.B.H., Scriven, F.M., and Greenfield, H. (1983). Nutrient composition of stone fruit *(Prunus* spp.) cultivars: apricot, cherry, nectarine, peach and plum. *Journal of the Science of Food and Agriculture,* **34**, 1383–1389.
Wills, R.B.H., Lim, J.S.K., and Greenfield, H. (1987). Composition of Australian foods. 40. Temperate fruits. *Food Technology in Australia,* **39**(11), 520–521, 530.
Wrolstad, R.E., and Shallenberger, R.S. (1981). Free sugars and sorbitol in fruits—A compilation from the literature. *Journal of the Association of Official Analytical Chemists* **64**(1), 91–103.

Chapter 4

Storage, Ripening, and Handling of Fruit

Brian Beattie and Neil Wade

4.1 INTRODUCTION

Fruits are valued as nutritive foods that are pleasing to eat. The fruits discussed here are the fleshy products of floral fertilization, which are eaten as a dessert, a salad vegetable, or as an ingredient of dishes such as stews and curries. Fruits are eaten fresh; processed into canned, frozen, and baked products; or converted into juice or jam. The raw material for processing is sometimes fruit that is unsuitable for the fresh market.

Fruits are classified by botanical or geographical relationships, similarities in fruit type or manner of cropping, or culinary use. Pome fruits (apple, nashi, pear), stone fruits (apricot, cherry, nectarine, peach, plum), and citrus fruits (grapefruit, lemon, mandarin, orange) represent botanically related species. Berry fruits (blueberry, boysenberry, loganberry, raspberry, strawberry) belong to different families but have in common a berry-type fruit that often grows on a vine. Tropical fruits (banana, durian, mango, pineapple, rambutan) have a common geographical origin. Salad vegetable fruits (capsicum, cucumber, egg plant, tomato, zucchini) are grouped by culinary use.

Fresh fruits are living plant organs, a fact that is the basis of correct handling procedures. Live produce respires; exchanges water with its environment; is subject to injury by mechanical means, insects, or toxic chemicals; and is subject to disease caused by fungi and bacteria and to metabolic disorders.

4.2 MATURITY AND RIPENESS

The terms *maturity* and *ripeness* have specific meanings to the postharvest biologist. Maturity refers to the stage of development of the fruit on the parent plant. There is a minimum period of development that must be undergone by any fruit before it is ready for harvest. A mature fruit is one that either has acceptable eating quality at the time of harvest or has the potential to ripen into a product of acceptable quality. A fruit can be mature but unripe, and indeed many fruits are harvested mature but unripe. Bananas, kiwi fruit, and avocados are examples of fruit that are picked unripe and that ripen (or are ripened) afterward.

Ripeness is attained when biochemical changes convert a mature but inedible fruit into an edible product. Softening, loss of astringency, biosynthesis of aroma volatiles, and conversion of starch to sugar are changes that commonly occur during ripening. The distinction between maturity and ripeness or ripening may be restated as follows: Maturation is a development process that occurs only while the fruit is attached to the parent plant. Ripening can occur on or off the plant and involves a physiological transformation of the fruit. The importance of the distinction may be illustrated by reference to the sugar content of a fruit. A fruit acquires all of its carbohydrate content from photosynthetic assimilates of the parent plant. The accumulation of carbohydrate is a developmental process, linked to maturation. The sugar content of fruits that do not undergo a rapid ripening phase cannot change appreciably after harvest. In fruits that ripen after harvest, starch or organic acids can be converted into sugars. The amount of starch or acid available for conversion is, however, determined during maturation.

4.2.1 Climacteric Behavior

Some fruits undergo a phase of rapid ripening, while others do not. All of the fleshy fruits that are the subject of this chapter fall into two distinct physiological classes with respect to their ripening behavior: climacteric and nonclimacteric fruit.

Climacteric Fruits

Climacteric fruits have a period of rapid ripening known as the climacteric. Respiration rate and heat evolution increase, sometimes dramatically; ethylene evolution often increases; and the fruit softens and develops flavor and aroma. Climacteric fruits often, although not always, have stored reserves of starch, and during the climacteric these reserves are hydrolyzed by starch-degrading enzymes to sugars. In fruits such as mango, both starch and carboxylic acids are converted into sugars. Climacteric fruits can be induced to ripen by treatment with ethylene. Examples of climacteric fruits are given in Table 4–1.

Nonclimacteric Fruits

Nonclimacteric fruits are fruits that do not undergo a rapid ripening phase. They mature slowly while attached to the parent plant, and their eating quality cannot improve after harvest. Nonclimacteric fruits have relatively low respiration rates that decline slowly after harvest. They produce ethylene at low rates. Application of ethylene to nonclimacteric fruits has been thought to have little effect other than the hastening of senescence changes, such as change in color from green to yellow, pedicel abscission, increased susceptibility to disease, and the development of off-flavors. Recent work has shown, however, that even the trace amounts of ethylene that accumulate around

Table 4–1 Recommended Conditions for Storage of Fruit Using Three Temperature Zones

Store at 0°C	Ripening Class*	Store at 7°C	Ripening Class	Store at 13°C	Ripening Class
Apple	C	Avocado (unripe)	C	Banana	C
Apricot	C	Capsicum (sweet bell pepper)	NC	Cucumber	NC
Avocado (ripe)	C	Honeydew melon	C	Custard apple	C
Berry fruit	C & NC	Mandarin†	NC	Grapefruit	NC
Cherry	NC	Olive	NC	Lemon	NC
Fig	C	Orange†	NC	Lime	NC
Grape	NC	Passion fruit	C	Mango	C
Kiwi fruit	C	Pineapple (ripe)	NC	Papaya	C
Lychee	NC	Tomato (colored)	C	Persimmon (ripe)	C
Nectarine	C			Pineapple (unripe)	NC
Peach	C			Tomato (mature green)	C
Pear	C			Watermelon	C & NC
Persimmon (unripe)	C				
Plum	C & NC				

*C, climacteric; NC, nonclimacteric. In some fruits the correct classification is arguable, and in any case different varieties of some fruits may be C or NC. When climacteric fruits ripen, they may evolve sufficient ethylene to initiate ripening in other climacteric fruits, or hasten senescence of nonclimacteric fruits.

†May also be stored satisfactorily at ambient (room) temperature.

produce during normal handling can reduce the shelf lives of nonclimacteric fruits. The postharvest lives of both climacteric and nonclimacteric fruits are reduced by the presence of ethylene. Some nonclimacteric fruits are listed in Table 4–1.

Use of Climacteric Class

Classification by climacteric class provides a simple guide to the general behavior of a fruit. Banana, a climacteric fruit, may ripen during transit with resultant self-overheating. Accidental exposure to ethylene may cause such an untoward event. Oranges, being nonclimacteric, cannot overheat like bananas, and exposure to ethylene, although not desirable, will usually have little noticeable effect on quality unless the fruit are stored for some time.

4.2.2 Ethylene

Ethylene (ethene) is a natural plant-growth regulator that is synthesized by all plants. Ethylene has many biological functions in growing plants, but in fruits it is particularly important as an agent of abscission (stem loosening), ripening, and senescence. Ethylene loosens many fruits from their stems and prepares the fruit to fall from the plant or to be picked. Ethylene initiates the ripening of climacteric fruits and hastens their ultimate senescence. In nonclimacteric fruits only the senescence effects of ethylene are apparent. The effects that ethylene has on fruits are exploited in their commercial handling. Ethylene treatments are used widely to ripen or degreen fruits.

Ripening

Many fruits, of which peach and nectarine may be cited as just two examples, begin to ripen on the tree if they are left to mature before harvest. Other fruit, such as the banana, is hard, green, and unripe when mature, and so can be picked in this condition and shipped to market with minimal wastage. Upon arrival at market and in response to prevailing market demand, the banana can then be induced to ripen with ethylene under controlled conditions that are optimal for good fruit quality. The technique of controlled ripening of bananas has been standard industry practice for many years, and it is possible for bananas to be supplied to a customer's order so that fruit of a prescribed color stage are delivered on a particular day. Controlled ripening is finding increasing application to other climacteric fruits that are picked unripe, such as avocados and kiwi fruit.

Controlled ripening is carried out in insulated rooms, constructed in the same way as coolrooms. Most fruits ripen best at a temperature of approximately 20°C, so ripening rooms are designed to operate at around this temperature. Banana ripening rooms are designed to run between 14° and 20°C. The room normally operates at high humidity, to prevent fruit shrivel, but in a "conventional" banana room where air movement around the fruit is poor, the humidity must be lowered once ripening has begun. Ripening is initiated by adding ethylene to the room atmosphere, either by periodic manual or automatic injections, or by the admission of a continuous bleed or "trickle" of the gas. Ethylene gas is both flammable and explosive, and strict safety precautions must accompany its use. The availability of ethylene as a compressed gas diluted in carbon dioxide has essentially eliminated the safety hazards that have been associated with ripening rooms.

The temperature and humidity of a ripening room are best controlled by the "pressure-ripening" system. Air is compelled to enter each package or bin stacked in the room and to pass around each fruit. The technique for doing this is exactly the same as that used for pressure cooling, which is described in more detail in a following section. Advantages of pressure ripening are uniform temperatures in all fruit anywhere in the room (a maximum temperature spread of less than 1°C is readily achieved) and uniform humidity.

Degreening

Although tomatoes are climacteric fruit, they are resistant to ethylene and their preclimacteric green life is only shortened a few days by ethylene treatment. This effect is nonetheless commercially important, especially because the market requires tomatoes presented at a uniform color stage. The differences in coloring time of a pick of mature-green to breaker tomatoes may be abolished by holding the fruit in an ethylene ripening or degreening room. Citrus fruits, which are nonclimacteric, may be similarly degreened with ethylene to improve their color for market. Degreening is an acceptable technological procedure if the fruits subjected to treatment are harvested when mature.

4.2.3 Maturity Standards

A most important, but often vexing, question in the fruit industry is the use of objective criteria to decide when a crop is ready to pick. Considerable effort has been expended in devising maturity standards for several major crops. Difficulties remain even with these major crops, and there is a lack of adequate maturity criteria for many other crops.

The application of maturity standards has been most successful with citrus and grapes, both of which are nonclimacteric fruits whose composition changes little after harvest. The ratio of soluble solids to titratable acidity is widely used as a maturity criterion for these fruits. Minimum acceptable values for the ratio, accompanied by minimum acceptable soluble solids contents, can be set for particular varieties. The soluble solids-to-acids ratio is an excellent criterion to use with fruits where the sugar-to-acid balance is the key to acceptability and where the concentrations of these constituents are fairly stable before and after harvest.

Maturity indices for climacteric fruits are more difficult to devise because until the fruit has ripened, only the potential quality of the fruit can be assessed. Since all the commercially important decisions about harvest, purchase, and payment are made before the fruit is ripe, maturity indices should be applicable to fruit whether ripe or unripe. The soluble solids content of a climacteric fruit is dependent on the stage of ripening. In kiwifruit, for example, the acceptability of the ripe fruit is highly correlated with the soluble solids content of the ripe fruit, but not of the unripe fruit. When unripe kiwifruit are tested for maturity, an estimate of the potential soluble solids content of the fruit when ripe would be a valid index. Starch content might be such a predictor, but according to research it is not. It has been suggested that the total dry matter content of the fruit at any stage of ripeness is an index of soluble solids content when ripe.

Maturity tests for fruits have been limited to a few attributes, such as soluble solids and acidity, that can be measured easily and inexpensively. The fruit is destroyed to prepare juice that is assayed by refractometry for soluble solids and titration for free acidity. Assays for other biochemical indicators of fruit development, such as specific proteins that are synthesized de novo at the onset of maturity, are being investigated. There is much current interest in nondestructive instrumental tests and especially in tests that could be applied to every fruit on a grading line and used as a sorting criterion. For example, near-infrared spectroscopy is being tested as a means to measure sugar and acid contents of intact fruits. The application of such technology is in its infancy. Some current uses do exist, however, for quality rather than maturity testing. When citrus fruits are damaged by frost, their density is affected and frost-damaged fruit can be graded out automatically by an on-line procedure for measuring the density of each fruit. Maturity and other quality criteria could be applied by such technology, which has as its aim the packing of more consistent lines of fruit that contain only product that satisfies objective quality specifications. The application of strict quality specifications is an important part of current efforts to introduce quality management principles to the fruit industry.

4.3   TEMPERATURE

Temperature is a vital tool in the postharvest management of fruit. The metabolic rate of a fruit slows down as the pulp temperature falls, and the rates of ripening and senescence are, therefore, decreased. The vapor pressure of water within the fruit tissues also decreases as the temperature falls; this reduces the rate of water loss by the fruit. Lowering the temperature also reduces infection of the fruit by microorganisms and slows the development of any existing infections. Fruit should be cooled quickly after harvest to a temperature appropriate for the particular commodity and held at this temperature subsequently, whether the product is destined for immediate retail sale, storage, or processing. The importance of correct temperature management cannot be overemphasized. High temperatures are more harmful to fruit quality after harvest than before the fruit is picked. At harvest, a fruit is disconnected from its water and respiratory substrate supplies. The rate of ripening will increase in fruits where ripening has begun on the tree or vine. The green life, or time taken to begin ripening, is often shortened when fruits are harvested unripe. All of these harvest effects, which reduce shelf life, can be moderated by correct temperature management.

Many temperate-climate fruits keep best at a temperature that is just above that at which the fruit tissues begin to freeze. The threshold temperature at which freezing begins (commonly called *freezing point*) depends particularly on the soluble solids content of the fruit, but most fruits will begin to freeze at or below −1°C. For practical reasons coolrooms holding such fruit are often run at 0°C, so that the spatial and temporal variation in air temperature throughout the room can be ± 1°C without any risk of freezing. Substantial benefits in product life can be achieved by a closer approach to the actual freezing point, and it is technically possible to do this in a well-designed coolroom.

Other fruits, particularly those grown in tropical and subtropical climates, are injured if they are held below a critical temperature that is above the freezing point of the fruit. The injury so caused is known as chilling injury; fruits subject to chilling injury are called chilling-sensitive fruits, while fruits that are not harmed by above-freezing temperatures are said to be chilling-insensitive. There is a discrete critical temperature for any particular fruit below which chilling occurs, although this fact may be obscured by varietal and climate effects. Variety and climate cause variations in the critical temperature for a particular commodity, just as the freezing point varies with the soluble solids content, which, in turn, is determined by the growing conditions. The critical temperature for chilling can be as high as 13°C, in the case of tropical fruits such as banana and mango, while other fruits, such as some varieties of avocado, may not be chilled until the temperature falls below 5–7°C. Some temperate fruits, including certain varieties of apple, may be injured by temperatures below 3–5°C, and a number of stone fruits (peach, nectarine) are injured by temperatures below about 7°C, and particularly by temperatures of 2–5°C. Symptoms of chilling injury are not expressed immediately and are often not seen at low temperature. After the fruit has been exposed to a chilling temperature for sufficient time, symptoms will appear when the fruit is transferred back to ambient temperature. In general, the lower the chilling temperature and the longer the time of exposure, the worse will be the ultimate injury.

A further complication to the response of fruits to chilling temperatures is the interaction of temperature with stage of ripeness. A common symptom of chilling is the inability of a chilled fruit to ripen properly. A fruit that is already ripe when it is chilled obviously cannot manifest symptoms of abnormal ripening, and if these are the main effects of chilling on quality, the ripe fruit can be said to be more chilling resistant than the unripe fruit. Chilled preclimacteric tomatoes do not ripen, while red-ripe tomatoes can tolerate moderate chilling treatment without a marked loss of eating quality. In contrast, moderate chilling of a ripe banana may also not markedly affect the eating quality of the pulp, but the fruit is usually rendered unacceptable by browning of the peel.

The distinction between chilling-sensitive and chilling-insensitive fruits is of the utmost importance in managing fruit temperature correctly. As a general rule, chilling-sensitive fruits should not be cooled below their critical temperature. In commercial handling, some compromises need to be made, and where limited storage facilities are available, it is not possible to hold each type of fruit in a mixed lot at its optimal temperature. Schemes have been devised whereby two, or preferably three, temperature regimens can be used to cater for most commodities, provided that storage for a short term only is intended. If only two storage areas are available, one can be run at 0°C and the other at 7°C. If three stores are available, the third area can be run at 13°C. Commodities can then be assigned to the two or three areas according to the dispositions suggested in Table 4–1. Particular care must be taken to clear quickly any commodities that are stored at 7°C that have a critical temperature for chilling above 7°C. If maximum postharvest life is sought, then storage must be at the optimal temperature for the particular commodity, and this temperature must be maintained within a narrow tolerance. Detailed tables of optimal storage temperatures are available, and particular attention should be paid to local knowledge and recommendations that take account of local circumstances.

There are two distinct stages in the cooling of fruit, which ideally are regarded as separate operations with dedicated facilities for each. The first operation is that of cooling the fruit by removal of field heat to the temperature at which it will be shipped or stored. The concept that field heat removal is a preliminary to refrigerated shipping or storage has resulted in the term *precooling* being used for this operation. The second refrigeration process is that of maintaining the precooled product at the desired temperature by removing respiratory heat and heat that has entered the cooled space from the outside.

4.3.1 Field-Heat Removal by Precooling

Effective precooling requires that large amounts of heat are removed from the fruit in a short time. Crops such as melons may have pulp temperatures of over 50°C because the fruit is substantially exposed to the sun and harvest is at the height of summer. At least 2.2 kW/tonne of heat must be removed from such fruit to reach a shipping temperature of 5°C in 24 hours. Apples, which grow in part shade and are picked in the autumn/fall, may be only 20°C at harvest, so that removal of 1.0 kW/tonne will prepare them for storage at 0°C in 24 hours. In fact, the summer melon crop would be better cooled in about 12 hours, whereas apples could be allowed 2–3 days to cool. Precooling must be done quickly so as to arrest deterioration of the fruit and to accommodate handling requirements such as shipment. Fruit has to be precooled to the carriage temperature before shipment because refrigerated vehicles and shipping containers cannot circulate sufficient cold air through tightly stacked loads to remove field heat. The refrigeration systems used during transportation are designed only to remove the heat of respiration of an already cooled product and heat that enters the refrigerated space from the outside.

The rate at which fruit can cool standing in a coolroom with adequate refrigeration and air circulation (at least 60 room volumes circulated per hour) is determined first by the stacking method and, ultimately, by the thermal properties of each individual piece of fruit. Fruit in a single, isolated carton will cool in several hours, especially if the lid is removed. Stacking the cartons one carton wide but several high will increase the cooling time of all but the top layer somewhat, but making the stack two or more cartons wide dramatically increases cooling time. If the cartons are stacked on a pallet of about 1 m² in area, it will take 2–3 days for fruit in the center of the stack to seven-eighths cool. The seven-eighths cooling time is the time required for fruit to cool to seven-eighths of the original difference in temperature between the fruit and the cooling medium. Seven-eighths cooling time (see Figure 4–1) may be taken as the "practical" or approximate cooling time for fruit exposed to a coolant that is a few degrees colder than the required shipment or storage temperature. Although fruit cools faster when in bulk bins than when packed in stacked cartons, it still takes a day or more for fruit in the

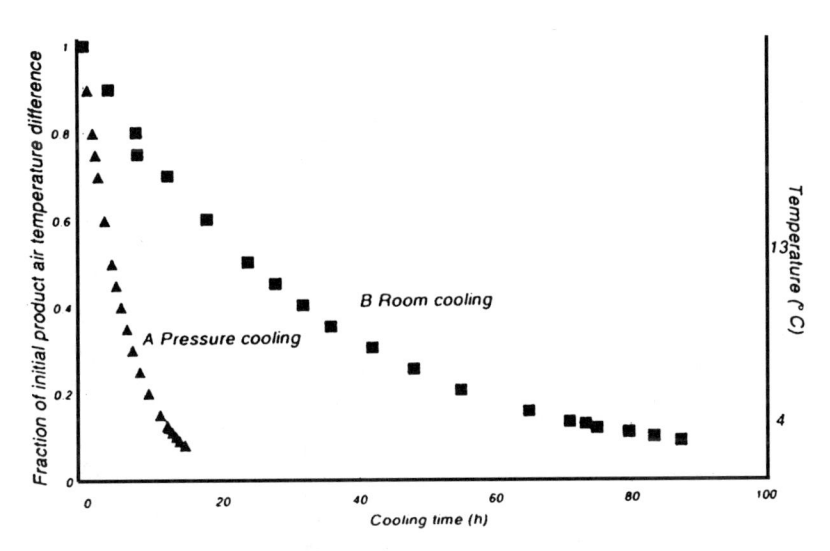

Figure 4–1 Effect of the cooling method on the way in which pulp temperature changes with time at the center of a pallet load of fruit. The initial pulp temperature of the fruit was 25°C, and it was cooled with air at 1°C. At the start of cooling, the fractional unaccomplished temperature change (shown on the left-hand axis) was unity. When seven-eighths cooled, the fraction was one-eighth or 0.125, and the pulp temperature was 4°C. Curve A is for a pallet of fruit precooled in a pressure cooler with airflow set to give a seven-eighths cooling time of 12 hours. Curve B shows the behavior at the center of a pallet standing in a conventional coolroom, such that the seven-eighths cooling time is 72 hours.

---

center of a 1-m² bulk bin to seven-eighths cool. Several methods are used to overcome the delays in cooling that are caused by packing and stacking.

Pressure or Forced-Air Cooling

Heat at the center of a stack of cartons escapes by conduction from fruit to fruit and through the layers of packaging material until it reaches the outer surface of the stack from where it is removed by convective heat transfer, which is a much faster process than conduction. If cold air can be admitted to each carton and brought into contact with the exposed surfaces of each piece of fruit, convective heat transfer occurs at the fruit surface, and cooling is hastened. In practice, cold air is brought into each carton or through each bulk bin by cutting ventilation slots in the ends of each carton or bin and slightly reducing the pressure on one face of the stack that is at right angles to the flow channels formed by the ventilation slots. Cold air is then drawn into the slots on the opposite face and this air passes through the width of the stack. Convective heat exchange occurs at the interface between the surface of each fruit and the turbulent air that now surrounds that surface. A number of stacking arrangements are used to achieve pressure cooling. The simplest, which is readily adapted to an existing coolroom, is effected by building parallel stacks of palletized cartons (Figure 4–2). The cartons are ventilated at right angles to the long axis of each stack, so that air can travel through

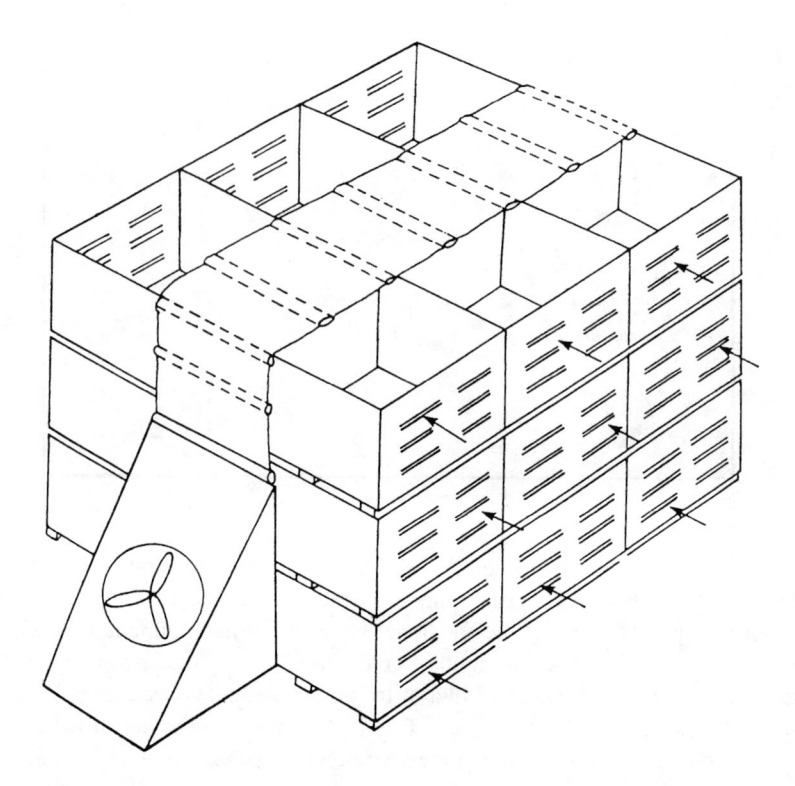

Figure 4-2 A simple system for pressure-cooling bulk bins or pallets of cartons. The bins or pallets are placed as shown inside a coolroom. The gap between the two parallel rows of bins or pallets is closed off completely with a flexible cover, and a portable fan is placed at one end. The fan exhausts air from the enclosed space, so that the pressure falls. Cold room air then flows through the ventilation slots in each bin or carton.

---

the short dimension. An air-return passageway is created between each pair of stacks by laying a cover sheet up from the floor and all the way along the gap between stacks. A small pressure-fan is fitted at one end of the air-return passage. The fan exhausts air from the passage and returns it to the room below the refrigeration evaporators.

Another arrangement is to construct air-collection chambers in which the fans can be mounted, either at the ends of the room for horizontal airflow, or under a false floor for vertical airflow. In theory, cooling should be faster if cold air is delivered directly through a false floor into the underside of the stacks, so that cold air flows vertically upward. The horizontal airflow system is, however, simple and readily adapted to existing practice.

For the improvement in heat exchange between the fruit and cold air to be of any benefit, there must, of course, be sufficient refrigeration capacity available so that the added heat load that is now applied to the evaporator can be removed by the refrigerant at least as quickly as it arrives. It is not necessary to match refrigeration capacity to the initial rate at which heat is evolved by the fruit, as long as balance is achieved by about the time that the initial temperature difference between the fruit and air has fallen by half. The penalty of such a design is a small (about 15%) increase in seven-eighths cooling time, but the refrigeration plant will be only half the size required to balance the two heat-exchange processes (fruit-air and air-refrigerant) at the commencement of cooling. Advantages of pressure cooling are its capital cost (which is little different from that of a well-designed conventional coolroom), flexibility (fruit can be cooled in bins or cartons, before or after packing), and the fact that the fruit is not wetted. The principal disadvantage (in some circumstances only) is that pressure cooling takes several hours.

Hydrocooling

Fruit are hydrocooled by flooding with cold water. Heat exchange at the surface of the fruit is very fast because of the high specific-heat capacity of water, and seven-eighths cooling times of 10 to 20 minutes can be achieved. The rate-limiting factor to cooling now becomes the thermal conductivity of the fruit itself, which determines the time taken for heat to travel from the center of the fruit to the surface. Fruit with smaller diameters, such as cherries, cool in 10 minutes in a hydrocooler. The particular advantage of hydrocooling in terms of managing a handling system is its speed. In terms of capital cost and energy use, however, this speed is achieved at the expense of a very large refrigeration plant, since the rate at which heat is evolved by the fruit has to be matched by the refrigeration capacity. The problems of insulating hydrocoolers often makes them less efficient than pressure coolers operating within a well-insulated coolroom.

It is almost essential that hydrocooled fruit be kept cool afterward, since the fruit, once wetted, will be very susceptible to disease unless treated with a protective chemical. The hydrocooling water itself acts as a reservoir of disease, as fungal spores washed from successive batches of fruit accumulate. The water in hydrocoolers is usually chlorinated to kill spores washed off the fruit, but chlorine treatments do not afford any residual disease-control benefit. Chlorine may corrode the plant, create unpleasant working conditions, and cause tainting of the fruit itself.

Other Cooling Methods

Vacuum cooling is used primarily for leafy vegetables, although it may have limited application to fruits. The pressure around produce is reduced until water in the tissues starts to boil and the latent heat of vaporization that is absorbed creates a cooling effect. Water ice or packets of gel ice may be placed in or around fruit packages to provide cooling during transportation. Boxes of dry ice are sometimes placed in air freight containers with fruit cartons. Dry ice must be well separated from the fruit so that freezing does not occur.

4.3.2 Cool Storage of Fruit

Produce that is to be stored for more than a few days should be kept in a coolroom specifically designed for storage. The temperature of a storage room should not fluctuate and it should be closely controlled at the optimum temperature for the particular fruit. The humidity of the room should also be optimal for the product. A survey of published values for the equilibrium relative humidities (ERH) of 39 species of dessert and salad fruits shows the mean ERH to be 98.4 ± 0.8%, with a range from 96.4% to 99.8%. When fruit have cooled to the temperature of their surroundings they will lose water to the surrounding air when the relative humidity of that air is less than the ERH of the fruit. Thus ripe banana fruit with an ERH of 96.4% will lose moisture to air with a lower humidity. Conversely, when fruit are held in air of higher relative humidity than the ERH of the fruit, they may absorb water vapor, gain turgor, and split or crack. The peel of a ripe banana splits in air of relative humidity greater than about 96%. Room humidity is important during storage because product water loss during storage is directly proportional to the vapor-pressure gradient between the fruit and the room air. As the humidity of the room air approaches the internal humidity of the fruit, water loss becomes less. It is usually not practicable or even desirable to match the internal humidity (ERH) of the fruit exactly because of problems with condensation, disease, and excessive turgor. An average room humidity of about 90% is satisfactory for most purposes. Room humidity is not as important during precooling (by room or pressure cooling) because at this stage the fruit is hotter than the room air, and the vapor-pressure gradient between the fruit and the air is primarily determined by this temperature difference. Once the fruit has cooled to the temperature of the room air, water loss is largely controlled by the humidity difference between the fruit and the air. Although the gradient is now smaller, the longer times associated with storage come into play, and harmful water losses can occur across small humidity gradients during a typical storage period.

Several techniques may be used to obtain high room humidities. Large evaporator coil surface area and high airflow over the fins of a forced-draught evaporator enable the evaporation temperature to be kept close to the temperature of the refrigerated space, so that the condensation of water vapor on the evaporator is kept low. A useful design criterion is that the difference in temperature between air entering and air leaving a "dry-coil" evaporator should not exceed 2.5°C. Dry-coil forced-draught evaporators are used widely because of their widespread applications in refrigeration generally. An alternative system that has particular application to fruit storage uses water as a secondary refrigerant. A fan circulates room air through a fine mist of chilled water so that the air is cooled to the temperature of the water and simultaneously saturated with water vapor at that temperature. A particular advantage of this system is that the air can be saturated at any operating temperature. Both the dry-coil and the water-mist systems can be designed so that air leaves the heat-exchange surface saturated with water vapor, but as the air gains heat (from the fruit and from the outside environment) the increase in air temperature will result in a decrease in relative humidity. It is simply not possible to transfer heat and keep the humidity at saturation, even if it were desirable to do so.

Another important aspect of room design that affects humidity is insulation. Heat that leaks into a coolroom from the outside raises the temperature of the air and lowers the relative humidity. This effect is minimized in a well-insulated room, which also requires less electricity to run. The walls, ceiling, and floor of a storage room should be well insulated. Various standards for permissible heat penetration have been suggested. A transfer rate of 10 W/m² should probably be considered the absolute maximum that can be tolerated.

Another method of obtaining a high humidity for storage is the "jacketed room" system. This method combines the principles that have been outlined previously. An uninsulated cubicle is built within the main coolroom and produce is cooled in this cubicle. The penetration of external heat into the cubicle is kept to a minimum because the ceiling and walls are jacketed with an envelope of cold room air. The nature of the refrigerant-to-air heat exchange is made irrelevant because the room air is now serving as a secondary refrigerant. The air inside the cubicle transfers the respiratory heat of the fruit and a very small amount of transmitted heat to the room air, with the exchange taking place on the exposed surfaces of the cubicle. Because this surface area is very large, the heat can be exchanged across a very small temperature gradient and so condensation onto the exchange surface is minimized and the humidity in the cubicle is kept high.

All of the methods for obtaining high humidity that have been described need to be incorporated into the original design. Where an existing room is running at too low a humidity, less satisfactory remedial measures can be taken. The most usual remedy is to install a spray humidifier that adds a mist of water to the room air. Humidifiers may cause wetness problems in the room, and the evaporator coils in a room operating below 2°C will ice up faster.

The maintenance of a uniform temperature and humidity throughout a storage room requires that there be good air circulation. The produce should be stacked with at least 100-mm clearance from the walls and floor and with gaps between adjacent pallets or bins so that air can move freely. The room air should be circulated continuously by fans. Less fan power is needed than during room precooling, and it is customary to switch off some of the forced-draught evaporator fans during storage, while leaving the rest to run continuously (rather than cycling with the room thermostat). In water-shower heat-exchange systems, the fan must run continuously anyway. An air circulation rate of 15 to 30 room volumes per hour is generally sufficient during storage.

4.3.3 General Requirements of a Coolroom

The full particulars of coolroom design are far too detailed to describe here, but some important general principles should be mentioned. Materials and methods of construction will not be dealt with since these vary considerably according to local requirements and procedures. The structure must be well insulated and sealed against the ingress of outside air. Doors must close tightly, and if they will be open for long periods during loading and unloading, the door entrances should be protected by plastic flaps, streamers, or air curtains to reduce the ingress of outside air. There must be sufficient working space within the room for machinery, such as forklift trucks, to operate freely. The floor must be strong enough to carry the mass of product to be loaded and the live load of vehicles. A most important element of the structure is the vapor barrier. This is a barrier against the permeation of water vapor into the wall insulation. In the absence of an efficient vapor barrier, water vapor will migrate into the wall on the hottest side and condense adjacent to the cool side. The condensate will reduce the efficiency of the insulating material and cause deterioration of the structural elements of the wall. Such deterioration can be quite rapid and very costly. The correct location of the vapor barrier depends on local climatic conditions. In warm climates, the outside temperature will almost always be higher than the temperature within the room, and so the barrier should be placed on the outer side of the insulation. If a room operating in a cold climate will usually be warmer inside than out, then the barrier should be placed on the inner side of the insulation. The vapor barrier is made out of material that is impermeable to water vapor, such as steel, plastic film, or bitumen-treated papers. Where coolrooms are built from panels made by laminating plastic foam insulation between two sheets of steel, one of the steel sheets will serve as the vapor barrier (being the outer or inner sheet, according to climate). All joints between panels must, of course, be well sealed with a durable and waterproof sealant. Once a secure vapor barrier has been constructed, it is important that the barrier be kept intact. Penetrations of the barrier should be minimized and properly sealed. It is good practice to seal plastic or metal conduit into any hole cut through a coolroom wall. The insulation exposed by the hole is thereby sealed off permanently. Pipework and wiring are then laid through the conduit.

4.3.4 Mixed Storage

The need to carry or store different commodities in the one place is a common problem. In evaluating the compatibility of different products, consideration must be given to their respective temperature requirements, relative sensitivity to ethylene, and the adsorption of undesirable odors. Detailed information on product compatibilities is available. The question of temperature compatibility has been discussed already and suitable courses of action outlined. The interactions between high- and low-ethylene–producing commodities are more complex, but the ethylene produced by apples will, for example, cause green bananas to start ripening, kiwi fruit to soften, cucumbers to turn yellow, and carrots to turn bitter. Produce such as onions, which have strong aromas, are likely to taint other produce, packing materials, and storage rooms. Potatoes are apt to cause an earthy taint and durians a complex taint.

4.4 STORAGE ATMOSPHERES

The storage life of a fruit can be affected quite substantially by the composition of the atmosphere in which the fruit is stored or carried. The storage atmosphere can be either beneficial or detrimental, and practical use is made of the beneficial effects. The harmful effects arise if the oxygen concentration approaches zero, as can happen at the center of a stack of hot fruit, or if carbon dioxide or ethylene accumulates because of poor ventilation. The storage lives of quite a number of fruits can be increased by reducing the O_2 concentration, often to about 2–3%. A smaller number of fruits keep better if the CO_2 concentration is increased, usually to just a small percentage, but sometimes up to about 20%. Most commodities that respond favorably to increased CO_2 will respond best if O_2 is simultaneously lowered. In some cases, such as peach and nectarine fruits, there is little or no benefit from lowering oxygen, but substantial benefit from relatively high concentrations of CO_2. It is most important to stress that too low a concentration of O_2 or too high a concentration of CO_2 will injure fruit. The threshold concentrations for injury vary with commodity and temperature. Fruits are a bulky mass of respiring cells. The internal atmosphere of a fruit is always different from that of air because respiration prevents the internal atmosphere from equilibrating with the outside atmosphere (air). Changes in the O_2 and CO_2 concentrations of the outside air make the internal atmosphere of the fruit even more extreme. If a fruit is placed in an atmosphere containing only 3% oxygen, the internal atmosphere of the fruit will contain only a fraction of a percent of O_2 at most. This small amount of O_2 will usually be sufficient to sustain normal aerobic respiration. If the outside O_2 concentration falls below about 1%, it is likely that the resultant trace of O_2 left in the fruit will be unable to sustain aerobic respiration, and anaerobic respiration will take over. Anaerobic respiration cannot completely oxidize sugars, and products such as ethanol and acetaldehyde accumulate. The accumulation of potentially toxic metabolites, the reduced yield of energy, or both, lead to cell injury and tissue death. Too high a concentration of CO_2 is also harmful. As little as 1%, CO_2 in the outside atmosphere may harm some varieties of citrus fruit, apple, and pear, whereas other apple varieties will benefit from 5% or even more of CO_2, and strawberries and a number of stone fruit will tolerate 20% CO_2. Carbon dioxide has many effects on cell metabolism, including inhibition of the respiratory pathway such that symptoms of injury similar to those caused by lack of O_2 are observed.

The beneficial effects of low O_2 and elevated CO_2 are varied, but where present they confer longer storage life and maintain quality. Apples can be kept crisp, juicy, and green for up to twice as long as in conventional cool storage. Kiwi fruit and bananas can be kept hard, green, and unripe. Peaches and nectarines can be kept firm, juicy, and free of mealy breakdown. Mangoes keep longer than otherwise in such altered atmospheres. Other fruits, particularly those of the nonclimacteric category, show less or no response to changes in the storage atmosphere. Citrus fruit, cherries, and strawberries, for example, do not respond dramatically to atmosphere changes. The beneficial effect that 20% CO_2 has on cherries and strawberries is probably caused by suppression of fungal growth.

The means by which changes in atmosphere composition improve storage life are not completely understood, but an important element of the response is control of the internal ethylene system. Molecular O_2 is required by the enzyme that produces ethylene from its amino acid substrate, and the lower O_2 concentrations that give beneficial storage responses inhibit ethylene synthesis by the fruit. Carbon dioxide competes with ethylene for its binding sites in the fruit cells, and the small percentage of CO_2 that has beneficial effects is often sufficient to greatly reduce ethylene binding. The concentrations of O_2 and CO_2 that improve storage life also reduce fruit respiration rates, in common with cool storage, and it is true that for a particular type of commodity reduced respiration rate is often associated with increased storage life.

Fruit may be stored under either a controlled or modified atmosphere. A controlled atmosphere is maintained within close tolerances by analysis and adjustment as required. Control may be manual or automatic. Modified atmospheres are not under close control, and they are usually generated by the respiratory activity of the fruit itself.

4.4.1 Controlled Atmosphere Technology

The first requirement for controlled atmosphere (CA) technology is a gas-tight storage room. It is difficult to build such a room, and the methods used vary widely. As with coolrooms, local building practices and codes determine the methods used. The principle of a CA store is that the room should be lined with material that is impermeable to gases. The requirements for such a gas barrier are similar to those of a vapor barrier, and local climate and construction method determine whether the gas and vapor barriers are separate or combined. In cold climates a combined gas and vapor barrier is placed inside the insulation. In warmer climates, the vapor barrier is still placed outside the insulation, while the gas barrier has often been inside where it can be repaired. Materials such as steel and plastics make good gas barriers, although coatings of various types are also used. Joins between sheet claddings are sealed with flexible sealants and covered with moldings. A technique for applying a plastic gas barrier is to spray the interior of the store with polyurethane foam. A sufficient thickness of foam forms a barrier to gas permeation and also serves as thermal insulation. This combination gas seal and insulation is ideal in a warm climate. Special doors with gas-tight seals are required. A common problem with CA rooms is that movement of the structure opens leaks in the gas barrier. Structural movement caused by pressure differences between the inside and outside atmospheres must be prevented by allowing the pressures to equilibrate. A water-sealed pressure relief pipe should be inserted through the wall of all CA stores. Movement caused by thermal expansion and contraction and by other factors, such as wind, is not easily accommodated.

4.4.2 Atmosphere Generation

Some common ways of generating the controlled atmosphere are listed below.

Using Fruit Respiration

Sealing the packed store allows fruit respiration to modify the atmosphere, which can then be controlled within the desired limits. It is only practicable in a refrigerated store to let the fruit generate the atmosphere once, immediately after the store is loaded, when the fruit is still warm and respiring rapidly. Cold fruit will not be able to restore an atmosphere that is lost through leakage or the need to open the room periodically to remove stock.

An Open-Flame Hydrocarbon Fuel Burner

Air is burnt in a gas flame and the combustion products (unburnt O_2, CO_2, and moisture) are discharged into the room. If fruit respiration causes O_2 to fall, outside air is bled in. If CO_2 increases, the room air is recirculated through a CO_2 scrubber.

If leakage causes O_2 to increase, more outside air is burnt and used to flush the room. Open-flame burners are economical only with small rooms because of their inability to burn O_2 that leaks into the room.

Catalytic O_2 Burner

These generators overcome the inability of open-flame units to use the O_2 in recirculated room air. They contain a bed of heated catalyst that allows fuel to burn in the low-O_2 concentration of the room atmosphere.

Air Separators

These take compressed air and fractionate the constituent gases by their molecular sizes. In one type of separator, air flows into hollow fibers made from a membrane in which O_2 permeates faster than N_2. By the time the air reaches the opposite end of the hollow fiber, much of the O_2 has diffused out, leaving an N_2-rich gas. In the pressure-swing adsorption separator, compressed air enters the bottom of one of two packed beds of adsorbent. The bed adsorbs O_2 so that an N_2-rich gas is discharged at the top of the bed. When the adsorptive capacity of the first bed nears saturation, the separation is switched to the second bed while the first is regenerated.

Liquid N_2

The O_2 concentration in a room can be reduced very quickly by flushing with N_2 that is injected into the room as liquid. Liquid N_2 is usually used only to establish the initial atmosphere and not for subsequent maintenance of low O_2.

Scrubbers

The use of scrubbers to control CO_2 in the atmosphere has been mentioned. Many different types of scrubbers are used to remove CO_2. The majority rely on the reaction between CO_2 and alkalies, and one type relies on the solubility of CO_2 in water. The air separators that have been described as generators also function as CO_2 scrubbers, since CO_2 partitions in a manner similar to O_2.

Safety Hazards

Hydrocarbon burners are a potential safety hazard, and they should be designed with safety devices and used by properly trained and experienced operators. It is possible for a storage room to be filled with flammable gas as a result of burner malfunction. Air separators pose no such safety risks and by running them in a flow-through mode, so that the store atmosphere is continuously flushed, it is possible that fruit quality may be improved by the removal of harmful fruit volatiles.

Analysis of Gases

The store atmosphere may be analyzed manually by the use of an apparatus such as the Orsat volumetric gas analyzer, or with instruments such as the infrared analyzer for CO_2 and a paramagnetic or fuel cell analyzer for O_2. Any of the electronic instruments can be set up to sample and read the room atmospheres regularly and record the results. The analyzer outputs may also be used to operate generators and scrubbers, so that the atmosphere is controlled automatically. Regardless of the technology used, accurate records should be kept of store atmospheres and store and fruit temperatures, so as to assist in the diagnosis of any quality problems that may arise in fruit after removal from storage.

4.4.3 Selection and Handling of Fruit for Storage

The selection and handling of fruit for CA storage is of the greatest importance in determining ultimate fruit quality. Unless fruits are mature when picked, they will never ripen into a product of acceptable quality. There are no exceptions to this rule, which applies whether the fruit goes to the domestic market or is stored. The stage of ripeness at which fruit enters storage is, however, critical. The efficacy of CA storage is reduced when fruits have entered the ethylene climacteric and have begun to ripen before storage. Fruits that are preclimacteric but mature will respond best to CA storage. It is common for harvest to be delayed until fruit size and color appear satisfactory. The fruit may well have begun to ripen before these attributes are judged satisfactory, with the result that the harvest will have a reduced storage life.

Speed of Cooling

The speed with which fruit is cooled and CA conditions are established is also critical if good results are to be obtained. Postharvest delays in attaining storage conditions allow the fruit to begin ripening, if it has not already done so, and for ripening changes to develop. Ripening usually begins sooner and develops faster once the fruit has been detached from the parent plant. The aim of good storage practice is, therefore, to pick the fruit and place it promptly in an optimal storage environment. With fruit such as bananas, storage conditions should be established within a day of harvest, while with apples the fruit should be cooled and under CA within a week of harvest. It is preferable to delay harvest if need be to achieve these time requirements, because ripening of attached fruit is slower. Too long a delay will result, however, in the fruit becoming too ripe for any useful storage.

Segregation of Fruit

Segregation of fruit within the store is another management tool. It is possible with at least some fruits, such as apples, to classify each line of fruit coming into the packing house as suitable for short-, medium-, or long-term storage. With apples, maturity and ripeness criteria such as pressure test, soluble solids, and iodine starch test may be used to assign lines of fruit to a storage class. Calcium analyses have also been used to predict the probability of storage disorders in apples. The store is then emptied according to the anticipated storage life of the several storage classes.

Evaluation of Shelf Life

Regular monitoring of the stored fruit by withdrawal of samples and evaluation of their shelf life is another useful tool. A small airlock can be installed in the room so that samples of fruit can be removed from outside without breaking the atmosphere.

The samples are allowed to ripen at room temperature and assessed for the incidence of disorders and other quality defects. At the first sign that quality is starting to deteriorate, the room should be opened and the affected fruit should be sent to market.

4.4.4 Modified Atmosphere Technology

Modified atmospheres have particular application as an in-package treatment for use during transport. Fruit is packed in a carton lined with a bag made of plastic film, and the top of the bag is either folded down or tied off. The plastic film restricts the outward movement of CO_2 and water vapor evolved by the fruit, and the inward movement of O_2. Provided that sufficient O_2 can cross the film barrier to support aerobic respiration, the atmosphere within the bag will acquire at constant temperature a steady-state composition such that $(21 - V_O)/V_C = (P_C/P_O)/R$, where V_O and V_C are the concentrations of O_2 and CO_2 respectively, as percentage by volume, P_C/P_O is the ratio of the permeability coefficients of CO_2 and O_2, respectively, and R is the ratio of CO_2 evolved by the fruit to O_2 consumed. The steady-state composition of the atmosphere inside a plastic film package is, therefore, a function of the respiratory gas ratio (a biological property of the fruit) and the selectivity factor of the film for CO_2 relative to O_2 (an attribute of the film). The selectivity ratio (P_C/P_O) of a hole is close to unity, whereas the ratio for polymeric films is invariably greater than unity. The selectivity ratio P_C/P_O for low-density polyethylene, for example, is commonly about 5. Suppose that a desired storage atmosphere must contain 5% O_2 and less than 10% CO_2. In a package where $P_C/P_O = 1$ and $R = 1$, a 5% O_2 atmosphere can only coexist with 16% CO_2. If $P_C/P_O = 3$, a possible atmosphere is 5% O_2 and 5.3% CO_2. This concept is shown diagrammatically in Figure 4–3. In a bag closed by folding, or made of a film that has been manufactured with micropores or small punctures, the selectivity ratio will be close to unity. If the bag is tightly closed and gas transfer is by permeation through the film, the selectivity ratio will approach that of the film, and CO_2 concentration will be reduced compared with a similar package of selectivity ratio unity. The selectivity ratio of a package rarely reaches that of the film from which it is made because of leaks at the bag closure and punctures in the bag.

A major limitation to the use of modified atmospheres is the effect of temperature on the atmosphere. As temperature increases, so does the respiration rate of the fruit and the permeability of the film to gases. The temperature coefficients of both processes are different, so that a more severe atmosphere invariably develops. Fruit that has been shipped under refrigeration in modified atmosphere packages is liable to injury if the consignment is allowed to warm up at its destination.

In the preceding discussion, controlled atmospheres have been treated as a technology appropriate to storage rooms, and modified atmospheres as having particular application to transport. This presentation reflects common commercial practice, but alternative applications do exist. Atmospheres in large storerooms have been generated by respiration and controlled by gas permeation units that use polymeric films. Similarly, controlled atmosphere equipment has been attached to shipping containers to control the container atmosphere during transit.

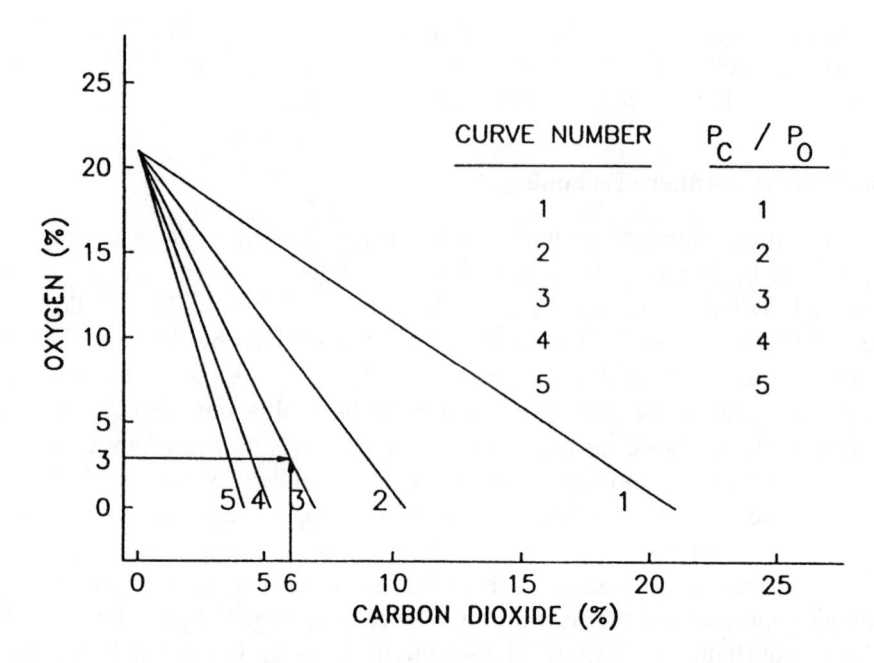

Figure 4–3 Principle of modified atmosphere packages that rely on film permeability. The graph refers to an ideal package without holes. Each line defines the combinations of package O_2 and CO_2 concentrations that can be attained with a particular film selectivity ratio (P_C/P_O), if the ratio R of CO_2 evolved by the fruit to O_2 consumed is unity. If, for example, $P_C/P_O = 3$, one of the possible atmospheres that may be attained is 3% O_2 and 6% CO_2. The actual atmosphere attained will depend on the mass and absolute respiration rate of the fruit, and the surface area of the package.

4.5 MAINTAINING QUALITY

The fulfillment of consumer demands for fresh, high-quality fruit and other horticultural produce has been possible only because of the use of the technology described here. Despite this, losses in produce still occur. The value of produce to the producer, wholesaler, retailer, or processor may be lowered or completely lost as a result of postharvest mismanagement. The losses to the community represented by consumer dissatisfaction with produce that is of poor quality and highly priced are probably greater. The world market is now calling, however, for one more requirement in addition to that for quality. Food safety must be documented and ensured. Food safety is now the prime market issue and a requirement of both public health authorities and the supermarkets that dominate the market. The management of both food safety and product quality contains the same system elements. In the next three sections technical aspects of quality management are discussed. The market situation and system elements for food safety and product quality are discussed in the final section.

Three important factors that cause quality loss during the harvesting and handling of fruit are as follows:

1. microorganisms that cause loss of product from disease or that compromise food safety
2. disorders caused by disturbances in the normal metabolism of the fruit (Stresses [associated with mineral nutrition, temperature, humidity, or atmosphere] are often implicated in fruit disorders.)
3. injuries caused by mechanical force (hail, dropping, squashing, rubbing, tearing, puncturing), insect attack, or chemicals

The principles that lie behind good postharvest management take into consideration the protection of fruit as a living organism. Losses and deterioration in quality can be greatly reduced by attention to field, packing house, storage, and retail management.

4.5.1 Disease

Disease is a major cause of loss in the harvested fruit crop. Most postharvest disease is the result of infection after harvest. Mechanical injuries inflicted on the fruit during and after harvest are probably the most important single event that allows invasion by wound-infecting organisms. Some important diseases arise from infection before harvest. Examples of fruit diseases that originate in the growing crop are anthracnose, a latent infection of fruits such as banana, and stem end rot in mango, which is the result of an initial symptomless endophytic colonization by certain fungi. All pathogenic diseases result from an interaction among a parasite, the environment, and a susceptible host. Most fruit parasites are fungi. The source of fungal inoculum (infective material) is usually spores. Spores may be conveyed in the orchard by wind, rain-splash, physical contact, or insects, and after harvest by contact with contaminated picking receptacles, boxes, bins, wash water, sorting equipment, and the like.

Environment and Host Susceptibility

Important elements of the postharvest environment are humidity, free water, and temperature. Most fungal spores require high humidity or free water to germinate. Fruit kept in a confined space (such as a carton) are surrounded by a humid microclimate, and spores resident at the site of an injury have the additional assistance of leaking cell sap. The rates of both spore germination and infection increase with increasing temperature up to a maximum, after which the rates decline, usually quite rapidly. Host susceptibility depends in the first instance on genotype. Most fruits are resistant to most pathogens, and a high degree of host specificity occurs. For example, *Penicillium* species are important causes of postharvest loss, but pathogenicity is quite specific. *P. expansum* is a pathogen of apples and grapes, but this species will not infect citrus. *P. digitatum* and *P. italicum* both infect citrus, but not grapes or apples. Susceptibility depends secondarily on the physiological state of the fruit, which varies according to stage of ripeness and handling history. Ripe fruits are more susceptible to infection than unripe fruits. Fruits that have been chilled are more susceptible to infection, and sometimes diseases are the most obvious symptom of chilling. Mechanical damage during handling provides wounds through which infection can occur. Latent infections occur when fungal spores form resting stages on the fruit surface in the field. Infection hyphae may penetrate the fruit surface, but no further development of the infection occurs until after the fruit has started to ripen. The anthracnose diseases of tropical fruits are examples of latent fruit diseases. Latent infections may occur in flowers before the fruit has developed. The gray mold rot (*Botrytis cinerea*) that affects many fruits can arise from latent floral infections.

Enzyme Maceration

Postharvest disease causes loss of fresh fruit, but the early stages of infection by postharvest pathogens can cause loss of fruit products. Fruit pathogens often secrete pectolytic enzymes that are used to secure entry to the host. Some of these enzymes are heat resistant. If apparently sound fruit in which fungal infection has just begun is canned, pectolytic enzyme activity left after thermal processing may macerate the product. This problem has been observed when apricots with an incipient infection of *Rhizopus stolonifer* were canned.

Disease Control

Control of postharvest fruit disease is achieved by integrated good practice both before and after harvest. Field disease-control measures reduce the carryover of disease and inoculum into the packing house. Cultural practices that let air and sun into the trees and prevent microclimates that are too warm and moist are helpful. Plant nutrition can affect disease. Excessive applications of nitrogen can increase pre- and postharvest disease, probably by lowering fruit calcium content. Fast cooling after harvest arrests infections that would otherwise form at wound sites and in ripening tissues. Good handling that minimizes mechanical damage reduces opportunities for infection. Pickers should know the correct method of handling fruit. They should have short fingernails and preferably wear gloves. Fruit should be picked with a swift movement that does not take wood or leaf. Picking bags should not be overloaded nor the fruit dropped when delivered to field lugs. The packing house and coolrooms should be kept clean. Floors should be swept regularly and at the end of each day's packing. Dirt and dust should be kept out of the building, and spilled or rejected fruit should be gathered and placed in a covered container that is replaced daily. Neglected moldy fruit can discharge clouds of spores. The equipment used to dump and sort fruit should be well maintained and free from defects that might injure fruit. Equipment should be steam cleaned each season. All belts and conveyors should be swabbed down each day with a sanitizing solution. Chlorine bleach preparations or alcohols can be used for sanitation, provided that care is taken against the corrosive properties of chlorine. Sometimes the interior of a coolroom has to be washed with a sanitizer because of mold growth. The primary cause of such problems (which are often associated with excessive condensation) should be investigated and remedied.

Fungicides

Fungicides can be applied to fruits after harvest in a dip before sorting, as a spray drench on the line, or as an additive to wax treatments. Fungicides may be progressively inactivated in dips or recirculating sprays by binding to soil or sap. Some formulations are unsuitable for application through spray nozzles, and mixing fungicides with wax reduces efficacy. Postharvest fungicides usually have to prevent infection, or resumption of infection, by an inoculum already resident on or in the fruit. The length of time between harvest and chemical treatment has a critical effect on efficacy of disease control, since infection is hastened by harvest wounds and ripening. Most fungicides should be applied within 24 hours of harvest when fruit are at ambient temperature (Figure 4–4). A longer delay may be permissible if the fruit is cooled quickly, since infection is slowed at low temperature. Treatment efficacy depends in part on uptake of fungicide at wound sites, where it prevents infection, often by interfering with spore germination or the growth of newly formed hyphae. Some postharvest fungicides, such as the benzimidazole group and imidazole ergosterol biosynthesis inhibitors, have an eradicant action whereby existing infections are controlled or suppressed. These chemicals can penetrate waxy cuticles and reach seats of infection within the fruit. Each postharvest fungicide has a specific spectrum of activity and is suitable only for the control of particular diseases. Postharvest fungicides present a food safety hazard, and their use must be in accordance with the applicable legislation.

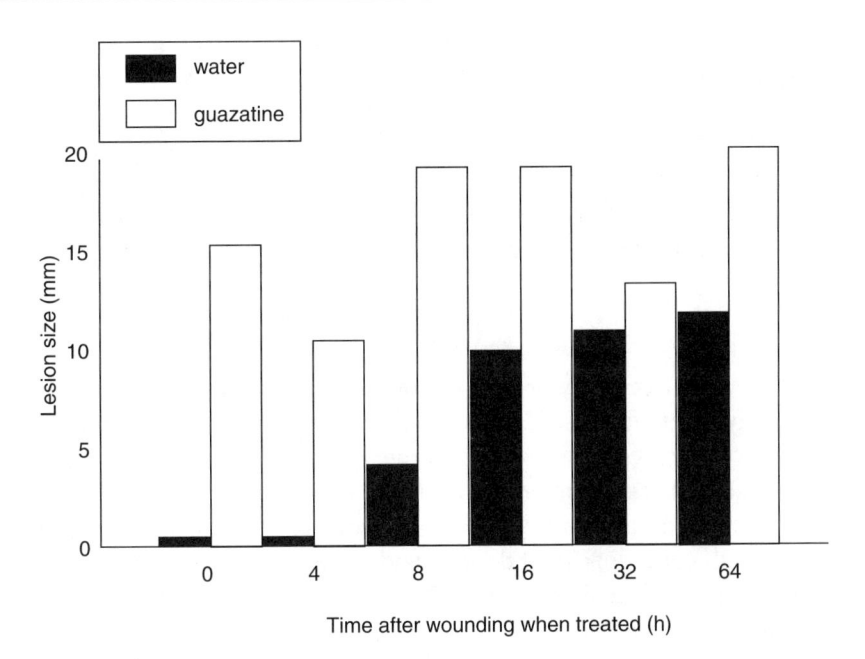

Figure 4–4 Effect on disease control of delays in application of a fungicide after fungal spores have been introduced at wound sites. Muskmelons were wound-inoculated with spores of the fungus *Geotrichum candidum*. Random samples of the fruit were dipped in either water (control) or a solution of the fungicide guazatine at subsequent intervals. The average size of sour rot lesions was measured 5 days later. The diagram shows that the efficacy of disease control was reduced when the delay between wounding and treatment exceeded about 8 hours. *Source:* Adapted from Wade, N.L., and Morris, S.C. (1982). Causes and control of cantaloupe postharvest wastage in Australia. *Plant Disease*, **66**, 549–552.

Fungi frequently develop resistance to postharvest fungicides. Populations of resistant strains can build up quite quickly, so that the fungicide in question suddenly loses efficacy in a particular packing house. The presence of resistance is diagnosed by laboratory testing of isolates of the suspected resistant strain. Strategies to avoid resistance rely on the use in rotation of several fungicides in both the field and packing house, and strict adherence to sanitation measures in the packing house. Waste fruits that go moldy or rot in the vicinity of the packing house may be multiplying a resistant strain that will then infect fruit entering the premises. It is best to prevent resistance developing in the first place, since the sudden appearance of resistance will cause loss, and withdrawal of the affected fungicide provides no assurance that the resistant strain will disappear.

Other Strategies

Several other disease-control strategies are currently being investigated to reduce fungicide use. Heat treatment can be effective, but it is often difficult to achieve satisfactory control without injuring the fruit. Heat can act by both inhibiting the pathogen and stimulating host resistance. Biological control has been achieved by dipping fruit in suspensions of bacteria or fungi that grow on the fruit surface and inhibit fungal pathogens. Two such "biofungicides," which contain inoculum of *Candida oleophila* or *Pseudomonas syringae,* have been approved for the control of postharvest diseases of citrus and pome fruits. The commercial utility of these pioneer biofungicides is still being assessed. The best results are being obtained where the biofungicide is used in conjunction with a reduced amount of synthetic fungicide, or is applied to fruit previously treated with a sanitizer. Treatments that increase host resistance are also being studied. Low doses of ultraviolet light between the wavelengths of 190 and 280 nm have reduced subsequent disease in apples, citrus, peaches, and tomatoes. It is probable that ultraviolet light acts by stimulating the defense mechanism of the fruit.

4.5.2 Disorders

Fruit may be affected by breakdown or blemish that is caused by an abnormality in the physiology of the fruit itself, rather than by the primary intervention of a pathogen. Such an injury is called a disorder. Disorders may have an external cause, but unlike diseases this cause is environmental rather than pathological. Although it is conventional to divide fruit breakdown into disease and disorder, abnormal fruit physiology is common to both conditions, and the distinction lies really with the predisposing cause, which may be pathological, environmental, or wholly internal. Some examples of fruit disorders are discussed below.

High-Temperature/Solar Injury Disorders

Most fruits develop sunburn if overexposed to solar radiation. Vein tract browning of melons is caused by preharvest solar ultraviolet radiation, followed by postharvest desiccation. The sun and other sources of heat may also cause thermal injury, manifested as abnormal softening and an inability to ripen normally.

Low-Temperature Injury

All fruits are liable to injury if their tissues begin to freeze. The resultant tissue injury becomes visible when the frozen tissue thaws, and the extent of injury depends on the rate of thawing. The phenomenon of chilling injury, which was discussed in the section on temperature, is another important disorder. The injuries that occur in tropical fruits stored below about 10–13°C are chilling injury disorders. Storage disorders that occur in temperate fruits such as apples and peaches below about 5°C have not yet been shown to arise from the same primary chilling event as the tropical fruit disorders, and it is not even clear if temperature causes the injuries observed. It may well prove, however, that temperate fruits are also subject to chilling injury, although at lower-threshold temperatures than tropical fruit. Breakdown, superficial scald, and soft scald of apples, and cool storage or mealy breakdown of peach and nectarine fruits are examples of storage disorders of temperate fruits.

Nutrient Deficiency or Imbalance

Fruits are peculiarly susceptible to disorders caused by low calcium or by a nutrient imbalance involving calcium. Calcium is carried into fruits and leaves by the transpiration (water) stream, and fruits, unlike leaves, have low transpiration rates. A classic calcium-deficiency disorder is bitter pit of apples. In extreme cases, corky pits develop in affected fruit on the tree, but symptoms usually appear after postharvest storage.

Atmosphere Stress

Fruit may be subjected to injurious atmospheres in improperly operated, controlled atmosphere stores, modified atmosphere packages, or if stowed in a poorly ventilated confined space. Low O_2 induces anaerobic fermentation in the fruit tissues, while high CO_2 may induce aerobic fermentation. Both conditions cause breakdown of the fruit. Specification of a "safe" atmosphere depends on the type, variety, and temperature of a fruit. Atmosphere stress is prevented by ensuring that fruit is either well-ventilated or well-monitored in controlled and modified atmosphere systems. It can be advisable to monitor controlled atmosphere stores not only for O_2 and CO_2, but also for ethanol that is often evolved by anaerobic fruits.

Control of Disorders

Control measures for disorders are suggested by the cause, where this is known. High-temperature injury is avoided by maximizing leaf cover (by pruning and disease control) and, in some cases, by erecting covers over the crop and even whitewashing fruit on the vine. Picked fruits are susceptible to heat injury and should be kept shaded. Postharvest low-temperature injury is generally avoided by holding fruit above the critical-threshold temperature for injury. The loss in shelf life that would otherwise occur at this higher temperature might be prevented by another technique, such as controlled atmosphere storage. The symptoms of chilling injury can sometimes be alleviated by prior treatment. Some examples of alleviation are the dipping of citrus fruits in a hot suspension of benzimidazole fungicide before exposure to cold, and the waxing of fruits before refrigeration. Apples are dipped in antioxidants to control the storage disorder superficial scald. Calcium-related disorders respond poorly to soil amendment with calcium, because little of this calcium can reach the fruit. Spraying fruit in the orchard can be effective, but coverage must be complete. Most success has been achieved with postharvest application of calcium salts by dipping. Atmosphere stress is preventable by good management.

4.5.3 Injury

Mechanical Injury

Mechanical injury is an important cause of downgrading and wastage of fruit. Damage may occur before harvest from wind rub, hail, pressure by adjacent fruit as they expand, and high turgor. Most damage occurs during and after harvest, from manual handling by pickers and packers, mechanical handling on dumping and sorting machinery, or during transport. Mechanically injured fruits are susceptible to infection and breakdown from disease.

Impact bruise injury occurs by fruit-to-fruit or fruit-to-hard surface contact when fruits are dropped or bounced. Falls of only several millimeters can be sufficient to cause bruising. Compression bruising is caused by a static load, such as the pressure of overlying fruit in a stack, or the pressure applied by the walls or lids of packages or bins. Bruise injury takes at least several hours to appear, and it may take weeks. Bruising is sometimes only apparent when a fruit is cut. Other types of mechanical injury include punctures (from fruit stalks, fingernails, or poorly maintained field boxes, bins, and handling equipment); abrasion injury, which occurs when fruit rub together when moving on a conveyor or during transport; and cracking, which occurs when the fruit skin fails under excessive internal turgor pressure.

The susceptibility of fruit to mechanical injury depends on several factors. Fruit pulp temperature affects susceptibility to bruising. The relation is often complex, and impact and compression bruising may each respond quite differently to temperature. Handling protocols may specify temperature limits for operations such as sorting, so as to minimize bruise injury. Some fruits should not be picked or handled when their turgor pressure is high, such as in the cooler times of day, or when wet with dew, condensation, or rain. Fruit may be too soft to handle. The decision of whether to grade fruit before or after storage will often depend on how soft it will become during storage.

Insect Injury

Insects infest fruit and cause damage in the field. Growth malformations, russetting, scars, and other blemishes are common symptoms. Some insects lay eggs in fruit, and when hatched the larvae consume the fleshy tissue. The best place to control insects is in the field, so that fruit is protected as it develops. Control is by integrated pest management, where biological and chemical measures are used to best effect. Postharvest disinfestation treatments are applied to fruits so as to satisfy quarantine requirements. The purpose of quarantine is to prevent the introduction of harmful pests, and high levels of security are required. Fumigation schedules based on halogenated hydrocarbons have been used as quarantine treatments for many years, but these fumigants are being withdrawn from use. Other treatments that are used include cold disinfestation, where the fruit is coolstored for about 2 weeks, and heat disinfestation, where it is heated to about 50°C, often for an hour or more. Gamma radiation is effective but unwelcome to the consumer. Alternative fumigants and controlled atmospheres are other possibilities that are currently under study.

Chemical Injury

Chemical injury typically appears as a surface blemish, such as a scald or circular spots around lenticels. Symptoms often take some time to develop or are made worse by storage, so that identification of the problem may be difficult. Chemical injuries that appear after harvest may have been caused by pesticide sprays applied in the field or by chemicals applied after harvest. Apples dipped after harvest in diphenylamine solutions for control of superficial scald may, for example, be injured if the fruit absorbs too much of the active ingredient. Excessive absorption occurs if the dip is prepared at too high a concentration, dipping time is too long, or the fruit is too warm. The fumigants used for postharvest insect disinfestation also cause injury if absorption is too high. Sometimes fruit are accidentally affected by chemicals. Fruit kept in coolstores in which ammonia is used as a refrigerant are liable to injury should ammonia gas leak from the refrigeration system.

Packaging and Mechanical Handling

Correct packaging and handling procedures prevent mechanical damage to fruit during transport and distribution. Fruit can be handled in bulk without packaging if a small amount of damage is acceptable. Bulk handling is appropriate where fruit comes straight from the field to a processing plant, but even fresh market fruit can be shipped in bulk if it is resorted before sale. The technology used to pack and handle depends on the cost and availability of materials and labor, the presence or absence of a crate-exchange system for reusable crates, and the means available to dispose of single-use packages. It is preferable that single-use packages be made of materials that can be converted for recycling. Molded liners may be used within cartons to reduce fruit-to-fruit contact and movement. Packages (cartons or crates) are best handled as unit loads on pallets or slip sheets. Various gluing or strapping systems are used to hold stacks of cartons together, so that they can be moved as a single unit on hand- or fork-lift trucks. Unit loads are much less vulnerable to mechanical damage than are individual packages. Unit loads cool slowly, however, unless they are ventilated for pressure cooling. The usefulness of unit loads is also diminished unless there is an integrated transport-and-materials-handling infrastructure designed for them. It is inefficient to have to break a unit load apart because vehicles will not accommodate it, or loading and unloading equipment is not available.

4.5.4 Managing the Storage, Ripening, and Handling System for Food Safety and Product Quality

The market for food, including fruit and fruit products, changed considerably in the late 1990s. Consumers are more demanding in both developed and developing countries, and supermarkets dominate food distribution. Newspaper headlines of food poisoning outbreaks have increased community concern about food safety. Many governments are reviewing food safety legislation. The food industry is fearful of litigation, and various forms of quality management are being introduced at each level of the industry. Market suppliers are required by supermarkets, market agents, merchants, and other retailers to have documented evidence that their produce is safe and of dependable quality. The major emphasis of these management systems is the control of food safety hazards. A food safety hazard is any microbial, chemical, or physical substance or property that can cause produce to be an unacceptable health risk to consumers.

Microbial Hazards

A limited number of foodborne illness outbreaks have been traced to fresh fruit or fruit products. Documented cases include the contamination of melons, tomatoes, and orange juice with *Salmonella,* apple cider and apple juice with *Escherichia coli,* strawberries with hepatitis A, and raspberries with the protozoan parasite *Cyclospora*. The implicated organisms are commonly of fecal origin. Likely sources of fruit contamination are, therefore, untreated animal manures and sewage, poor sanitary facilities for workers, poor worker hygiene, and contaminated water. Water quality may be especially important in the fruit industry, where it is used in the field for irrigation and spraying, and in the packing house for operations such as washing, conveying, dipping or drenching fruit, and ice making. The important foodborne bacterium *Listeria monocytogenes* is, however, endemic in soil and decaying vegetation. *Listeria* may be a common contaminant of processing plants. The control of microbiological hazards should include ensuring that the water supply is free from contamination, and that hygiene and cleanliness programs for the premises and staff include the control of pests.

Another food safety risk is that of mycotoxins. The main such problem known in fruits is the mycotoxin patulin that is produced by such fungi as *Penicillium expansum* when it infects pome and stone fruits to cause blue mold rot. Patulin is a very toxic substance isolated from the rot-affected and adjacent healthy tissue. The possible risk of mycotoxin contamination is a compelling reason why decayed or moldy fruit should be culled out and discarded.

Chemical Hazards

The product can be chemically contaminated by use of nonapproved or incorrectly applied pesticides or the failure to observe withholding periods. Other sources include water and soil contaminated from previous use by fertilizers and pesticides, and oil, grease, and hydraulic fluid from machinery. Control is by staff training and use of correct procedures, including calibration of spray equipment.

Physical Hazards

Physical contamination includes foreign objects such as weed seeds, sticks, stones, splinters, glass, metal, and rubbish. Control of the hazard is again by staff training and the use of prescribed cleaning and equipment maintenance procedures.

Management Systems and Market Requirements

Marketing fresh and processed fruit to export and domestic destinations now requires documented evidence that food safety and product quality systems are in place. In the United States, the Department of Health and Human Services, the US Department of Agriculture, and the Environmental Protection Agency have identified horticultural produce as an area of food safety concern. Consequently, a plan entitled the president's food safety initiative: "Initiative to Ensure the Safety of Imported and Domestic Fruits and Vegetables" (produce safety initiative) has been introduced to provide further assurance that fruits and vegetables consumed by Americans, whether grown domestically or imported, meet the highest health and safety standards.

The Council of the European Union (EU) has a system of directives concerning food safety issues. All participants in the supply, importation, or marketing of fresh produce must ensure that their product is safe, wholesome, and legal. General rules govern food hygiene and the procedures for verification of compliance with these rules. Member states of the EU must encourage and participate in the development of industry guides to good hygiene practice. Food hygiene measures that ensure the safety and wholesomeness of foodstuffs apply to all stages after harvest, including preparation, processing, packaging, storing, transportation, distribution, handling, and retailing.

In Australia the major supermarket chains require their suppliers to have full food safety documentation based on third-party audit. Any enterprise preparing fruit or fruit products for the market must have a fully documented food safety and product quality system. There are many such systems, but all are based on the ISO9000 standard, which is the internationally agreed model for quality assurance in the production, installation, and servicing practices of a business. The methodology most commonly required by customers and legislation for use in supplier quality management systems is HACCP (hazard analysis critical control point). Public health authorities are reviewing food safety legislation to accommodate these new developments.

Quality management HACCP-based practices for suppliers of fruit and fruit products are a combination of good hygiene practices (GHP) and best management practice (BMP) together with hazard identification and control. These practices include the following:

- working to a customer-approved specification for a product and its handling
- following GHP and BMP in the use of technology for control of hazards
- ensuring that staff are appropriately trained for the food safety and quality tasks they are doing
- ensuring that areas, premises, and equipment where product is handled, processed, packed, and stored are appropriately maintained and kept clean and pest free
- using reliable transport operators who maintain product safety and quality during transport to customers
- maintaining essential documentation and records of safety and quality practices
- maintaining sufficient records to identify and trace product from receipt at the packing or processing premises back to harvest and forward through processing

to dispatch (so that any problems identified at a later stage can be traced back to their origin)
- maintaining regular communication with customers about product performance in the marketplace
- continually reviewing records and practices to identify safety and quality hazards

In this millennium it will no longer be sufficient for fruit and fruit processing businesses to rely on postharvest technology alone to produce a quality product. They will be compelled by the market and legislation to implement HACCP-based documented systems, or to go out of business.

SUGGESTED READING

Beattie, B.B., and Revelant, L.J., eds. (1992). *Guide to Quality Management in the Citrus Industry.* Australian Horticultural Corporation, Sydney, Australia.

Beattie, B.B., McGlasson, W.B., and Wade, N.L., eds. (1989). *Postharvest Diseases of Horticultural Produce,* Vol. 1, *Temperate Fruit.* CSIRO Publications, Melbourne, Australia.

Beattie, B.B., McGlasson, W.B., and Wade, N.L., eds. (1995). *Postharvest Diseases of Horticultural Produce,* Vol. 2, *Tropical Fruit.* Queensland Department of Primary Industries, Brisbane, Australia.

Centers for Disease Control and Prevention. (1998). *Guide to Minimize Microbial Food Safety Hazards for Fresh Fruits and Vegetables.* Food and Drug Administration, US Department of Agriculture, Washington, DC.

Hardenburg, R.E., Watada, A.E., and Wang, C.Y. (1986). *The Commercial Storage of Fruits, Vegetables and Florist and Nursery Stocks,* rev. ed. Agriculture Handbook No. 66. US Department of Agriculture, Washington, DC.

Kader, A.A., ed. (1992). *Postharvest Technology of Horticultural Crops.* No. 3311, ANR Publications, University of California, Oakland, CA.

Lipton, W.J., and Harvey, J.M. (1977). *Compatibility of Fruits and Vegetables during Transport in Mixed Loads.* Marketing Research Report No. 1070. US Department of Agriculture, Washington, DC.

Peleg, K. (1985). *Produce Handling, Packaging and Distribution.* AVI, Westport, CT.

Ryall, A.L., and Pentzer, W.T. (1982). *Handling, Transportation and Storage of Fruits and Vegetables,* Vol. 2, *Fruits and Tree Nuts,* rev. ed. AVI, Westport, CT.

Shewfelt, R.L., and Prussia, S.E., eds. (1993). *Postharvest Handling. A Systems Approach.* Academic Press, San Diego, CA.

Snowdon, A.L. (1990). *A Colour Atlas of Post-harvest Diseases and Disorders of Fruits and Vegetables,* Vol. 1, *General Introduction and Fruits.* Wolfe Scientific Ltd., London.

Snowdon, A.L. (1991). *A Colour Atlas of Post-harvest Diseases and Disorders of Fruits and Vegetables,* Vol. 2, *Vegetables.* Wolfe Scientific Ltd., London.

Snowdon, A.L., and Ahmed, A.H.M. (1981). *The Storage and Transport of Fresh Fruit and Vegetables.* National Institute of Fresh Produce, London.

Thompson, J.F., Mitchell, F.G., Rumsey, T.R., Kasmire, R.F., and Crisosto, C.H. (1998). *Commercial Cooling of Fruits, Vegetables, and Flowers.* No. 21567. ANR Publications, University of California, Oakland, CA.

Wade, N.L., and Graham, D. (1987). A model to describe the modified atmospheres developed during the storage of fruit in plastic films. *ASEAN Food Journal,* **3**, 105–111.

Wade, N.L., and Morris, S.C. (1982). Causes and control of cantaloupe postharvest wastage in Australia. *Plant Disease,* **66**, 549–552.

Watkins, J.B., and Ledger, S. (1990). *Forced-air Cooling,* 2nd rev. ed. Information Series QI 88027. Queensland Department of Primary Industries, Brisbane, Australia.

Wills, R.B.H., and Ben-Yehoshua, S. (1998). Effect of low ethylene levels on senescence of fruit and vegetables. *14th International Congress on Plastics in Agriculture, Tel Aviv, Israel, March 1997.* Laser Pages Publishing, Jerusalem, Israel, pp. 593–599.

Wills, R.B.H., McGlasson, W.B., Graham, D., and Joyce, D.C. (1998). *Postharvest: An Introduction to the Physiology and Handling of Fruit, Vegetables and Ornamentals*, 4th ed. University of NSW Press, Sydney, Australia, and CAB International, Wallingford, UK.

Wills, R.B.H., Ku, V.V.V., Shohet, D., and Kim, G.H. (1999). Importance of low ethylene levels to delay senescence of non-climacteric fruit and vegetables. *Australian Journal of Experimental Agriculture*, **39**, 221–224.

Chapter 5

Production of Nonfermented Fruit Products

Peter Rutledge

5.1 INTRODUCTION

The extraction of juice from fruit is an ancient art dating from the earliest of records, where wine is often mentioned. Fermentation of fruit juice so that the alcohol content preserved the fermented juice was one of the earliest forms of food preservation by the human species. Although fermented beverages are dealt with in another chapter, the extraction of fruit juice must be considered a mature technology. With rapid changes taking place in most technologies during the past century, the manufacture of fruit juice has progressed from the farm or cottage industry into the efficient technology of modern food processing.

Throughout the temperate areas of the world, fruits used for the major quantities of juices are citrus (predominantly orange), pome and grape, or vine fruits. Some production of stone fruit and berry juices is carried out but only in small quantities. Pineapple dominates tropical fruit juice production, with highly flavored fruits such as mango, passion fruit, and guava becoming more popular as blending juices.

Methods of extracting fruit juices are dependent on the structure and edible portion of the fruit. Preservation methods include thermal treatments, freezing, chilling, concentration (drying), and, for some clear juices, fine filtration. Juices may be taken apart by removing volatile flavor components, water, bitterness, and acidity and then recombined to produce a consistent product. Fruit-derived drink bases may be manufactured from the remaining fruit material after the juice has been extracted.

5.2 FRUIT QUALITY

Fruits used for juice manufacture are often those rejected because of the high specifications for the fresh market, or they may be off-cuts from other fruit processes or fruit that is specifically grown for juicing. Juicing is near the bottom of the fruit usage chain, so care must be taken to ensure that only sound material is used (Figure 5–1). Fruit that is infected with molds, starting to ferment with yeasts, or is rotten is not suitable for juicing and must be removed from the processing line, preferably before washing so that microbial and off-flavor contamination of the juice is prevented.

Fruit maturity is important for optimum flavor because flavor volatiles are produced near the fully ripe stage. However, if the fruit is allowed to ripen to the stage where senescence begins, the structure of the fruit is degraded, and this can cause problems during extraction. For example, pears that are overripe form a "porridge" from which it is almost impossible to extract the juice.

Variety also plays an important part in the quality of the extracted juice. Some varieties are more suitable than others for processing. The classic example of this is the difference between processed Valencia and navel oranges, with the production of bitter limonin in the navel juice making it unpalatable. Juices from apples of different varieties may be blended to give the required flavor characteristics.

Figure 5–1 Unloading of juice fruit from bulk transport.

5.3 TEMPERATE FRUIT JUICES

The climate in Europe, the northern United States, and smaller areas in the southern hemisphere is suitable for growing temperate fruit. Most significant of these temperate fruits are orange, apple, and pear, which make up the bulk of processed juice used for nonalcoholic purposes. Smaller volumes of juice from grape, berry, and stone fruits are produced.

5.3.1 Orange Juice

Valencia and navel are the two generic varieties of oranges produced in most countries. Other smaller plantings are available for juice, but again these can be separated into the two principal types above. Valencias have a 6-month season, which is followed by a similar season for navel oranges. Valencia types produce excellent orange juice that is slightly sour at the start of the season, changing through to slightly bland at the end of that season. This is a reflection of the maturity of the fruit and the requirements of processors to extend their season for as long as possible.

Navel orange juice, however, is not only sour at the start of the season but also produces bitter processed juice because of the production of limonin. This bitterness is not apparent in the market fruit because it takes time or thermal processing to produce the bitter limonin component. As the season progresses, the amount of bitterness in the juice diminishes and is masked by the increasing sweetness of the fruit. Blending preserved Valencia juice with navel juice during the navel season does extend the use of bitter navel juice. The demand for orange juice exceeds the supply of Valencia orange juice alone, making it necessary to juice navel oranges.

The manufacture of orange juice also includes the production of several by-products. Before the juice is extracted, the fruit is washed and the skin oil is removed by passing the fruit over an oil extraction machine, which inflicts hundreds of small cuts on the skin (Figure 5–2). The oil sacs in the surface of the skin are ruptured and the oil is removed by washing. The oil and washwater are centrifuged so that the emulsion is removed from the water, which is returned to the machine. The emulsion can be "broken" in a high-speed centrifuge to yield the peel oil. In other systems, the oil is removed from the juice after juice extraction, during the finishing operation.

Citrus juice is extracted by either a reaming action (Figure 5–3) or a crushing action, as in extractors manufactured by the Food Machinery Corporation (FMC). The reaming operation involves cutting the fruit in half, followed by the automatic presentation of each half to the reamers to remove the contents. The FMC system uses a cup-shaped set of fingers top and bottom that mesh, as shown in Figure 5–4. As the fruit is crushed between the cups, a small piece of skin is cut from the center of the bottom cup and the contents are squeezed through the cut hole into a tube. The tube is perforated and, via a pumping action, the juice is forced through the perforations into a manifold for removal from the machine. If the skin oil has not been removed prior to extraction, the juice from the crushing operation has a higher oil content than that from the reaming method, assuming correct adjustment of the extractors.

Figure 5–2 Machinery for removing citrus skin oil. Courtesy of Brown Machinery Australia Ltd.

Figure 5–3 Brown reaming-type citrus juice extractor. Courtesy of Brown Machinery Australia Ltd.

Figure 5–4 FMC-type citrus extractor: left, fingers meshed; right, fingers released.

Because the citrus juice from the FMC extractors has had to pass through 1-mm diameter holes in the perforated tube, the screening process to remove unwanted material has been partially carried out. Juice cells and excess pulp that remain in the juice are easily removed during the following screening operations. The reaming method results in an excess of white skin lining (albedo), segment walls, cell tissue, pieces of skin, and seeds, which must be removed from the juice. A small screw press, paddle, or brush finisher using screens with 0.5-mm diameter holes is often used to fine-screen the juice. In some plants, the juice is centrifuged to remove pulp, which can then be used in by-products.

Orange juice is heat treated as soon as possible after extraction to inactivate pectinases (enzymes) naturally occurring in the juice. The pectin in the juice is responsible for holding the "cloud" or suspended solids in the juice. If the pectin is degraded by the pectin esterase, then the attractive cloud in the juice will fall, leaving a serum (clear liquid) and a deposit on the bottom of the container. The required heat treatment is 95°C for 30 seconds, which is more severe than that required for microbial stability.

Methods of preservation and marketing of orange juice will depend on the local food laws and customer expectation and acceptance. Orange juice that has been fully pasteurized and received further heat treatments during concentration may undergo detrimental changes in flavor. Customers in the Asian region find thermally induced flavors unpleasant and prefer to purchase orange juice without these flavors.

Orange juice that has been extracted and pasteurized at 65–70°C for a few seconds to inactivate yeasts and molds will have a shelf life of up to 21 days when stored at 4°C or less. Juice treated in this way does not exhibit undesirable flavors. A shelf life of up to 3 months at chill temperatures can be attained with orange juice treated with chemical preservatives. Preservatives used, when allowed by law, are typically sulfur dioxide at 100 ppm and a combination of sorbic and benzoic acids at up to 400 ppm. Thermal treatment at 65–70°C does not inactivate pectinases, and some separation of the orange juice cloud is evident during storage. Because the Valencia season lasts for only half the year, the storage of "fresh-like flavored" juice has become an interesting problem. Some manufacturers have found that storage of untreated orange juice at temperatures close to the freezing point extends the storage time without noticeable loss in orange flavor. There is a temptation to package this juice and give it a short shelf life under refrigeration. In practice this has resulted in several cases of food poisoning in the United States and Australia because there has been no effective treatment given to eliminate vegetative microbial cells. Although the market potential of this untreated juice may be inviting, the risks to customers and businesses is extremely high.

Orange juice may be hot filled into cans from the pasteurization step, inverted, and spin cooled to give a high-quality product. Aseptic processing into Tetra Pak or bag-in-the-box systems is common.

Problems associated with sour and bitter juice may be addressed by modifying the juice using adsorbents that remove the unwanted characteristic. These adsorbents are in the form of polymer beads. Johnson and Chandler (1985) investigated systems for removing the bitterness from orange juice, and in 1986 they published work on the removal of sourness from citrus juices. Their method is to remove the fruit pulp by centrifuging so the pulp will not clog the bed of beads. The juice is then pumped through the beads and the citric acid or the limonin is removed, depending on the bead used. The pulp is returned to the juice after it has been treated. Beads may be regenerated by passing caustic soda solutions through the bed of beads.

Concentration of orange juice is used to store juice for blending and for transport to reconstitution plants. The soluble solids of orange juice contain mainly sugar and acid. The percentage of soluble solids is called °Brix (°B) and in orange juice will vary from about 9 to 14°B. Orange juice is passed through multiple-effect evaporators under vacuum, which produce concentrate in excess of 65°B. Concentrate is stored chilled or frozen to reduce the non-enzymatic browning reactions and attack from *Saccharomyces rouxii*. Highest-quality concentrate is produced by removing the flavor volatiles from the fresh juice, at low temperature, in a spinning cone column (see Figure 5–10) followed by concentration. The volatiles are returned to the concentrate, which, when diluted, will have the full orange flavor. Other recovery systems remove the volatile components from the distillate during concentration. These volatiles will have undergone some thermally induced changes during this process. When the volatiles are added back to the concentrate, it results in an inferior flavor compared with concentrate using the spinning cone column method.

Other products produced from oranges include pulp wash, where the orange pulp from the finisher or centrifuge is water washed and then refinished. Pulp wash is a water extract of orange pulp and may be used in other products, such as drinks and cordials, where their use is allowed by food regulations. Some manufacturers use a countercurrent extractor to extract useful material from the skins, which is then suitable for the manufacture of cordial bases. Skins that have been steamed, milled, stone milled, and dried also make a comminuted drink base.

5.3.2 Citrus Juices

Lemon, lime, grapefruit, tangerine, and mandarin are other important citrus fruits. The processing of these fruits into juices is similar to the manufacture of orange juice.

5.3.3 Apple Juice

The process used for producing apple juice from apples depends on the volume of production and the amount of capital to be invested. There is a distinction between the processing methods used by a small two- or three-person operation run by a grower and the larger factory operation run by a company. However, the following processing steps generally apply to both situations.

Before washing, the fruit is inspected to remove rotten or moldy apples. This is conveniently carried out on an inspection belt in small plants, but with developments in technology such fruit can be automatically removed by scanning systems. Cleaning operations depend on the condition of the fruit entering the factory. If the apples have been handpicked and are free from extraneous matter they will require washing only. If the apples have been machine harvested, which is possible for juice production, then leaves and twigs will require removal by a dry cleaning operation using a high-velocity fan. Washing removes dirt and water-soluble agricultural spray residues, and it is important to rinse the apples with fresh water as they are elevated from an immersion tank.

Apples are milled for most extraction systems, but for diffusion extraction they are sliced into 3-mm ripple-cut slices. Hammer mills using 1- to 1.5-cm screens are suitable for producing apple pulp for pressing. This apple pulp is not too finely divided and has the required structure for the pressing operation. To prevent oxidative or enzymatic browning, some ascorbic acid is often added at the milling stage, especially if there are likely to be any delays in further processing operations.

Juice may be extracted from the apple pulp or slices by hand-operated rack and cloth pressing, automatic batch pressing, continuous machine pressing, centrifugal separation, or diffusion extraction.

The capital cost of the rack and cloth method of extraction is relatively cheap, but it is very labor intensive. A coarse-weave nylon cloth is spread over the top of a frame placed on a slatted wooden divider. Apple pulp is poured into the cloth until the frame is full. The edges of the cloth are folded over the pulp, forming a cloth-bound bed of apple pulp called a "cheese" because it resembles the European-style bound cheese. The frame is removed, a divider is placed on the cheese, and another cheese is built on top of the first, and so on. A stack of several cheeses is made, depending on the capacity of the press. The stack is placed in a hydraulic press with a collection tray located under the stack. Pressure is gradually applied to the stack until the free-run juice is expressed, and then pressure is applied up to about 1500 kg in some presses. This process normally takes about 30 minutes. Juice from the rack and cloth press is low in suspended solids and is often processed as cloudy juice without further treatment. Some difficulty is experienced when using old fruit, and a filter aid, such as diatomaceous earth, is put on the cloth before adding the pulp. Juice yield for this method is about 70% of the weight of pulp when using fresh apples but can drop to below 60% with old, mushy apples.

There are several designs of automatic batch presses available, with Willmes, Bucher, Bucher-Guyer, or Atlas Pacific commonly used for apple pressing. These presses usually need press aid and pectinase added to the apple pulp to aid the pressing operation. The press aid, which is normally shredded paper or rice hulls, forms channels through the press cake for the juice to be expressed. Care must be taken with press aid to ensure that unwanted flavors are not added to the juice.

Automatic batch presses are the most varied in the method used to press the pulp. One uses a large piston to force the pulp into a perforated cylinder, while another has a solid cylinder with perforated flexible nylon tubes between the piston and the end of the cylinder to extract the juice. Another design has the pulp spun onto the walls of a perforated cylinder, and an inflatable bag pushes the juice through the cylinder walls. These types of presses give juice yields from good apples of 70% to 80%, but the yields from old fruit can drop dramatically. The advantage of these machines is that the labor input is small and they can juice at least 5 tonne/hour of fruit.

Common commercial continuous presses are the screw press and the belt press. The most common is the screw press, which is also used for extracting other kinds of fruit for juice. The screw press consists of a tapered screw surrounded by a close-fitting screen. The end cap of the machine, which resists the exit of the pumice, is held in place by springs or air pressure. The pulp travels along the screw with the pressure being built up by the resistance of the end cap. The juice is expressed through the screen and the press cake is pushed out from the end of the screw when sufficient pressure is applied to overcome the resistance of the end cap. Press aid may be added to the pulp when the apples are old and mushy. A yield of 70% to 75% can be expected when using good, crisp fruit. Juice from the screw press often contains excessive suspended solids that have to be removed before a cloudy apple juice can be processed.

Belt presses are continuous presses having two belts through which juice may pass (Swientek, 1985). Apple pulp is spread between the belts, which enclose the pulp by coming together before traveling between a series of rollers. The rollers exert pressure on the pulp and juice is expressed through the belts. In some cases, about 1% wood fiber is mixed with the pulp as a press aid. A correctly maintained belt press will yield about the same as a screw press.

Centrifugal continuous extraction of pulp is possible using a screening centrifuge such as the Sharples Conejector. In this machine, the pulp is pumped onto the center of a fast-rotating screen. The juice is forced through the screen and the pumice is centrifugally removed. The juice is unavoidably aerated during this operation, giving a darker juice than other methods of extraction. The yield is slightly less than continuous pressing, and the suspended solids are slightly higher.

Continuous diffusion extraction can be used when the juice is to be manufactured into apple juice concentrate (Figure 5–5). Countercurrent extraction on an inclined single or twin screw conveyor is the method normally used. A screen is fitted in the lower end of the machine so that liquid can be separated from the solids and removed. Apples are sliced into 2-mm ripple-cut slices and dropped into the lower end of the extractor just above the screen. A small stream of water or distillate from the evaporator

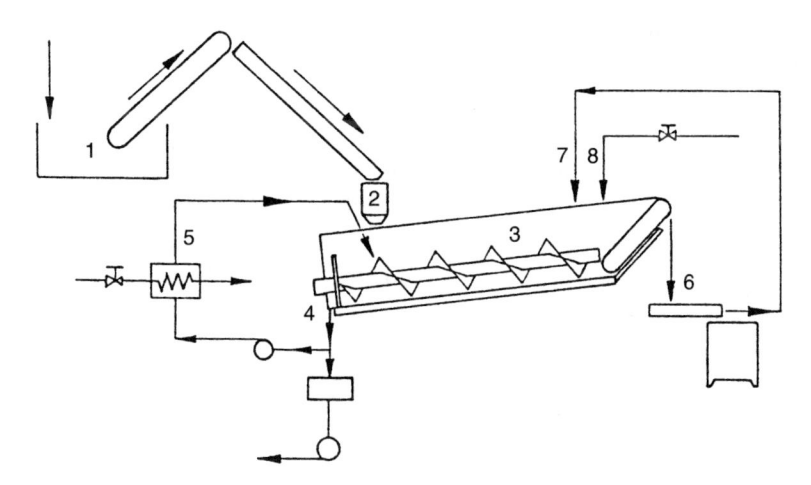

Figure 5–5 Countercurrent extraction of apples. 1, Wash tank; 2, slicer; 3, screw conveyor; 4, juice exits through screens; 5, beat exchanger; 6, screw press; 7, returning liquid from screw press; 8, water inlet.

is added at the top of the conveyor and runs countercurrent to the apple slices. The liquid level is adjusted by changing the level of the take-off tube behind the screen. A small stream of the juice is heated in a heat exchanger to about 70°C and sprayed onto the incoming fruit to inactivate enzymes and to reduce the thermal load on the hot water jacket of the conveyor. The slices that exit the top of the conveyor are pressed, and the liquid is returned to the extractor at the point on the conveyor where the soluble solids are similar. The extraction time is on the order of 1 hour. Casimir (1983) measured yields of soluble solids of 89–91% from Granny Smith apples, compared with 68–69% from Bucher-Guyer presses using the same batch of fruit.

5.3.4 Pear Juice

Extraction of pear juice is similar to that for apple juice. The use of pear juice as a fruit syrup for canned fruit requires it to be clarified, including the brown color, by amino acid removal. Color in pear juice may be removed by polymer bead systems (Cornwell and Wrolstad, 1981), although aeration is required to complete the phenolic reactions before removal. If aeration is not carried out, further darkening of the juice will occur on storage. Darkening problems in pear juice concentrate manufactured from treated juice are the result of browning reactions more than phenolics.

5.3.5 Stone Fruit Juices

Plums, peaches, apricots, and cherries are the stone fruits most commonly processed into pulps and juices. With these fruits, the normal crushing operation does not give a reasonable yield because of the structure of the fruit flesh. Plum and cherry juice browns rapidly because of enzymatic action once the fruit is broken. To prevent enzymatic action, the whole fruit is heated to a temperature that will inactivate the enzymes before it is crushed. This heating is called a hot-break process. This can be carried out using a tomato hot-break machine or by steaming the fruit for several minutes (small plums need 7 minutes) to inactivate the enzymes and to break the structure of the fruit. The fruit is passed through a paddle finisher to remove the skins and seeds from the pulp. The pulp is cooled to about 50°C and then pectinase is added to break the pulp structure further. A decanting centrifuge is used to remove a partially clarified juice from the pulp. Prune plums that are too small for drying give a clarified yield of between 50% and 60% juice containing soluble solids in excess of 20°B. The juice when concentrated to 75°B has a deep maroon color. Concentrate with these color characteristics is ideal for blending with lighter-colored juices.

Peach pieces from canning lines, where the peach has been peeled and destoned, that are unsuitable for canning may be pulped mechanically and pasteurized in a scraped-surface heat exchanger. The pulp is concentrated to 30°B without enzyming and aseptically filled into bulk systems for further processing as pie filling or for drying for confectionery.

5.3.6 Berry Juices

Berry juices are usually highly flavored and colored and as such are ideal for blending with other juices such as clarified apple or pear juice. Importance must be placed on retaining these characteristics during processing into juice. Berries can be broken between rollers or in mills, but the juice may still be held in the cellular structure of the fruit flesh. Enzymes may be added to break down the structure and the juice recovered in a decanting centrifuge. Heat may be used instead of enzymes, but there might be some loss of flavor and color as a result.

Preservation of the juice will depend on its ultimate use, but freezing is not uncommon if flavor is important. Thermal processing will allow long-term storage. Concentration of berry juices and subsequent storage of the concentrate, chilled or frozen, is an option if the flavor volatiles are removed before concentration and returned to the concentrate.

Grapes may be considered as berries, and some grape juice is clarified for use in blended juices. (The extraction of grapes can be found in the chapter on fermented beverages.) The removal of tartrate or the prevention of tartrate crystallization, sometimes known as "wine stone," is a problem for long-term storage of the juice. Exchanging sodium for potassium in the juice by ion exchange will change the relatively insoluble potassium hydrogen tartrate for the more soluble sodium salt. Concentrated grape juice can have a sugar precipitate if it is concentrated above about 55°B, but this will quickly redissolve on dilution.

5.4 TROPICAL FRUIT JUICES

Tropical fruits require some care in processing if the delicate flavors are to be retained. In some tropical fruits, the acidity normally associated with temperate fruits is lower (with a corresponding rise in pH). Acidification of the juice from these fruits to a pH of about 4 will be required to enable them to be processed at pasteurization temperatures.

5.4.1 Pineapple Juice

Pineapple juice is often recovered from the ejected skins and cores from the Janaka machines that cut the pineapple in preparation for canning. Other sources of fruit for juice are small pineapples, physically damaged fruit that is unsuitable for canning, and off-cuts from the canning line. This waste material is utilized by using a screw press to recover the juice. In some cases, a finisher with a fine screen or centrifuge is used to ensure that a light, opalescent juice is collected. Pineapple juice is a high-acid product that is preserved in a manner similar to that for normal fruit juices.

5.4.2 Papaya Purée

Papaya is one of the bland tropical fruits that may require acidification to ensure a safe product when processed by normal methods. Fruit is chopped into pieces large enough to fit into a paddle finisher, where the seeds and skin are removed from the flesh. Only ripe fruit should be used because the finisher will not extract efficiently if hard pieces of fruit are fed into the machine. Pulp is blended with citric acid to give the required pH, which should be below 4.2 but is often as low as 3.5. Blended pulp is then fed through a scraped-surface heat exchanger to heat the pulp to 94°C. It is held for 2 minutes to ensure microbial stability, followed by cooling in a scraped-surface heat exchanger prior to aseptic packaging (Morris, 1982). The heated pulp may also be hot filled into cans, inverted, and cooled. In some countries, the papaya fruit is comparatively high in nitrates, which will corrode tinplate cans very quickly and unduly shorten the shelf life of this product.

Freezing papaya pulp has been used by some American processors for the preservation of unacidified pulp. This is a relatively expensive method to use, but it is necessary if unacidified pulp is required by the customer. The other alternative would be to give a botulinum-type thermal process, but the pulp would be badly damaged by the heat treatment, with the loss of the delicate flavor.

Treatment of papaya pulp with pectinases will reduce the pulp to a solid and a serum after separation in a decanting centrifuge. The serum may be pasteurized if acidified and concentrated, but this product does not have the appeal of the pulp.

5.4.3 Mango Pulp

Mango pulp is a popular blending material for many types of mixed fruit juice product. The mango has a large stone; stringy, sometimes fibrous flesh; and a skin that, if included in the pulp, has a characteristic terpene flavor that can be unpleasant. Pulp is extracted by steaming the fruit for 2 to 3 minutes, followed by a rough chop to break the fruit but not the stone. Chopped mango is extracted with a paddle finisher, using a screen fine enough to eliminate the fibrous material in the flesh. This produces a pulp that may need acidification for preservation, like the papaya pulp.

Mango is also extracted using countercurrent systems, which have been described in the extraction of apple juice (section 5.3.3). The mangoes are washed and then squashed to break the fruit without breaking the stone. The squashed fruit is dropped into the lower end of a countercurrent extractor (Figure 5–6) with water at 65°C running countercurrent to the fruit. Reheated juice is sprayed onto the incoming fruit to inactivate enzymes and reduce the thermal load on the system. A wiper is required on the bottom screen to prevent the build-up of fibrous material, which would eventually block the screen. The juice from the extractor has been diluted up to 10% by water and requires concentration to bring it back to the original strength. Mango from the countercurrent extractor may be concentrated to 30°B without enzyme treatment. Excellent-quality, light yellow-orange-colored mango juice concentrate is produced from countercurrent-extracted mangoes.

5.4.4 Passion Fruit Juice

Passion fruit has two main varieties, the purple and the yellow types. The purple variety has a flavor superior to the more prolific and larger yellow variety. Passion fruit

Figure 5–6 Extraction of mango using a countercurrent extractor.

has a delightful flavor (Whitfield and Last, 1986), which is used for flavoring drinks or for blending with other juices.

Extraction of passion fruit pulp is accomplished by dropping the washed fruit between two rotating converging cones. Fruit is caught in the nip of the cones and burst as the cones rotate toward the bottom of the machine, where the clearance is reduced to the thickness of the skin of the fruit (see Figure 5–7). The skins carry through in the cones and the pulp drops into a finisher that removes pieces of skin and, depending on the finisher screen size, also the seeds.

Passion fruit juice has a higher starch content compared with other fruit juices. If the starch is not removed, thermal processing produces a highly viscous product that is ideal for some uses. However, the starch may be removed in a decanting centrifuge, leaving a juice that can be concentrated to about 50°B. Casimir et al. (1981) have discussed the types of evaporation equipment suitable for concentrating passion fruit juice. The important requirement for the equipment is a very short residence time in the evaporator to prevent thermal damage, which will destroy the flavor of the juice.

High-quality passion fruit juice is often frozen in preference to thermal processing so the flavor remains intact. Freezing is an expensive option because the cost of storage

Figure 5–7 Twin-cone passion fruit extractor with the top cover removed.

increases with time, but the flavor of passion fruit is its main attribute and worth preserving. Short-time pasteurization processes from spin-cooking cans or for aseptic processing also produce a high-quality passion fruit juice.

5.4.5 Guava Pulp

Guavas are similar to pears for processing because both fruits have stone cells (sclereids) in the flesh. Stone cells in juices give an unpleasant "gritty" mouthfeel that is unacceptable. The guavas are washed and chopped for presentation to a paddle finisher fitted with a screen containing holes of about 1-mm diameter. A second finisher with a finer screen of about 0.5-mm holes is used to remove stone cells. An alternative method to remove the stone cells is to grind the juice in a stone mill; this will make the stone cell small enough not to have an unpleasant mouthfeel in the juice. Stone cells may also be removed by centrifuging the juice, but this may also remove some of the wanted pulp material.

Guava pulp may be preserved by freezing without pasteurization. Aseptic processing is also a popular method of preserving the pulp (Figure 5–8), which requires a temperature of 95°C for 30 seconds for microbial stability.

Figure 5–8 Aseptic processing of fruit pulp. Courtesy of Tetra Pak, Ltd.

Pectinase-treated guava pulp will produce a clear guava juice after pressing or centrifuging in a decanter centrifuge. If a very clear juice is required, then filtration will give a polished fruit product. This juice can be used for clear jellies or blending with other clarified juices such as apple juice. Guava pulp may be concentrated to about 30°B and the clarified juice to 70°B.

5.4.6 Universal Extraction of Tropical Fruit Juices

In larger factories, specialized extraction equipment for specific fruit is used and this has been mentioned during the discussion on the various types of fruit. However, for a small-scale operation, which occurs in many Pacific Islands, a more universal extraction machine is required. The Polyfruit extractor is a machine that has been shown to extract the pulp from bananas, pineapples, papaya, citrus, custard apples, and guava. To change from one fruit to another involves, in most cases, a simple adjustment to the clearance in the machine.

Fruits do not require size grading for Polyfruit extraction. Fruit falls between two rotating drums, which convey the fruit onto a vertical blade. The fruit is cut in half, and each half still being conveyed by the drum is pushed against perforated screens. Extracted pulp passes through the screens into a closed chamber for pumping to a finisher. Skin or peel is expelled through inclined chutes at either end of the machine. Polyfruit might not be as efficient as some of the specialized equipment, but the versatility more than compensates for any small losses.

5.5 CLARIFICATION OF FRUIT JUICES

The initial stage of juice clarification is the removal of excess pulp, which can be carried out by centrifugation in a decanting centrifuge or by finishers with fine screens. Juice is clarified by removing pectin, starch, gums, proteins, polyphenolics, metal cations, and lipids, which may cause hazes before or after preservation. Specialized enzyme preparations are available for specific fruit juices; for example, a mixture of pectinase and amylase might be used for early season apples. If an enzyme is used in the mash (Schmitt, 1985), then this mixture might include some arabinase.

The traditional method of fining is to heat the juice to the required temperature and stir in the enzyme or enzyme mixture, wait until the enzyme has acted, and then add agents that will bring down tannins and other unwanted material (Heatherbell, 1976a,b). These agents are gelatin, bentonite, and silica-sol, and flocculation usually occurs rapidly after their addition. This is carried out at temperatures near ambient to get good flocculation; however, Grampp (1977) describes a system for hot clarification that is faster than the traditional method. The juice is decanted or racked from the lees and filtered using a filterpress or a diatomaceous earth filter such as a rotary vacuum filter.

Control of the traditional method of fining was often difficult, with occasional unexplained additional flocculation occurring after filtration. This often occurred when pectin tests, using acidified alcohol, were carried out by inexperienced personnel who were not able to distinguish a fine pectin haze. Membrane filter systems have become a practical reality since the cross-flow system started to be used industrially in the 1970s (Pepper, 1987) and these have eliminated many of the haze problems.

Ultrafiltration of fruit juice has become the industry standard because of the simplicity and effectiveness of the system. The juice is enzymed to reduce clogging of the filter and then passed across the filter under pressure. The pores in the filter are small enough to hold back tannins and other compounds that cause haze formation in the clarified juice. Yields of juice from ultrafiltration are typically in the 95% to 97% range, compared with 90% to 93% from the traditional method. Chemicals used in the traditional fining are no longer required, although the ultrafilter cartridges have to be replaced.

5.6 METHODS OF PRESERVATION

Prior to any preservation steps, the juice should be deaerated to remove dissolved or entrapped oxygen, which will react with ascorbic acid and darken the juice. This entails spraying the juice into a chamber that has been partially evacuated and then pumping the juice from the bottom of the chamber.

5.6.1 Thermal Treatment of the Juice

Because fruit juice is a liquid food, the easiest method of heating is by heat exchanger. There are several types of heat exchangers, and the choice of type will depend on the amount of insoluble solids in the fruit juice. Clear juices and those with only a small amount of insoluble solids can be heated in a plate heat exchanger, preferably with a regeneration section.

Spiroflo-type heat exchangers have more widely separated internal surfaces than the plate type and, therefore, will accept juices with a higher proportion of solids. These can be used for quite viscous products and could be used for single-strength pulp. Shell in-tube heat exchangers are similar to Spiroflo but do not have the advantage of turbulent flow.

Scraped-surface heat exchangers are used for pulp with high levels of solids and for concentrated pulps such as 30°B peach pulp concentrate. Scraped-surface heat exchangers have a jacketed drum that can be heated or cooled. The product is wiped with a scraper along and around the inside of the drum. This equipment will quickly heat or cool products with high viscosity and poor thermal characteristics.

5.6.2 Canning

With the exception of some tropical fruits, fruit juices and pulps are acid products with a pH of less than 4.2 and often in the 3.5 to 4.0 range. To inactivate microbial growth at these pHs, the juice requires heating to 80 to 93°C for a few seconds only. Plain tinplate cans are used for fruit juice because the tin has a reducing effect on the juice. Juice that has darkened as a result of oxidation will return to its natural color under the reducing influence of the tin–acid reaction of corroding tinplate cans. This is the major difference between packing juice in cans and in plastic or plastic–cardboard containers. Juice packed in plastic will darken with time.

Preserving juice in tinplate cans can be carried out by hot filling, closing, and cooling the cans at speeds in excess of 500 cans per minute. The juice is heated in a heat exchanger to the required temperature, filled into the can, steam flow closed, inverted to pasteurize the end and, ideally, spin cooled under water sprays. The ideal rate of spin for 74-mm diameter cans is 180 rpm for single-strength thin fruit juice.

Spin cooking juice in cans that have been vacuum closed is another efficient method for canning juice. Vacuum closers are slow, up to 200 cans per minute, but the cans are thermally processed by spinning them in atmospheric steam and then cooling them under water sprays. Processing time for orange juice is between 60 and 90 seconds, and cooling takes about 45 seconds. Canned juices with superior quality are produced from spin cooling or spin cooking–cooling.

Cooking cans in boiling water is another method, but the product is usually heat affected and the marketplace will not accept this poor-quality product.

5.6.3 Aseptic Processing

Technically, aseptic processing is a thermal process like canning, where the product is pasteurized and then filled without contamination into sterilized containers and hermetically sealed. The Tetra Pak system seen in Figure 5–9 uses a heat exchanger to pasteurize and cool the juice, which is filled into a cardboard–aluminium–plastic laminated box. The box material enters the machine as a sheet that is sterilized with hot peroxide solution, formed into a tube, filled, sealed, cut, and then folded into the box shape.

Figure 5–9 Tetra Pak aseptic packaging machinery. Courtesy of Tetra Pak, Ltd.

Other aseptic systems are designed for filling bags that will fit into 200-liter drums or into boxes about the size of 0.5-tonne fruit bins. It is possible to fill road and rail tankers using the aseptic system. In the case of tankers, they are steam cleaned and then filled with sterile air as they cool before filling with product.

5.6.4 Bottling

Bottling fruit juice is similar to hot filling cans but uses glass bottles. The bottles pass under hot-water sprays after capping to pasteurize the cap, or the cap can be put on the bottle under steam. Cooling is achieved by passing the bottles under water sprays, using warm water initially to prevent breakage through thermal shock on the glass.

Plastic containers are more difficult to fill because of the thermal instability of the plastic. Bottles made from polypropylene will withstand the temperature of normal hot filling, but the material is unsuitable for storing fruit juice because of its high permeability to O_2. Early plastics used for storing fruit juices were unstable above 72°C. These bottles may be used if they can be held at 70°C for 10 minutes after filling. Juice is heated in a plate heat exchanger to 85°C and then cooled in a regeneration section to between 71 and 72°C before filling the bottles. The bottles are squeezed during filling to remove any headspace and are then held for 10 minutes at 70°C before water cooling. Polyester materials such as polyester terephthalate (PET) can be hot filled like glass bottles.

The shelf life of plastic containers with fruit juice is in the order of 6 months before the permeation of oxygen through the plastic either darkens the juice or oxidizes the constituents such as ascorbic acid. Although plastic containers have technical difficulties with filling and pasteurization, they are favored by marketing because of the different shapes available and the good presentation of the product.

5.6.5 Chemical Preservatives

Juice treated with chemical preservatives such as sulfur dioxide and sorbic and benzoic acids will have a shelf life of many months at 4°C. In some countries, the use of preservatives in juices is banned. The levels required to preserve juice at 4°C are about 100 ppm sulfur dioxide (usually added as sodium metabisulfite) and 400 ppm of a combination of sorbic and benzoic acids (usually added as the sodium or potassium salts).

5.6.6 Freezing

Freezing of juice is carried out where the pH of the pulp is high or the preservation of a particular flavor component is important. The juice is chilled in a heat exchanger and then filled into plastic containers that are frozen in a blast freezer using air temperatures of about –40°C. The juice should be frozen to –18°C and stored at that temperature.

5.6.7 Filtration Sterilization

Membrane filters are able to filter clarified juices such as apple and grape juice so finely that the yeasts and molds that normally spoil juices are eliminated. Mulvany

(1966) describes some early filter systems. Grape juice stored for future wine fermentation is commonly sterile filtered and stored in sterile tanks. Sterile filtering replaces the treatment of the grape juice with sulfur dioxide and eliminates the problem associated with removing the preservative before fermentation can take place.

5.6.8 High-Pressure Processing

Emerging technologies for the preservation of fruit juices in particular are high-pressure processing, thermosonication, and pulsed electric field treatment. High-pressure processing has been commercialized in Japan, Mexico, France, Spain, and the United States, while the other techniques are still in the research stage. High-pressure processing subjects the juice to an isostatic pressure of about 400 to 500 MPa for commercial operation, although pressures vary between 100 and 1000 MPa for various applications. When one considers that 1 atm is about 100 kPa, then these pressures are extremely high.

Machinery required for high-pressure processing is a pressure vessel, a hydraulic pressure-generation system, temperature control of the hydraulic liquid (usually water), and a materials-handling system. Product packed in flexible containers is loaded into the vessel, the vessel sealed, and then filled with water. Pressure is applied with the hydraulic pump to a set pressure and time and then released, the vessel drained, and the product removed. Fruit juice processed in this manner is usually processed in retail containers. If bulk processing is used, then aseptic filling of the product is required.

Thakur and Nelson (1998) outline the effect of high-pressure processing on the microflora of food. Vegetative cells are destroyed but spores remain active, which is an effect similar to normal thermal pasteurization of fruit juice. Rovere and Gola (1999) claim a six-fold reduction in lactic acid bacteria, yeasts, and molds with a five-minute treatment at 500 MPa. They make the point that while yeasts and molds are eliminated, there is only a partial elimination of enzymes, which may cause loss of shelf life due to gradual chemical degradation. On the positive side, the natural flavor and constituents such as vitamin C are fully retained during processing, giving a high-quality product.

5.7 CONCENTRATION OF FRUIT JUICE

5.7.1 Essence Recovery

Fruit is prized for the esters and flavors that are produced during ripening. Importance must be placed on the retention of these flavors during processing of the fruit into juice and through any subsequent treatments. Most of these flavor components have a boiling point less than that of water and are considered to be "volatile" during concentration. Volatile flavors are lost into the distillate stream during concentration. In many of the older concentrators there was an essence recovery column for the distillate, which removed the volatile flavors for inclusion into the concentrate. These volatiles had received the heat treatment given to the juice during concentration and then another heat treatment to remove them from the distillate. Heat treatments for concentration are normally carried out under vacuum and, therefore, the temperatures are not high, but some degradation of the flavor does occur through each cycle.

A better method is to remove the volatiles prior to concentration, and a simple way to do this is in a spinning cone column. The column consists of a series of cones with every second cone attached to a spinning shaft in the center and every other cone attached to the wall of the column, as in Figure 5–10. The juice passes through a heat exchanger to warm the juice to just below its boiling point at the vacuum applied in the column. The temperature used is normally between 40 and 50°C. The juice enters the column at the top and runs down the first cone by gravity. It falls into the gap between the cone and the shaft onto the spinning cone. The juice is spun to the wall under the centrifugal force on the spinning cone, and so on down the column. This presents a large surface area to the gaseous phase, which enters the bottom of the column and travels countercurrent to the juice. The gaseous phase may be air, nitrogen, or steam under vacuum. The volatiles are picked up by the gaseous phase and removed from the top of the column for condensation and removal. The juice passes out of the bottom of the column and is removed by a bottoms pump (Figure 5–11).

Resident time for juice in the spinning cone column is on the order of a minute, so little thermal damage is done to the juice, which is then pumped to the evaporator for concentration.

Other types of columns are available for essence recovery before concentration. These columns act as a multiple-effect evaporator, as shown in Figure 5–12. Juice is heated in a heat exchanger and then flashed into a cyclone with the vapor entering a fractionating column. The juice is pumped from the bottom of the cyclone into the first stage of the evaporator, as it retains some heat from the flashing process. The vapor passes up the column and into a condenser to remove the essence fraction.

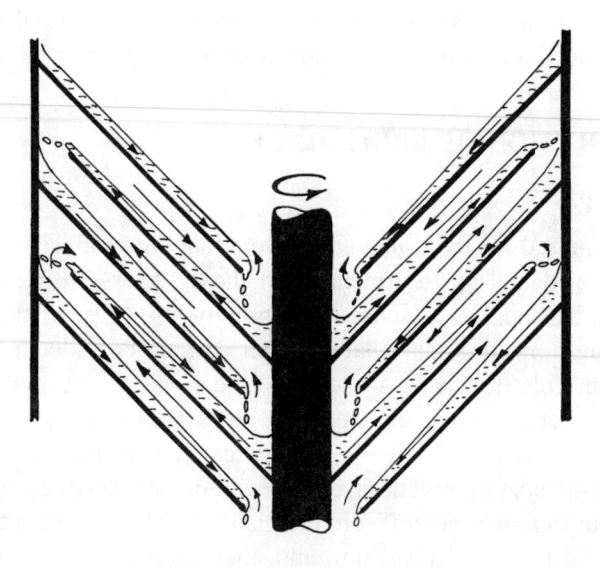

Figure 5–10 Extraction of volatiles by the spinning cone column.

Figure 5–11 A commercial spinning cone installation.

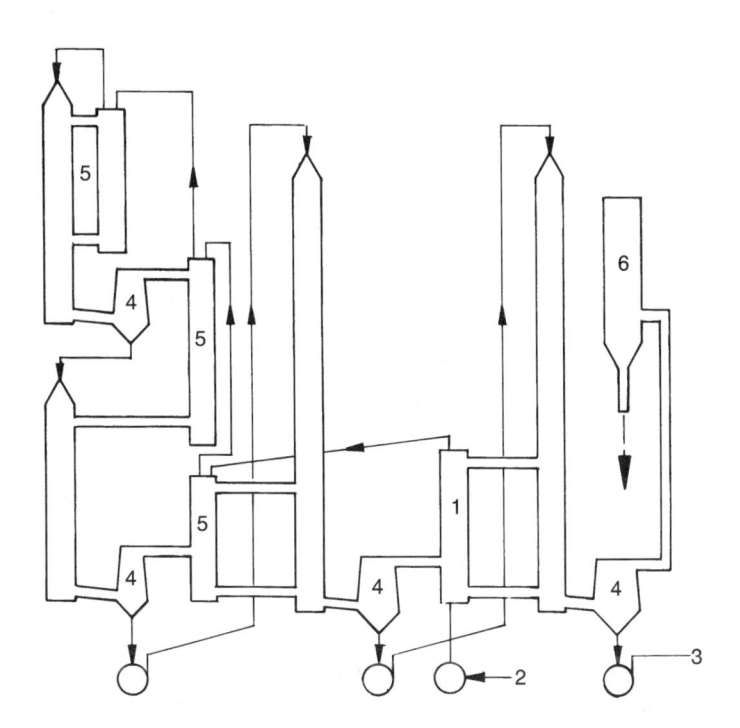

Figure 5–12 Multiple-effect evaporator. 1, heaters; 2, entry of product; 3, exit of product; 4, cyclone; 5, tube-bundle heat exchanger; 6, condenser.

5.7.2 Concentration

Concentration of fruit juice involves boiling the juice under vacuum and removing the steam by condensation. There are several types of concentrators but the most efficient are the multiple-effect evaporators. Pollard and Beech (1966) outline methods of vacuum-concentrating fruit juices, but since that time, multistage, multieffect, thermal-accelerated-short-time-evaporators (TASTE) have evolved. These evaporators are described by Nagy et al. (1993). A stage-and-effect has a preheater, distribution cone, tube bundle, and liquid vapor separator. The tube bundle has juice traveling down the inside of the tubes, while the vapor from the previous stage is traveling up the outside of the tubes. A typical layout is shown in Figure 5–12.

Centrifugal evaporators such as the Alfa-Laval Centritherm have very short holdup times of about a second. Conical design of the heat-transfer surface in the Centritherm increases the *g* forces on the concentrate as the viscosity increases through concentration. Although these machines are only a single effect and, therefore, expensive to operate compared with multiple-effect evaporators, they produce excellent-quality juice concentrate (Robe, 1983).

Juices with delicate flavors that are affected by the heat treatment during concentration may be freeze-concentrated. As ice forms in the juice it may be removed, leaving behind a concentrate. This process is explained by Tannous and Lawn (1981), but they also found that on freezing apple juice some constituents apart from water were removed with the ice.

Combinations of ultrafiltration and reverse osmosis have been used to concentrate fruit juices without heating (Mans, 1988; Harrison, 1970). Although membrane technology has been available for some time, it has not gained the degree of acceptance by the industry as other forms of juice concentration have done.

5.8 PRODUCTS DERIVED FROM FRUIT JUICE

Fruit juice in most countries is defined as 100% fruit content. Products derived from fruit, such as nectar, fruit juice drinks, cordials, and carbonated beverages, have definitions as to their fruit content that will depend on national food regulations. Australian regulations, for example, have definitions for diluted fruit juice that contain at least 50%, 35%, and 5% fruit juice. Also for a cordial, syrup, or topping, the fruit juice content must be at least 25% and the soluble solids a minimum of 25°B. Therefore, fruit nectar under these regulations is a fruit juice syrup containing at least 25% fruit juice. Regulations will vary from country to country, so fruit juice drink products will also vary.

5.8.1 Fruit Juice Drink

Fruit juice drink will contain fruit juice that has been diluted with a sugar–acid syrup. In most cases, the acidity is on the order of up to 1% and the soluble solids will be between 10 and 12%. Because the pH is between 3.0 and 4.0, the drink can be preserved in a manner similar to that for fruit juice.

5.8.2 Fruit Nectars

Fruit nectars are a mixture of fruit juice, water, and sugar. The soluble solids of nectars can be between 25 and 50% depending on country and customers. These products are acidic and can be preserved as for fruit juices, with an allowance for the increased viscosity. The juice used in nectars is usually a full pulp juice, and additional pulp may be added in some cases.

5.8.3 Carbonated Beverages

Carbonated beverages may or may not contain fruit juice. In general, the carbonated beverages that are sold as soft drinks are a mixture of flavored, colored, and acidified syrup, which is stabilized with benzoic acid (sodium benzoate). This syrup is diluted with soda water and filled into bottles or plastic containers under cool conditions to preserve the carbonation before capping. Adcock (1968) explains the conditions for carbonation and the removal of air from the carbonator. Figure 5–13 shows the bottle pressure required to achieve various volumes of carbonation with CO_2.

Carbonated fruit juice is manufactured in some countries. The fruit juice is cooled and sparged with CO_2 before entering the carbonator, where the juice is carbonated under pressure with CO_2. Carbonated juice is filled into glass bottles, capped, and passed through a bottle pasteurizer so the product reaches 70°C, which is held for 10 minutes. Bottles are cooled in the pasteurizer under water sprays. The bottles for this process must be strong enough to withstand the pressure of the carbonation at 70°C. In practice, there is some breakage in the pasteurizer because of variation in the wall thickness of bottles.

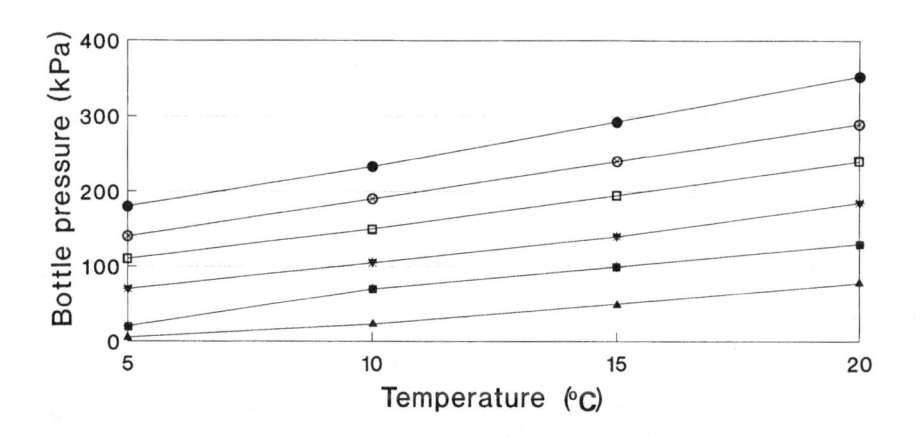

Figure 5–13 Carbonation pressures. -▲- 1.5 volumes; -■- 2 volumes; -▼- 2.5 volumes; -□- 3 volumes; -○- 3.5 volumes; -●- 4 volumes.

5.9 ADULTERATION OF FRUIT JUICE

For various reasons, the manufacturers of modern fruit juices have adopted systems where fruit juice is taken apart, modified, and then put back together again. Concentration is a typical example of this; for example, citrus fruit juice has had the pulp, skin oil, and essence removed prior to concentration. It could have been debittered or deacidified and a skin extract could have been made. There is a great temptation for a manufacturer with all these fractions at hand to manufacture some pulp wash, add some comminuted pulp for body, some of the essence, sugar, acid, juice, and some water to make a product similar to fruit juice.

Manufacture of "artificial" fruit juices prompted the importing countries in Europe and Germany in particular to look into ways of detecting adulteration of fruit juice. Germany's solution was to look at many detectable attributes of the particular juice and determine RSK values. RSK stands for Richtwert (guide value) Schwankungsbreite (range) and Kennzahl (reference number) (VdF, 1987).

The United States also has problems with adulteration of fruit juices, and Brause (1993) recounts the Food and Drug Administration's success in convicting companies by using different analytical determinations. Referring to the American situation, he states: "Despite all the less than honest people who have been caught at it, adulteration continues to occur at about a 10% incidence."

The Campden & Chorleywood Food Research Association in the United Kingdom has developed a near-infrared (NIR) spectroscopy system for verifying fruit juices (Scotter et al., 1992). Using NIR, a wide variety of juice characteristics can be predicted: for example, country of origin, several kinds of adulteration, the type of process that the sample came from, as well as the fruit variety from which the juice was derived. The spectral database currently contains 1000 samples from seven species, and over half are orange juices. In a blind test of 46 samples, including authentic and adulterated samples of seven juice types, 85% were correctly identified. The results prove NIR to be a valuable screening technique for fruit juice verification, and, as the database expands, the predictive ability of the technique will become more accurate.

Other countries also have their systems for determining whether fruit juice has been adulterated. There does not appear to be any universal foolproof system, or this would have been adopted by all countries. Unfortunately, while the rewards for adulteration are so high the unscrupulous few will attempt it without regard for the good name of the industry.

REFERENCES

Adcock, G.I. (1968). *The Jusfrute Book—A Text Book of Soft Drink Manufacture*. Jusfrute Ltd, Gosford, NSW, Australia.

Brause, A.R. (1993). Detection of juice adulteration. *Journal of the Association of Food and Drug Officials*, **57**(4), 6–25.

Casimir, D.J., Kefford, J.F., and Whitfield, F.B. (1981). Technology and flavour chemistry of passionfruit juices and concentrates. *Advances in Food Research*, **27**, 243.

Casimir, D.J. (1983). Countercurrent extraction of soluble solids from foods. *CSIRO Food Research Quarterly*, **43**, 38–43.
Cornwell, C.J., and Wrolstad, R.E. (1981). Causes of browning in pear juice concentrate during storage. *Journal of Food Science*, **46**, 515–518.
Grampp, E. (1977). Hot clarification process improves production of apple juice concentrate. *Food Technology*, November, 38.
Harrison, P.S. (1970). Reverse osmosis and its applications to the food industry. *Food Trade Review*, 33.
Heatherbell, D.A. (1976a). Haze and sediment formation in clarified apple juice and apple wine—I: The role of pectin and starch. *Food Technology in New Zealand*, May, 9.
Heatherbell, D.A. (1976b). Haze and sediment formation in apple clarified apple juice and apple wine—II: The role of polyvalent cations, polyphenolics and proteins. *Food Technology in New Zealand*, June, 17.
Johnson, R.L., and Chandler, B.V. (1985). Ion exchange and absorbant resins for removal of acids and bitter principles from citrus juices. *Journal of the Science of Food and Agriculture*, **36**, 480–484.
Johnson, R.L., and Chandler, B.V. (1986). Debittering and de-acidification of fruit juices. *Food Technology in Australia*, **38**, 294–297.
Mans, J. (1988). Next generation membrane systems find new applications. *Prepared Foods*, April, 69–72.
Morris, C.E. (1982). Aseptic papaya puree. *Food Engineering*, April, 84.
Mulvany, J. (1966). Filtration-sterilisation of beverages. *Process Biochemistry*, December, 470.
Nagy, S., Chen, C.S., and Shaw, P.E. (1993). *Fruit Juice Processing Technology*. Agscience, Auburndale, FL.
Pepper, D. (1987). From cloudy to clear. *Food*, October, 65.
Pollard, A., and Beech, F.W. (1966). Vacuum concentration of fruit juices. *Process Biochemistry*, July, 229–233, 238.
Robe, K. (1983). Evaporator concentrates juices to 70°B in single pass vs. 2 to 3 passes before. *Food Processing*, 92–94.
Rovere, P., and Gola, S. (1999). High pressure processing. A safe technique to process foods whilst maintaining their natural characteristics. Presented at the 10th World Congress of Food Science and Technology, Sydney, Australia, 3–8 October 1999.
Schmitt R. (1985). Turbidity in apple juices-arabanes. *Confructa-Studien*, **29**(1), 22–24, 26.
Scotter, C.N.G., Hall, M.N., Day, L., and Evans, D.G. (1992). The authentication of orange juice and other fruit juices. In *Near Infra-Red Spectroscopy*, ed. K.I. Hildrum et al. Ellis Horwood, Chichester, England, pp. 309–314.
Swientek, R.J. (1985). Expression-type belt press delivers high juice yields. *Food Processing*, April, 110.
Tannous, R.I., and Lawn, A.K. (1981). Effects of freeze-concentration on chemical and sensory qualities of apple juice. *Journal of Food Science Technology*, **18**, 27–29.
Thakur, B.R., and Nelson, P.E. (1998). High-pressure processing and preservation of food. *Food Reviews International*, **14**(4), 427–447.
VdF. (1987). *RSK values. The Complete Manual*. VdF Verband der deutschen Fruchtsaftindustrie e.V., Bonn, Germany.
Whitfield, F.B., and Last, J.H. (1986). The flavour of passionfruit—a review. *Progress in Essential Oil Research*. Walter de Gruyter, Berlin, Germany.

Chapter 6

Cider, Perry, Fruit Wines, and Other Fermented Fruit Beverages

Basil Jarvis

6.1 INTRODUCTION

The fermentation of fruit to produce wines, as well as the brewing of beer, is recorded in ancient Egyptian and Greek writings. Although production was based largely on the fermentation of grape juice, there is no doubt that fermentation of fruits other than grape had been practiced widely, although, because of the lower alcohol content, such wines did not store well. Over the years the production of grape wine became dominant, except in those areas where cultivation of vines was limited by climatic conditions; in such areas wine was produced by the fermentation of juice from other fruits.

Only the product of fermentation of grape juice can be ascribed the unqualified name of "wine." Fermented fruit juices are, therefore, given the generic name "fruit wines" or, in some places, "country wines." Techniques for the production of fruit wines, and also for cider and perry, which are special categories of fruit wines, closely resemble those used for fermentation of grape wine, although there are often significant differences of detail.

The development of distillation practices to produce stable spiritous beverages led to the development of distilled products from wines of all types. In addition, some distilled products were subsequently flavored by percolation through macerated fruit or by steeping the fruit in a distilled spirit. Further variations have included the blending of fruit juices, sugar, and other ingredients, such as herbs and vegetative plant extracts, with distilled spirits to produce liqueurs and aperitifs.

This chapter considers the characteristics of a range of fruit-based alcoholic beverages (other than grape wine, on which many books have been written) and the manner of their production. The principles of the production of cider are described in some detail. Shorter accounts are given of the production and properties of perry, fruit wines, fruit liqueurs, and fruit spirits. Fruits that are used to produce alcoholic fruit-based beverages include both cultivated and "wild" species. Some characteristics of alcoholic fruit-based products are summarized in Table 6–1.

6.2 CIDER

Cider (*synonyms*: cyder, hard cider, applejack, cidre, apfelwein) is the alcoholic product of fermentation of apple juice and differs from the nonalcoholic product sold in the United States. In recent years, cider has become an increasingly important commercial product in both domestic and export trade, with over 60% of European cider produced in England. Both commercial and "farmhouse" ciders are produced in many countries. This account reviews the current state of knowledge of cider fermentation and the factors that affect its production.

6.2.1 A Brief History

The fermentation of apple juice to produce an alcoholic beverage is believed to have been practiced for over 2,000 years. The fermented juice of apples is recorded as being a common drink at the time of the Roman invasion of England in 55 BC. Celtic mythology revered the "sacred apple," and in his famous history (AD 77) Pliny the Elder refers to a drink made from the juice of the apple. Cider was drunk throughout Europe in the third century AD, and in the fourth century St. Jerome used the term *sicera* (whence the name cider was possibly derived) to describe drinks made from apples.

Cider was reputedly a more popular drink than beer in the eleventh and twelfth centuries in Europe. Traditionally, cider has been produced throughout the temperate regions of the world where apple trees flourish; such localities include the northern coastal area of Spain, primarily Asturias and the Basque region; France, especially Normandy and Brittany; Ireland; Belgium; Austria; Switzerland; Germany; the west, southwest, and southeast regions of England; Finland; Sweden; and, more recently, Argentina, Australia, Canada, New Zealand, South Africa, and the United States.

Table 6–1 Characteristics of Fruit-Based Alcoholic Beverages

Type	Fruit Source	Alcohol Content (g/100 mL)
Cider	Apples	1.2–8.4
Perry	Pears	1.2–8.4
Fruit wines	Various	1.2–14
Fortified fruit wines	Various	6–22
Fruit liqueurs	Various	20–30
Fruit spirits	Various	35–40

There are references to cider in many writings from the Middle Ages. The popularity of cider in fourteenth-century England was such that William of Shoreham reflected the Church's concern for the niceties of sacramental rites by stating that "young children were not to be baptised in cider"! William Langdon in *Piers Plowman* and Shakespeare in *A Midsummer Night's Dream* refer to the consumption of cider; Daniel Defoe observed that Hereford people "boaft the richeft cider in all Britain," and Samuel Pepys noted in his Diary that on 1st May 1666 he "drank a cup of Syder." From the seventeenth century onward, cider was praised in numerous poems and other literary works as an aid to good cheer and a homely cure for almost every ailment known to man.

Up to the twentieth century, cider was a popular rural drink, cheaper than beer and often more potent at about 7% alcohol by volume (ABV). In England, farm workers often had part of their wages paid as truck (i.e., in kind), and every farmer would make his own cider for consumption by his workers and his own family and guests. Most farms in the West of England and the West Country had their own cider press, or used the services of a traveling cider press that was hauled by horse from farm to farm.

Commercial cider production commenced during the latter part of the nineteenth century in England, although a few farms had sold cider from early in the eighteenth century. Cider production in England was estimated at 55 million gallons in 1900. By 1920 the level of cider production had decreased significantly; although some 16 million gallons were still produced on farms, only 5 million gallons came from factory producers. In the early 1980s cider sales had risen to over 60 million gallons per year and by 1994 cider sales had risen to 92 million gallons. In the 12 months to September 1999, sales of UK-produced cider reached 109 million gallons, and a further 5 million gallons of cider were exported.

The growth in cider production in the United Kingdom, which had been inhibited by the imposition of excise duty in 1976, results both from the introduction of new concepts in cider products and from changes in drinking habits, especially of younger adults. In 1999, it was possible to obtain cider on draught (both keg and cask conditioned), prepackaged in glass and polyethylene terephthalate (PET) bottles and in cans. Products range in alcoholic strength from less than 0.5% ABV to 8.4% ABV, and in sweetness from totally dry (i.e., no residual sweetness) to sweet. In addition, cider products now range from "white" ciders (fermented from decolorized apple juice or produced by decolorization of fermented cider) to "black" cider (a blend of cider and fermented malted barley). Cider made from apples of a single crop may be sold as a defined-year Vintage, while others may be made from a single apple cultivar (e.g., Kingston Black).

6.2.2 Cider and Culinary Apples

Traditional European ciders are made from the juice of cider apples, believed to have been imported into Britain by the Normans, although it is recorded in Gaulmier's *Traite du Sidre* (1573) that a Spaniard named Dursus de l'Etre brought apple trees into France in 1486 (Jarvis, 1993).

Traditional cider apples are of four main types: bittersweet, low in acidity but high in tannin; bitter sharp, high acidity and high tannin; sharp, high acidity but low tannin;

and sweet varieties, low acidity and low tannin. Details of some typical cultivars and the composition of their juices are given in Table 6–2.

In some countries, cider is made primarily from culinary varieties of apple such as the Granny Smith and the Bramley. The juice of such varieties contains low levels of tannin but has a higher acidity than do the more traditional cider apples. Hence ciders fermented from culinary juices have a softer, less astringent flavor. In Germany, where most cider is made from culinary or dessert apples, an extremely astringent Frankfürter *apfelwein* known as *Speierling* is produced by suspending linen bags, containing berries of *Sorbus domestica*, in the fermentation vat in order to extract astringent polyphenols. In commercial cider making, blends of juices from different types of apples may be fermented to develop particular flavor profiles, or fermented ciders made from different types of juice may themselves be blended.

Cider Orchards

The traditional farm orchard of standard trees still exists and, although generally declining in acreage, many hundreds of such orchards provide both an apple crop and grazing for livestock. Modern cider apple orchards are largely intensive bush orchards with trees frequently planted as closely as 2.5 meters apart in rows some 5–6 meters apart. After planting, staking, and protecting using a wire guard against rabbits, the grass under the trees is treated with a suitable herbicide to reduce competition for nutrients and water. Some growers cover the herbicide strips with a mulch of apple

Table 6–2 Characteristics of Selected Cider Apple Varieties

Type	Typical Varieties	Acidity (g/100 mL)	Tannin (g/100 mL)
Sweet	Sweet Coppin	0.20	0.14
	Taylor's Sweet	0.18	0.14
	Court Royal	0.21	0.11
Bittersweet	Dabinett	0.18	0.29
	Michelin	0.25	0.23
	Yarlington Mill	0.22	0.32
	Néhou	0.17	0.60
	Médaille D'Or	0.27	0.64
Bitter sharp	Kingston Black	0.58	0.79
	Breakwell's Seedling	0.64	0.23
	Bulmer's Foxwhelp	1.91	0.22
Sharp	Brown's Apple	0.72	0.13
	Reinette O'bry	0.63	0.13
	Frederick	1.02	0.09

pomace, woodchips, straw, or other suitable material to ensure maximum retention of moisture. Although it has become standard practice to retain the herbicide strip into the productive years of an orchard, this practice is currently the subject of research to assess whether a sward of slow-growing grass varieties may be better in relation to the quality of harvested fruit.

During the growing season, it may be necessary to spray the trees against pests (e.g., red spider mite), mildew, apple scab, canker, and other conditions. Many cider apple varieties suffer from biennialism; that is, they produce a very large crop of fruit in one year and a small crop in the following year. It is frequently recommended that in years of heavy potential cropping, the trees should be sprayed during blossom with a growth regulator such as paclobutrazol (Cultar). The objective of such chemical thinning is to control stress in the tree by reducing the crop of apples in an "on" year and increasing the crop in an "off" year. However, not all cider or culinary apple trees are sprayed, and there is a growing demand for fruit from orchards that have not been treated with agrochemicals and that conform with the legislative requirements for "organic fruit."

Harvesting and Pressing

Traditionally the apples, falling naturally or shaken by hand from the trees, would be raked into piles (Figure 6–1) or filled into sacks that were stored under the trees until the fruit was in a suitably ripe condition to be milled and pressed. In modern intensive bush orchards (Figure 6–2), it is normal to shake the trees mechanically in order to

Figure 6–1 Traditional harvesting of cider apples from standard trees. Apples were shaken to the floor of the orchard, where they were gathered by hand into baskets and then tipped into piles to mature.

Figure 6–2 A modern bush cider apple orchard showing the herbicide strip along the row under the tree canopy.

cause the fruit to fall. Such shaking does not harm the tree and permits fruits to be harvested mechanically immediately after falling, so reducing the risk of the rot that may occur if fruit are left for any length of time on an herbicide strip. The collected fruit normally will be washed mechanically in the orchard and then transferred by road to the cider mill where, after weighing, the fruit is tipped into a fruit canal (Figure 6–3) or onto a concrete pad prior to further washing, milling, and pressing.

Traditionally the fruit was crushed between stone rollers and the fruit mash was then pressed in a screw press after being built into a "cheese" consisting of layers of fruit in hessian cloths held on ash slats (Figure 6–4). In modern processing, the fruit is milled using either a hammer or knife mill; the mash is then transferred into a batch press such as a Bücher-Guyer hydraulic ram press or a continuous belt press such as a Belmer (Figures 6–5 and 6–6). The pomace from the primary pressing may be extracted with warm water and again pressed to give a secondary juice with a lower sugar content; alternatively a continuous-diffusion process may be used to increase the total yield of tannin and sugar extracted from the apple pulp. Residual pomace is frequently used for extraction of food-grade pectin, or it may be used as animal feed or for mulching.

Great care is needed to ensure that the starch content of the fruit has been converted to sugar before milling; otherwise significant quantities of starch in the juice will affect both the economics and the practical aspects of fermentation and subsequent clarification. However, overripe fruit can give rise to significant problems during milling and pressing by modern equipment, especially in a continuous-belt press.

Figure 6–3 An "apple canal" into which fruit is tipped from bulk carriers, farm trailers, or sacks. Above the canal is a conveyor carrying fruit from a bulk discharge area to the appropriate part of the canal.

It is normal practice to treat freshly pressed juice with about 50–75 mg of sulfur dioxide per kilogram to inhibit both oxidative browning and the growth of wild yeasts (qv). Since much of the sulfur dioxide will become bound (qv), it is normal to adjust the level of sulfur dioxide in the juice after 12–24 hours to give a residual of 10–30 mg of "free" sulfur dioxide per kilogram. Increasingly, much juice produced by large commercial cider makers will be evaporated to produce a juice concentrate that can be stored for fermentation throughout the year, rather than making cider only during the harvest period. Prior to concentration, the juice will normally be treated with a preparation of pectinases and amylases to remove soluble pectin and starch, which would otherwise cause formation of a gel in the concentrated juice.

6.2.3 Fermentation of Cider

Fermentation Vessels

Traditionally, the apple juice was fermented in oak barrels (Figure 6–7) or vats (Figure 6–8); although many such vats are still in use, vats of concrete or mild steel with a ceramic or resin lining, glass-reinforced plastic, and, more recently, stainless steel are used. Most cider vats are typically of 10,000- to 200,000-gallon capacity, but

118   Fruit Processing

Figure 6–4 Traditional pressing of cider apple pulp using a hydraulic screw press. The "cheese" of apple pulp contained in a hessian cloth is held on a series of wooden slats.

Figure 6–5 Installation of a modern continuous-belt press at H P Bulmer Ltd., Hereford, UK.

Figure 6–6 Continuous pressing of apple pulp in a belt press.

Figure 6–7 Traditional cider making in oak barrels, Herefordshire, UK, 1935.

much larger vats exist. The largest fermentation/storage vat in the world is located at the premises of H P Bulmer Ltd. in Hereford (UK); this vat holds some 1.6 million gallons of cider. A few cider makers use modern conico-cylindrical vats (Figure 6–9). However, tall, narrow conico-cylindrical vats, such as those used by brewers, are not

120   Fruit Processing

Figure 6–8 Traditional oak cider vats at H P Bulmer Ltd., Hereford, UK.

Figure 6–9 A modern cider fermentation hall at H P Bulmer Ltd., Hereford: the photograph shows the base cones on a number of 150,000-gallon stainless steel fermentation vessels

ideal. Jarvis, Forster, and Kinsella (1995) demonstrated an effect of hydrostatic pressure on the fermentative ability of *Saccharomyces cerevisiae* in cider making. At pressures in excess of 1.5 bars (equivalent to a fermenter height of 48 feet [15 meters]), the fermentation rate is slowed significantly, and the flavor of the fermentation product also changes. The fermentation of cider is more akin to the fermentation of wine, and vats of limited height, such as the Unitank-style vessel, minimize the effect of hydrostatic pressure stress on the fermentation yeast.

Fermentation Substrate

Traditional cider making used whole juice, often with much of the apple solids remaining. Such solids included the pips, which contain cyanogenic glucosides; hence the cider could contain small quantities of cyanide derivatives. Folklore has it that farm cider making was improved by the addition of dead rats or other animal corpses to the cider vat; such apocryphal stories abound, but there is evidence that farmers would often suspend a leg of ham in a cider vat to "enhance the curing process." One of the consequences would be to increase the available nutrients and thereby stimulate the fermentation process.

In some countries, where the traditional production processes are prescribed in law (see AICV, 1998), only fresh apple juice can be fermented into cider (e.g., production of *cidre* in France). However, in many other countries (e.g., United Kingdom, Ireland, Belgium, Australia, South Africa) the process of chaptalization has become increasingly common. Chaptalization is the process of supplementation of the base juice with a suitable fermentation sugar (e.g., glucose syrup), which enables the production of a strong cider with an alcohol content up to about 12% ABV that can be broken back to a commercial-strength product during blending; without chaptalization, the maximum alcohol yield will be about 6% to 7% ABV, dependent upon the natural sugar content of the juice. Recently, some French cider makers have started to use an existing definition, intended originally for products derived from the second-pressing juice of apples, to develop English-style ciders fermented from chaptalized apple juice; the product must be described as *boisson alcoolisée à base de pomme*, an alcoholic beverage derived from apples, to differentiate it from traditional *cidre*.

In the preparation of cider, the apple juice will be fresh or reconstituted from an apple juice concentrate; in many countries the apple juice can be blended with up to 25% by volume of pear juice (either perry pear [qv] or culinary pear juice); such blending produces a juice with distinctly different properties because of the high sorbitol content of pear juice. Other than fermentation sugars, the only other primary addition will be a suitable yeast culture. However, in fermentation brews originating wholly or largely from juice concentrates, it is normal to add yeast nutrients, such as diammonium phosphate, sodium pantothenate, and thiamine.

The Fermentation Process

The fermentation of apple juice to cider can occur naturally through the metabolic activity of the yeasts and bacteria present on the fruit at harvest, which are then transferred into the apple juice on pressing. Other organisms, arising from the milling and pressing equipment and the general environment, can also contaminate the juice at this stage. Unless such organisms are inhibited (e.g., by maintaining 10–30 mg of free sulfur dioxide per kilogram), the mixed fermentation will yield a product that varies considerably from batch to batch, even if the composition of the apple juice is constant. Hence, control of the indigenous and adventitious microorganisms followed by deliberate inoculation with a selected strain of yeast is the preferred commercial route for the production of cider. However, there is a tendency for addition of an excess of sulfur dioxide to the juice such that many fermentation cultures are themselves partially inhibited, thus nullifying the effects of using selected culture strains. Indeed, in such circumstances, sulfur dioxide–resistant wild yeasts can then grow unchecked and produce undesirable flavor and aroma changes in the cider.

The process consists of partially filling the vat with juice followed by inoculation with a specific starter culture of yeast (frequently a wine yeast capable of growth at elevated alcohol levels). Once fermentation has started, additional substrate (apple juice, concentrate, or fermentation sugar) is added from time to time until the vat is filled. The fermentation process typically continues without control of temperature, pH, or other parameters until all the fermentable sugar has been metabolized into alcohol. Depending upon conditions, this process can take 2–12 weeks! There is no doubt that control of temperature can give more consistent fermentation profiles and products. While ambient temperature fermentation occurs widely, increasingly, commercial cider makers control temperature within a range of 20–25°C. Often the fermentation will be started at about 18°C to avoid any excessive increase in temperature resulting from the initial exothermic aerobic growth of the yeast. In France and Belgium, cider is often fermented at a temperature of 15–18°C.

Secondary Fermentation and Maturation

When fermented to dryness, the cider is frequently left for a few days on the lees to permit the yeasts to autolyze, thereby adding cell constituents such as enzymes, amino acids, and nucleic acids to the brew. The cider will be racked (separated) from the lees and transferred either directly or after clarification into storage vats (traditionally made of oak). A secondary malo-lactic fermentation (qv) process occurs either in parallel with the primary yeast fermentation (typical of the process of cider making in northern Spain) or following completion of the primary fermentation. In addition, storage of the cider following fermentation results in a number of chemical changes that affect significantly the flavor of the cider. Judgment as to the extent of maturation and the suitability of the cider for use is still an art vested in the craft skill of the cider maker.

Final Processing

When required for use, different batches of cider, generally made from mixtures of different juices, will be blended by the cider maker to provide specific aroma and flavor attributes. Traditionally, the raw cider will be fined using agents such as bentonite, gelatin, or chitin and filtered to give a bright product with no haze. Modern processing refinements include the use of cross-flow or other microfiltration systems to obviate the need for fining and to speed the process.

If the cider has a high alcohol content (e.g., 9–12% ABV) it may be "broken back" to the final product strength using water or dilute apple juice. Other ingredients, such as sugars, intense sweeteners (e.g., saccharin), colors, and/or additional preservative (in the European Union [EU] this is restricted to sulfur dioxide or sorbic acid) may be added (Anon, 1994a, b; 1995). The finished blend may be carbonated or subjected to further processing prior to packaging.

The Association of the Cider and Fruit Wine Industry of the EU has published a Code of Practice, which consolidates information on the permitted cider-, perry-, and fruit-wine–making practices of all member Associations within Europe (AICV, 1998). This code defines cider (and its national synonyms such as *apfelwein*, *cidre*, *sidra*, etc.) as

> An alcoholic beverage obtained only by the complete or partial fermentation of the juice of fresh apples, or the reconstituted juice of concentrate made from the juice of apples, or a mixture of such juices, with or without the addition of a limited volume of the juice of pears, with or without the addition of potable water and sugars, before or after fermentation, with or without the addition of fresh or concentrated apple juice after fermentation and with or without the addition of food additives permitted by EU legislation. The product may be uncarbonated (i.e., still) or carbonated by secondary fermentation or by injection of carbon dioxide. The product will contain an actual alcohol content with the range 1.2% to less than 8.5% alcohol by volume (ABV). *Fortification of cider by addition of distilled alcohol is not permitted.* A *low alcohol cider* is defined as one that is produced by the practices described for cider, but which contains less than 1.2% ABV. This definition excludes all fruit wines, fruit spirits, alcohol fortified fruit wine-based products and "alcopops" (i.e., blends of alcohol and fruit juice).

In the United Kingdom, the only specific legislation controlling cider making (other than the general requirements of food legislation) are the Official Notices produced by the Commissioners for Excise. In the absence of official quality criteria, the National Association of Cider Makers (NACM) has operated a voluntary Code of Practice for British cider and perry for many years. The code defines, *inter alia*, approved ingredients and processes for cider (and perry) and lays down a minimum quality criteria with which members of the NACM must comply. A regular monitoring exercise is undertaken by the NACM to ensure compliance with the Code of Practice, which requires that products contain the following:

actual alcohol	>1.2 and <8.5 g/100 mL (ABV)
volatile acidity (as acetic acid)	not more than 1.4 g/L
iron	not more than 7 mg/L
copper	not more than 2 mg/L
arsenic	not more than 0.2 mg/L
lead	not more than 0.2 mg/L

These criteria are not dissimilar to those laid down by legislation in other countries. The Guild of Craft Cider Makers was formed in the United Kingdom in 1998; its members are required to comply with more restrictive manufacturing and compositional criteria than those designated by the NACM. Ciders that conform to the defined analytical and organoleptic criteria are permitted to display a special quality mark on the product label.

6.2.4 Special Types of Cider

Low-Alcohol Cider (<1.2% ABV)

Low-alcohol ciders have been developed either by dealcoholization of strong cider or by using a "stopped fermentation." Dealcoholization uses processes such as falling film evaporation or reverse osmosis. The dealcoholized base lacks flavor and body and is unsuitable for sale without considerable improvement. One such product, which has been marketed for several years, uses the dealcoholized cider as a base, which has then been fortified with cider, apple juice flavors, and selected alcohols to rebuild both the flavor and body of the product. In the stopped-fermentation process, the cider is chilled, racked off from the yeast cells, and clarified or filtered once the alcohol level reaches about 2% ABV. The cider is then diluted with fresh unfermented apple juice to give a product having some cider flavor.

"White" Cider (Generally 6.5–8.4% ABV)

White ciders are prepared by either fermentation of decolorized apple juice or by decolorization of the fermented cider. Decolorization of apple juice also removes many juice flavors and flavor precursors such that the fermented product has a softer, less cidery flavor than does a product made by decolorization of fermented cider. By the careful selection of decolorizing agents (e.g., polyvinylpolypyrrolidone (PVPP), charcoal), it is possible to minimize the loss of flavor compounds.

"Black" Cider

Black cider was introduced to the UK market on a trial basis in 1992; it consists of a cofermentation of apple juice with hops and a heavily malted barley. The flavor is very complex, having overtones of both cider and stout. Although not widely developed, it demonstrates the flexibility of cider as a base material for the development of a range of new alcoholic beverages. An alternative approach is to add caramel to the final product in order to develop a reddish-brown cider. One interesting process is to sweeten the final cider by addition of a deliberately caramelized apple juice concentrate; however, great care is required in producing the caramelized concentrate if a consistent color is to be achieved.

"Ice" Cider

Following the successful development of various processes for the production of "ice beers," the concept has been extended to cider. The details of the processes used are currently confidential, but cooling of cider to a temperature at which ice crystals form and the subsequent separation of the cider from the ice results in subtle flavor changes. However, the chemical nature of the changes is not yet fully understood.

"Vintage" and "Single Cultivar" Ciders

A cider fermented from the fresh juice of fruit harvested in a named year can be designated as a vintage cider. Similarly, cider prepared from a single apple cultivar (e.g., the highly astringent Kingston Black apple) can be designated to be a single cultivar cider. Combination of the two can thus generate a single cultivar vintage cider.

Cidre Nouveau

A recent marketing development in both England and France has been the introduction in early November of a freshly fermented (i.e., nonmatured) cider given the designation *cidre nouveau*. Such products are available for sale for a few months only (in France until the following March).

Traditional Ciders

Many traditional ciders are produced in France, where reserved designations and production practices are defined by law (AICV, 1998). Table 6–3 summarizes some of the compositional requirements of different ciders and perries. The term *cidre bouché* is reserved for naturally carbonated cider with >3 g of CO_2 per liter or artificially carbonated cider with >4 g of CO_2 per liter, packaged in a champagne-style bottle, and closed with a mushroom stopper and a wire muzelet. The product must contain at least 1.5% alcohol and have a potential alcohol greater than 5.5% ABV. French ciders are designated brut (dry), demi-sec (semisweet), or doux (sweet) in accordance with the residual sugar level and density of the cider. In Belgium a traditional product similar to cidre bouché is known as *cidre fermette* (farmhouse cider).

In Germany, *apfelwein* must have a minimum alcohol content of 5% ABV, a sugar-free dry extract of at least 18 g/L, and a maximum volatile acidity of 1 g/L, measured as acetic acid. Product descriptors include *Frankfürter Apfelwein* (cider made in the Frankfürt area), *Speierling* (cider with a high acidity and astringency), and *Most* (a product from the Baden-Würtenberg area, which contains up to 30% pear juice and has an alcohol content of 4% ABV). Variants include *Apfel-Schaumwein*, a cider carbonated

---

Table 6–3 Analytical Criteria for French Cider and Perry

Analytical Parameter	Units	Cidre (Poiré)	Cidre (Poiré) Bouché	Fermenté de Pomme (de Poire)	Pétillant de Pomme (de Poire)
Potential alcohol—minimum	% ABV	5.0	5.5	4.0	4.0
Actual alcohol —minimum	% ABV	1.5	1.5	1.5	1.5
—maximum		—	—	—	3.0
Total dry extract	g/L	16.0	18.0	14.0	14.0
Volatile acidity*—maximum	g/L	1.0	1.0	1.0	1.0
Ash—minimum	g/L	1.4	1.4	1.4	1.4
Iron —maximum for cidre	mg/L	10.0	10.0	10.0	10.0
—maximum for poiré		17.0	17.0	17.0	17.0
Ethanal (acetaldehyde)					
—maximum	mg/L	120	100	120	100
Sulfite—maximum	mg/L	175	150	175	150
Carbon dioxide—minimum	g/L	—	4.0†	—	2.5

* Measured as sulfuric acid.
† 3.0 g/L for naturally carbonated products.

Source: Reprinted with permission *AICV Code of Practice 1998: Definitions of Cider, Perry and Fruit Wines,* International Food and Beverage Consultancy.

to a pressure of at least 3 bars of CO_2 at 20°C and packaged in a champagne-style bottle with a mushroom stopper, *Apfeldessertwein* (minimum alcohol content 12% ABV), and *Apfeltischwein* (8–11% ABV). Spanish cider (*sidra*) must have an actual alcohol content of at least 4% ABV (4.5% for *Sidra Natural*). As in France, the production of cider is controlled by legislation that defines a range of compositional parameters (AICV, 1998). *Sidra natural* must be made only by traditional practices without addition of chaptalization sugars, and the carbonation must be developed exclusively by fermentation. Information on other national cider specialities is given in AICV (1998).

Organic Cider

There is a small, but growing, market in the United Kingdom for organic cider that is produced from fruit grown only in orchards that have received no treatment with agrochemicals or fertilizers, other than those deemed acceptable under the EU regulations (European Council Regulation, 1991). The use of additives and the labeling of the product is also constrained by the EU regulation.

6.2.5 The Microbiology of Apple Juice and Cider

Freshly pressed apple juice contains a variety of yeasts and bacteria, many of which will be incapable of growth at the acidity of the juice. Examples of organisms often present in juice are shown in Table 6–4, together with an indication of their susceptibility to sulfur dioxide.

The Role of Sulfur Dioxide in Apple Juice

The use of sulfur dioxide as a preservative in cider making is controlled by legislation; in the EU the maximum level permitted in the final product is 200 mg/kg (Anon, 1995). The addition of sulfur dioxide to apple juice results in the formation of so-called sulfite addition compounds through the binding of sulfur dioxide to carbonyl compounds (q.v.). When dissolved in water, sulfur dioxide and its salts set up a pH-dependent equilibrium mixture of "molecular sulfur dioxide," bisulfite, and sulfite ions (Beech and Carr, 1977; Beech and Davenport, 1983). The antimicrobial activity of sulfur dioxide is believed to be caused by the molecular sulfur dioxide moiety of the part that remains unbound (the so-called "free" sulfur dioxide). Less sulfur dioxide is needed in juices of high acidity; for instance, 15 mg of free sulfur dioxide per kilogram at pH 3.0 has the same antimicrobial effect as 150 mg/kg at pH 4.0.

The binding of sulfur dioxide is dependent upon the nature and origin of the carbonyl compounds present in the juice. Naturally occurring compounds that bind sulfur dioxide include glucose, xylose, and xylosone. If the fruit has undergone some degree of rotting, other binding compounds will be present, including 2,5-dioxogluconic acid and 5-oxofructose (2.5-D-*threo*-hexodiulose). Such juices will require increased additions of sulfur dioxide if wild yeasts and other microorganisms are to be controlled effectively.

The addition of sulfur dioxide to a fermenting juice results in rapid combination with acetaldehyde, pyruvate, and α-oxoglutarate, produced by the fermenting yeasts. Consequently, all such additions must be completed immediately after pressing the

Table 6–4 Typical Microorganisms of Fresh Apple Juice

Type	Typical Species	Sensitivity to Sulfur Dioxide*
Yeast	*Saccharomyces cerevisiae*	+ or –
	Saccharomyces uvarum	+ or –
	Saccharomycodes ludwigii	–
	Kloeckera apiculata	+++
	Candida pulcheriima	++++
	Pichia spp.	++++
	Torulopsis famata	++
	Aureobasidium pullulans	+++
	Rhodotorula spp.	++++
Bacteria	*Acetobacter xylinum*	++
	Pseudomonas spp.	++++
	Escherichia coli	++++
	Salmonella spp.	++++
	Micrococcus spp.	++++
	Staphylococcus spp.	++++
	Bacillus spp.	– (spores)
	Clostridium spp.	– (spores)

*Sensitivity: –, insensitive; +, relatively insensitive; ++, +++, ++++ increasingly more sensitive.
Source: Reprinted from R. Macrae, et al., eds., *Encyclopedia of Food Science, Food Technology and Nutrition*, B. Jarvis, et al., Cider: Chemistry and Microbiology of Cidermaking, pp. 984–989, Copyright 1993, by permission of the publisher Academic Press London.

---

juice, although, provided the initial fermentation is inhibited, further additions to give the desired level of free sulfur dioxide to inhibit wild yeasts can be made during the following 24 hours.

Fermentation Yeasts

The fermentation process is carried out by strains of *Saccharomyces* spp., especially *S. cerevisiae*, *S. bayanus*, and *S. uvarum*, which are added to the sulfited juice as a pure culture. Traditionally, the starter culture is prepared in the laboratory, although, increasingly, commercially produced dried yeast cell preparations are used, either for direct vat inoculation or as inocula for the yeast propagation plant.

The choice of culture is dependent upon many criteria (Table 6–5), such as flocculation characteristics, ability to ferment efficiently at subambient temperatures, alcohol and sulfur dioxide tolerance, and lack of ability to produce hydrogen sulfide (Beech, 1993). One desirable characteristic is the ability to produce fusel oils (e.g., higher alcohols), which affect both the flavor and aroma of the cider (Table 6–6).

The fermentation process typically takes some 2 to 12 weeks to proceed to dryness (i.e., to a specific gravity of 0.990–1.000), at which time all fermentable sugars have been converted to alcohol, CO_2, and other metabolites. After inoculation, the starter yeast together with those sulfur dioxide–resistant wild yeasts selected from the juice

Table 6–5 Desirable Properties for Yeast Strains Used in Cider Making

Produces polygalacturonase to break down soluble pectins	Ferments to "dryness" (i.e., no residual fermentable sugar)
Relatively resistant to SO_2, low pH value, and high ethanol level	Does not produce excessive foam
	Good flocculation characteristics
Low requirement for vitamins, fatty acids, and oxygen	Produces good levels of desirable aroma compounds, organic acids, and glycerol
Produces rapid onset of fermentation	Minimal production of SO_2 and SO_2-binding compounds
Good rate of fermentation at chosen temperature	Does not produce H_2S or acetic acid
Efficient utilization of a range of sugars	Compatible with malolactic bacteria

Source: Reprinted with permission from B. Jarvis, M.J. Forster, and W. Kinsella, Factors Affecting the Development of Cider Flavour, in *Microbial Fermentations: Beverages, Foods and Feeds*, R.G. Board, D. Jones and B. Jarvis, *Journal of Applied Bacterial Symp Supplement*, Vol. 79, pp. 155–185, © 1995, Blackwell Sciences, Ltd.

will increase in numbers from an initial level of about $(0.2–1) \times 10^6$ cfu/mL to $(2–5) \times 10^7$ cfu/mL. Following an initial aerobic growth phase, the resulting oxygen limitation and high carbohydrate levels in the media trigger the onset of the anaerobic fermentation process.

In controlled fermentations, a maximum temperature of 25°C will generally be permitted, although fermentations controlled at 15–18°C are not uncommon, especially in France, where long, slow fermentations are often preferred. Temperatures higher than 25°C are generally undesirable, since metabolism by the desired yeast strain may be inhibited, leading to "stuck" fermentations and the growth of undesirable thermoduric yeasts and spoilage bacteria. Stuck fermentations can sometimes be restarted by addition of nitrogen (10–50 mg/L), usually as ammonium sulfate or diammonium phosphate, sterols (usually in the form of yeast hulls), and/or thiamine (0.1–0.2 mg/L).

Table 6–6 Higher Alcohols in Apple Juices and Ciders

	Concentration Range (mg/L)	
Constituent	Apple Juices	Ciders
Propan-1-ol	0.2–9	4–200
Butan-1-ol	0.5–24	4–32
Butan-2-ol	<0.1	14–74
Heptan-1-ol	0.1	42–196
Octan-1-ol	0.1–2	16–39
Hexan-1-ol	1–2	2–17
2-Phenylethanol	<0.1	7–260

Source: Reprinted from R. Macrae, et al., eds., *Encyclopedia of Food Science, Food Technology and Nutrition*, B. Jarvis, et al., Cider: Chemistry and Microbiology of Cidermaking, pp. 984–989, Copyright 1993, by permission of the publisher Academic Press London.

As with any fermentation process, the viability and vitality of the fermentation yeasts is critical. Measurements of plasma membrane potential and acidification power of a yeast during the 3- to 6-week cider fermentation process have shown that although significant cell membrane changes do occur, the yeasts are extremely robust (Dinsdale, Lloyd, and Jarvis, 1995). Studies of yeast strains grown with added ethanol and mixtures of higher alcohols and other metabolites confirmed the high tolerance of the cider yeasts studied (Seward et al., 1996; Willetts et al., 1997). Prolonged anaerobiosis, together with accumulation of toxicants, might be expected to compromise viability and vitality (Lloyd et al., 1996). However, studies have shown unequivocally that, despite a gradual decline in respiratory and fermentative capacities and in the adenylate energy status of the organisms, sufficient viable numbers survive even after 16 weeks in the fermentation liquor to enable reinitiation of fermentation within a few days of transferring into a fresh fermentation base (Dinsdale et al., 1999). Thus, despite loss of vitality during the long cider fermentation process, viability is extensively retained within the yeast population.

This has several implications for the cider maker: first, if a cider is sweetened by addition of sugar or apple juice, then residual fermentative yeasts may reinitiate an in-pack fermentation unless the cider has been sterile filtered or pasteurized; second, yeast collected at the end of fermentation could be washed and stored for subsequent initiation of new fermentations, as is commonly practiced in the brewing industry; third, the extent of cell autolysis at the end of fermentation is probably much less than has been generally supposed to occur, although it is still good practice to remove the cider from the flocculated yeast as soon as possible.

Maturation and Secondary Fermentation

The maturation vats are filled with the racked-off cider and either provided with an overblanket of CO_2 or otherwise sealed to prevent ingress of air, which would stimulate the growth of film-forming yeasts (e.g., *Brettanomyces* spp., *Pichia membranefaciens*, *Candida mycoderma*) and aerobic bacteria (e.g., *Acetobacter xylinum*). Growth of the former will produce precursors for the development of defects such as "mousy" flavor (e.g., 1,4,5,6-tetrahydro-2-acetopyridine), while the latter will acetify the cider through the production of acetic and other volatile acids, which impart a vinegary note to the product. Of course, deliberate acetification of cider can be used to produce cider vinegar.

During the maturation process, growth of lactic acid bacteria (*Lactobacillus pastorianus* var. *quinicus*, *L. mali*, *L. plantarum*, *Leuconostoc mesenteroides*, *Leuc. oenus*, etc.) can occur extensively, especially in wooden vats. The control of the malolactic fermentation processes in cider is only now beginning to be understood (see Jarvis et al., 1995). The malolactic fermentation process results in the conversion of malic and quinic acids to lactic and dihydroshikimic acids, respectively. Such secondary fermentation reduces the acidity of the cider and imparts subtle changes that are generally considered to improve the flavor of the product. However, in certain circumstances, some metabolites of the lactic acid bacteria may damage the flavor and result in spoilage (e.g., excessive production of diacetyl [and its vicinal-diketone precursors]) (Table 6–7), the butterscotch-like taste of which can be detected in cider at a threshold level of about 0.4 mg/L.

Table 6–7 Volatiles in Normal and Diacetyl-Spoiled Cider*

	Normalized Peak Area	
Compound	Normal	Spoiled
Alcohols		
2-Methyl propanol	35.4	97.3
Iso-amyl alcohol	305	213
2- and 3-Methyl-butan-1-ol	503	456
Heptan-1-ol	1.26	0.75
Hexan-1-ol	35.7	29.5
Nonan-1-ol	0.79	0.86
Methionol	1.29	0.55
2-Phenylethanol	69.0	61.6
Esters		
Ethyl acetate	86.1	89.2
Ethyl lactate	45.9	37.9
Ethyl-2-methylbutyrate	10.3	12.9
Ethyl-2-hydroxy-4-methyl pentanoate	4.19	7.03
Ethyl benzoate	4.70	5.66
Ethyl hexanoate	233	179
Ethyl octanoate	280	226
Ethyl decanoate	57.6	53.4
Ethyl dodecanoate	2.10	1.04
Diethyl succinate	5.23	2.88
Hexyl acetate	1.95	6.25
Hexyl octanoate	3.39	11.5
2-Phenylethyl acetate	11.4	6.42
Aldehydes		
Benzaldehyde	1.42	1.41
Octanal	0.67	1.19
Decenal	7.52	0.36
Undecanal	2.16	1.23
Ketones		
Decan-2-one	3.58	0.90
Acids		
Hexanoic acid	12.6	12.6
Heptanoic acid	0.00	3.39
Octanoic acid	34.6	29.4
Nonanoic acid	1.01	1.21
Miscellaneous		
Diacetyl	0.00	3.42
γ-Decalactone	4.39	3.39
Ethyl guaiacol	2.47	4.53
Eugenol + 4-ethylphenol	1.86	1.52

* Based on gas chromatography-linked mass spectroscopy (GC-MS) analysis of headspace volatiles.
Source: Reprinted from R. Macrae, et al., eds., *Encyclopedia of Food Science, Food Technology and Nutrition*, B. Jarvis, et al., Cider: Chemistry and Microbiology of Cidermaking, pp. 984–989, Copyright 1993, by permission of the publisher Academic Press London.

Swaffield, Scott, and Jarvis (1994) studied the microbial colonization of cubes of sterilized old vat oak suspended in a vat of maturing cider. Over a period of several weeks it was found that organisms are able to penetrate into and colonize to a depth of at least 1 cm from the surface of the timber. The organisms were diverse strains of both yeasts and bacteria, including many that exude mucilaginous materials. This is of some significance in relation to the manner in which entrapped microorganisms cause changes in cider during maturation; it suggests that the secondary maturation process in wooden vats may be a natural immobilized microbial fermentation. This opens up possibilities for introducing controlled immobilized fermentation of cider in the future.

Spoilage and Other Microorganisms in Cider

Bacterial pathogens such as *Salmonella* spp., *Escherichia coli,* and *Staphylococcus aureus* may occasionally occur in apple juice, where they will have been derived from the orchard soil, farm and process equipment, or human sources. However, the acidity of the product prevents growth, and such organisms do not survive for long in the fermenting product. Report of outbreaks of food poisoning in the United States due to *E.coli* O157:H7 in cider refer not to fermented (or hard cider) but to apple juice. Unpublished studies in the United Kingdom (F. Campbell, personal communication) and Senancheck and Golden (1996) clearly established that although the high-acid-tolerant strains of *E.coli* associated with the US outbreaks can survive for long periods in apple juice, they are extremely sensitive to alcohol and die within a few hours in fermenting or fermented apple juice. Similarly, the report of a serious outbreak of cryptosporidiosis in the United States (Millard et al., 1994) from consumption of apple cider refers to consumption of freshly pressed apple juice not fermented cider. However, the report does highlight a potential risk of contamination of apples and juice by oocysts of *Cryptosporidium* spp. if apples are harvested from the orchard sward following grazing by cattle. Unpublished studies in the United Kingdom have shown that *Crytosporidium* oocysts are sensitive to pasteurization; in addition, filtration processes such as those used in commercial production of cider would remove any oocysts if they did contaminate the juice. Bacterial spores from species of *Bacillus* and *Clostridium* can survive for long periods and are frequently found in cider but are of no concern in relation to either spoilage or public health because of the low pH value of cider; nonetheless, their presence may be indicative of poor plant hygiene.

The juice from unsound fruits and from contamination within the pressing plant may show extensive contamination by microfungi, such as *Pencillium expansum, P. crustosum, Aspergillus niger, A. nidulans, A. fumigatus, Paecilomyces varioti, Byssochlamys fulva, Monascus ruber,* and *Phialophora mustea* and by species of *Alternaria, Cladosporium, Botrytis, Oospora,* and *Fusarium.* None is of particular concern in cider making, except that thermoresistant spores of organisms such as *Byssochlamys* spp. can survive pasteurization and may grow in cider if it is not adequately carbonated.

The occurrence of the mycotoxin patulin in apples infected with *Penicillium expansum* and other fungi may result in carryover of patulin in the apple juice base used for cider fermentation. Investigations by Moss and his colleagues (unpublished data) show that although patulin initially inhibits growth of the yeasts used in cider production, the organisms rapidly become tolerant to patulin and, once growth is initiated, the patulin is rapidly metabolized to form ascaldiol and smaller amounts of other metabolites. Hence, if apple juice is contaminated with patulin the fermentation may be slow to start, but the patulin will be destroyed within a few hours. Claims from France that patulin has been found in cider are believed to be associated with postfermentation sweetening of the cider by addition of patulin-contaminated apple juice.

Reference has been made to the role of organisms such as *Brettanomyces* spp. and *Acetobacter xylinum* in the spoilage of ciders during the later stages of fermentation and maturation. Of equal concern is the yeast *Saccharomycodes ludwigii*, which is often resistant to sulfur dioxide levels as high as 1000–1500 mg/L. Maloney (1993) showed that isolates of the yeast *S. ludwigii* were very tolerant of sulfur dioxide because of the production of an excess of acetaldehyde, which was able to minimize the accumulation of intracellular sulfite. He also found that sulfur dioxide–resistant strains had relatively low intracellular pH values and a low intracellular buffering capacity.

S. ludwigii can grow slowly during all stages of fermentation and maturation and is often an indigenous contaminant of cider-making premises. Its presence in bulk stocks of cider does not cause overt problems. However, if it is able to contaminate "bright" cider at bottling, its growth will result in a butyric flavor and the presence of flaky particles that spoil the appearance of the product. Although single cells of the organism are sensitive to pasteurization conditions, clumps of organisms can survive pasteurization, thereby contaminating the final product at the packaging stage; contamination can also arise from the packaging plant and the environment.

Environmental contamination of final products can also occur from other yeasts such as *S. cerevisiae*, *S. bailii*, and *S. uvarum*, which will metabolize any residual or added sugar to generate further alcohol and, more important, to increase the concentration of CO_2. Strains of such environmental contaminants are frequently resistant to sulfur dioxide. Bottles of cider inoculated with such fermentative organisms may develop carbonation pressures up to 9 bar. For this reason, it is essential to minimize any risk from contamination of packaged product by fermentative yeasts and to maintain an adequate level of free sulfur dioxide in the final product. This is particularly important in multiserve containers, which may be opened and then stored with a reduced volume of cider. This precaution is less important for product packaged in single-serve cans and small bottles, which receive a terminal pasteurization process after filling.

Special Secondary Processes

Traditional "conditioned" draught cider results from a live secondary fermentation process. After filling into barrels, a small quantity of fermentable carbohydrate is added to the cider followed by an inoculum of active alcohol-resistant yeasts. The subsequent growth is accompanied by a low-level fermentation during which sufficient CO_2 is generated to produce a *pettilance* in the cider together with a haze of yeast cells. Such products have a relatively short shelf life in the barrel.

Double-fermented cider is initially fermented to a lower-than-normal alcohol content (e.g., 6% ABV) by restricting the amount of chaptalization sugar added. The liquor is racked off as soon as the cider has fermented to dryness and is either sterile filtered or pasteurized prior to transfer to a second sterile fermentation vat. Additional sugar or apple juice is added, and a secondary fermentation is induced by inoculation with a different yeast strain (e.g., an alcohol-tolerant strain of *Saccharomyces* spp.). Such a process permits the development of very complex flavors in the cider.

Sparkling ciders are normally prepared nowadays by artificial carbonation to a level of 3.5–4 vol of CO_2. Traditionally, sparkling ciders were prepared according to the "méthode champenoise." After bright filtration of the fully fermented dry cider, it would be filled into bottles containing a small amount of sugar and an appropriate champagne yeast culture. The bottles were corked and wired and laid on their side for the fermentation process, which lasts 1–2 months at 15–18°C. Following this stage, the bottles were placed in special racks with the neck in a downward position. The bottles were gently shaken each day to move the deposit down toward the cork, a process that can take up to 2 months. The disgorging process involved careful removal of the cork and yeast floc but without loss of any liquid (sometimes the neck of the bottle would be frozen to aid this process). The disgorged product would then be topped up using a syrup of alcohol, cider, and sugar prior to final corking, wiring, and labeling. It is not difficult to understand why this process is rarely used nowadays! An alternative procedure, which gives a gentle pettilance similar to that of the traditional méthode champenoise, is to undertake a secondary "cuvée close" fermentation in a sealed pressurized vat so that the CO_2 generated is retained within the cider.

6.2.6 The Chemistry of Cider

The chemical composition of cider is dependent upon the composition of the apple juice, the nature of the fermentation yeasts, microbial contaminants, and their metabolites, and on the nature of any additives used in the final product.

The Composition of Cider Apple Juice

Apple juice is a mixture of sugars (primarily fructose, glucose, and sucrose); oligosaccharides and polysaccharides (e.g., starch); malic, quinic, and citromalic acids; tannins (e.g., polyphenols such as chalcones, flavonols, proanthocyanidins, hydroxycinnamic acids, and flavonol gyclosides); amides and other nitrogenous compounds; soluble pectin; vitamin C; minerals; and a diverse range of alcohols, aldehydes, ketones, esters, fatty acids, and hydrocarbons. Some esters (e.g., ethyl 2-methylbutanoate) give the juice a typical applelike aroma (Paillard, 1990). The relative proportions are dependent upon the variety of apple, the cultural conditions under which it was grown, the state of maturity of the fruit at the time of pressing, the extent of physical and biological damage (i.e., mold rots), and, to a lesser extent, the efficiency with which the juice was pressed from the fruit.

Much of the sulfur dioxide used during cider making is bound to constituents originating from the juice itself (e.g., galacturonic acid); from mold damage to the fruit (e.g., 5-keto-fructose); from ascorbic acid (L-xylosone); and from metabolic carbonyls, especially acetaldehyde, pyruvate, and α-ketoglutarate, arising from the yeast fermentation. Jarvis and Lea (2000) showed significant differences in the production of sulfite-binding compounds by strains of yeast used in UK commercial cider production, one strain producing more than twice the amount of sulfite-binding compounds than the lowest-producing strain. It is not surprising, therefore, that in commercial ciders there is a considerable difference in the levels of sulfite that is bound. Also important as a cause of sulfite binding is the pretreatment of the juice with pectinase, which results in the formation of elevated levels of galactose and galacturonic acid in the juice.

Products of the Fermentation Process

The primary objective of fermentation is the production of ethyl alcohol from simple sugars. Various intermediates in the Embden–Meyerhof–Parnass pathway can also be converted to form a diverse range of other metabolites, including glycerol (up to 0.5%). Diacetyl and acetaldehyde may also occur, particularly if the process is inhibited by excess sulfite and/or if uncontrolled malolactic fermentation occurs. Other metabolic pathways will operate simultaneously with the formation of long- and short-chain fatty acids, esters, lactones, etc. Methanol will be produced in small quantities (10–100 mg/L) as a result of demethylation of pectin in the juice. Table 6–7 illustrates some of the volatile compounds found in a normal and a spoiled cider.

If other organisms are also present in the fermentation (e.g., lactic acid bacteria), these can convert the fruit acids malic and quinic acids to lactic and dihydroshikimic acids, respectively, thereby reducing the acidity of the cider by about 50%. These reactions are accompanied by further diverse but not widely understood chemical changes that result in subtle, yet important, flavor changes in the final product. Lactic and acetic acids can also be formed directly from residual sugars, and great care needs to be taken to avoid excessive production of volatile acids in cider.

The levels of tannins in cider are reduced significantly during fermentation, and by adsorption on the surface of a stainless steel process plant (Mitchell, 1999). The level of tannins such as 5-caffeoyl quinic acid is reduced by more than 90% during the fermentation process; others, such as the chlorogenic, caffeic, and *p*-coumaryl quinic acids, are reduced to dihydroshikimic acid and ethyl catechol.

The nitrogen content of cider juice comprises a range of amino acids, the most important of which are asparagine, aspartic acid, glutamine, and glutamic acid; smaller amounts of proline and 4-hydroxymethylproline also occur. Aromatic amino acids are virtually absent from apple juices. With the exception of proline and 4-hydroxymethylproline, the amino acids will be largely assimilated by the yeasts during fermentation. However, leaving the cider on the lees for an appreciable length of time will significantly increase the amino nitrogen content as a consequence of release of cell constituents during autolysis.

Inorganic compounds in cider are derived largely from the fruit and will depend upon the conditions prevailing in the orchard. These levels will not change significantly during fermentation. Small amounts of iron and copper may also occur naturally, but the presence of more than a few milligrams per liter will result in significant black or green discolorations and flavor deterioration. Such discolorations are caused by the formation of iron or copper tannates from traces of metal ions derived from equipment and/or from use of rotten fruit.

Cider Maturation and Cider Flavor

Changes in the composition of the cider occur during maturation, but the nature of the changes is but poorly understood. The flavor of cider is very complex and is dependent upon the source, type, and pretreatment of the juice, the nature of the yeast strain(s) used, the type and conditions of the fermentation processes (including the occurrence or otherwise of the malolactic fermentation), and many other variables. The primary attributes for assessing cider flavor are related to descriptive notes concerned primarily with astringency, bitterness, fruitiness, sweetness, sourness, and acidity; other related attributes include the bouquet, the alcohol content, and sulfur notes. Off-flavors include those described as mousy, musty, yeasty, diacetylic, sulfury, "rotting vegetable," acetic, and metallic (oxidized).

Unlike wines, and spirits such as whisky, the wooden vats traditionally used for maturation contribute few, if any, flavor compounds to the cider; however, they are an important source of organisms for the secondary fermentation and maturation processes, which can produce a diversity of flavor characteristics. Analysis of green, fermented, and matured ciders has so far led to the identification of over 200 volatile and nonvolatile compounds that individually and in combination affect the flavor of the product. Some of the factors affecting the occurrence of cider flavor compounds have been reported (Jarvis et al., 1995) and this is an area of on-going investigation.

6.3 PERRY (*POIRÉ*)

The production of perry parallels that of cider in most respects. Perry production is traditional in certain parts of the United Kingdom (especially Gloucestershire and Somerset) and in France (Normandy). The fermentation base can be blended with up to 25% of apple juice. Perry is defined by AICV (1998) as an alcoholic beverage obtained *only* by the complete or partial fermentation of the juice of fresh pears, or the reconstituted juice of concentrate made from the juice of pears, or a mixture of such juices, with or without a limited volume of the juice of apples. As with cider, potable water and sugars may be added before or after fermentation, fresh or concentrated pear juice may be added after fermentation, and additives may be used as permitted by EU legislation. The product may be uncarbonated (i.e., still) or carbonated by secondary fermentation or by injection of carbon dioxide. The product will contain between 1.2 and 8.5% ABV. Fortification of perry by the addition of distilled alcohol is not permitted.

As with the production of cider, special varieties of perry pears are grown traditionally (Table 6–8), especially in Gloucestershire. Perry pears have a low to moderate content of tannin and introduce a high degree of astringency to the perry that is not found in fermented culinary pear juice. Although fewer studies have been made on pears than on apples, it has been suggested by Paillard (1990) that pear aromas contain significantly fewer volatile constituents (79) than does the aroma of apple (212). Pear juice contains a higher proportion of acetate and other short-chain aliphatic esters than does apple juice, with fewer alcohols and higher esters or lactones. However, in many varieties of pear, the aroma is characterized by a range of deca-2,4-dienoate esters, all of which possess a powerful pear aroma. The methyl- and ethyl-deca-2,4-dienoates have high boiling points and are retained even in concentrated pear juices.

Because it was feared that the genetic base of traditional perry pears would be lost, a national collection of perry pear varieties has been established at the Three Counties Agricultural Showground at Malvern in Worcestershire, UK.

Details of the production of perry are given by Pollard and Beech (1963). The major difference between perry and cider is in the flavor: perry has a much softer and

Table 6–8 Characteristics of Some Perry Pear Varieties

| | | Typical Composition ||
Type	Typical Varieties	Acidity (g/100 mL)	Tannin (g/100 mL)
Low acid, low tannin	Red Pear	0.29	0.09
Low acid, medium tannin	Late Treacle	0.33	0.15
	Nailer	0.37	0.23
Medium acid, low tannin	Brown Bess	0.55	0.10
	Gregg's Pit	0.57	0.11
Medium acid, medium tannin	Blakeney Red	0.42	0.13
	Moorcroft	0.50	0.17
	Tumper	0.53	0.15
High acid, low tannin	Yellow Huffcap	0.62	0.10
High acid, medium tannin	Holmer	0.84	0.30

Source: Data from Williams, Faulkner, and Burroughs (1963).

less astringent character, and the occurrence of a high level of nonfermentable sorbitol gives the product increased body. Perry stored for a long time will tend to harden and lose the smooth flavor characteristics; the reasons for such changes are unknown, but they are probably related to instability of the polyphenols.

6.4 FRUIT WINES

The production of fruit wines is undertaken commercially in Europe and in many other parts of the world, especially those unsuited to the cultivation of grape vines. In Europe, the AICV (1998) has defined two types of fruit wine: fruit wines produced with and without fortification with alcohol.

Fruit wine, without addition of alcohol, is made only by the complete or partial fermentation of the juice or pulp of fresh edible domestic and/or tropical fruits except grapes, or the reconstituted juice of concentrates from these fruits, with or without the addition of water and sugar or honey. Juice and/or concentrated juice of such fruits may be added after fermentation, together with, or without, the addition of food additives permitted by EU legislation. The product may be uncarbonated (i.e., still) or carbonated by secondary fermentation or by injection of carbon dioxide. The strength of fruit wine without the addition of alcohol will range from 1.2% to 14% alcohol (ABV).

Fruit wine with added alcohol is made only by the complete or partial fermentation of the juice or pulp of fresh edible domestic and/or tropical fruits except grapes, or the reconstituted juice of concentrates from these fruits, with or without the addition of water, sugar, or honey, and distilled *agricultural* alcohol. Juice and/or concentrated juice of such fruits, plants, and/or flavorings may also be added after fermentation, together with, or without, the addition of food additives permitted by EU legislation. The product may be uncarbonated (i.e., still) or carbonated by secondary fermentation or by injection of carbon dioxide. The strength of fruit wine with the addition of alcohol will range from 6 to 22% alcohol (ABV), but the product must not have the characteristics of a spirit.

Fruit wines produced in accordance with these definitions will be colored and/or flavored by the juices used in their manufacture; they may also be further colored by the addition of permitted food colors (Anon, 1994b). By contrast, fruit wines manufactured in Eastern European countries such as Poland will not necessarily be perceived as being different from grape wines, being generally white, rosé, or red in color and classified according to their alcohol and sugar contents. Jarczyk and Wzorek (1977) noted that legislative definitions exist in Poland, the former Czechoslovakia, and the former USSR for different styles of fruit wine. Some examples of fruit wine compositional data are presented in Table 6–9.

As with cider, compositional standards relating to maximum volatile acidity and sugar-free dry extract are prescribed to prevent excessive dilution of the products. For example, the maximum permitted level of volatile acidity (measured as acetic acid) should not exceed 1.2 g/L for white and 1.6 g/L for red wines (Jarczyk and Wzorek, 1977).

6.4.1 Fruits Used in Fruit Wine Manufacture

In Western and Central Europe, fruit wines are prepared by the fermentation of cultivated fruits, such as apples; cherries; white, red, or black currants; pears; plums; strawberries; and wild fruits such as the arctic bramble, bilberry, blackberry, cloudberry, elderberry, lingonberry, and rose hips. Wild fruits frequently contain higher levels of acidity and tannins; they also frequently have stronger aromas than cultivated varieties of the same plants. Similar fruits have been used in other countries also. In recent years, considerable attention has been paid to the use of subtropical and tropical fruits for the

Table 6–9 Proximate Composition of Some Fruit Wines

Style of Wine	Country of Origin	Alcohol (g/100 mL)	Sugar (g/100 mL)	Acidity (g/L)	SFDE* (g/L)
Dry	Former Czechoslovakia	10.0	<2.0	ND	ND
	Poland	12.5	6.6	5.8	ND
	Former USSR	12.5	2.5	8.0	22.0–30.1
Medium sweet	Former Czechoslovakia	12.0	3.8	ND	ND
	Poland	13.0	10.4	6.6	ND
	Former USSR	13.0	6.3	7.8	25.0–31.5
Sweet	Former Czechoslovakia	14.0	10.0	ND	ND
	Poland	13.8	13.3	6.9	ND
	Former USSR	13.5	10.2	8.2	30.6–35.1

* SFDE, sugar-free dry extract; ND: no data.
Source: Data from Jarczyk and Wzorek (1977).

production of wines. These include apricots (Joshi et al., 1990), banana (Mosso et al., 1990), carambola (Poullet, 1994), kiwi (Broussous and Ferrari, 1990; Tjomb, 1991), mango (di Baggi et al., 1986; Adesina, Oguntimein, and Obisanya, 1993), muskmelon (Teotia et al., 1991), orange (Rose, 1991), papaw and plantain (Mosso et al., 1990), and persimmon (Gorinstein et al., 1993).

Compositional data for a variety of fruits used in wine making are summarized in Table 6–10.

6.4.2 Processing of the Fruit

Fruits for wine making are taken to the winery, where they are sorted, washed and macerated, or milled prior to pressing. Components that will damage the flavor and aroma (e.g., stalks) are removed prior to maceration. After maceration, the fruit pulp will be pressed either in a rack-and-cloth hydraulic press (cf. cider making) or in a modified Bücher–Guyer type press. Continuous belt-and-screw presses are used less extensively. The macerated fruit will be treated with a suitable enzyme preparation to break down the pectin and thereby increase the juice yield. In some instances, the pulp will be heated to 80–85°C during treatment with pectolytic enzymes prior to pressing. Such elevated temperatures will increase the color of the juice, stabilize the anthocyanin pigments, and destroy wild yeasts and other microorganisms (Rommel, Wrolstad, and Heatherbell, 1992; Withy, Heatherbell, and Fisher, 1993).

Juice not required immediately for pressing may be stored following the addition of sulfur dioxide at levels of up to 1200 mg/L, usually in the form of gaseous sulfur dioxide

Table 6–10 Proximate Compositional Analysis of Edible Parts of Fruits Used in Wine Making

Fruit	Total Sugar (g/100 mL)	Total Acidity (g/L)	Tannin (g/100 mL)
Apple (*Pirus malus*)	10.0–11.2	6.0–7.0*	0.1–0.50
Blackberry (*Rubus fructicosus*)	5.5–12.8	9.0–12.0†	0.3
Cherry, sweet (*Prunus avium*)	9.5–16.6	6.0–12.0‡	ND
Cherry, sour (*Prunus cerasus*)	10.0–12.2	13.0–18.0‡	0.14
Black currant (*Ribes nigrum*)	7.0–15.4	10.0–21.0†	0.39
Plum (*Prunus domestica*)	9.3–12.0	9.0–12.0*	0.07
Raspberry (*Rubus idaeus*)	4.7–11.6	13.5–16.0†	0.26
Apricot (*Armeniaca vulgaris*)	6.7–8.5	7.0–13.0*	0.07
Peach (*Persica vulgaris*)	7.8–12.8	8.0–9.0†	0.10
Pear (*Pirus communis*)	9.5–13.0	3.0–5.0*	0.03–0.44
Mango (*Mangifera indica*)	11.3–15.4	2.4–6.8*	ND

* Acidity expressed as malic acid.
† Acidity expressed as citric acid.
‡ Acidity expressed as tartaric acid.
ND, not determined.
Source: Data from Nagy, Chen, and Shaw (1993); Jarczyk and Wzorek (1977).

or as potassium metabisufite. Alternatively, and preferably, the juice will be concentrated some sevenfold following depectinization and clarification; the concentrate can then be stored at chill temperature (7–10°C) until required. Berry fruits can also be deep-frozen and stored at −18°C until required for pressing.

6.4.3 Fermentation of Fruit Wines

The Fermentation Process

As in the fermentation of cider, fruit wine fermentation is done largely in vats made of concrete, steel, or glass-reinforced plastic (GRP), alternatively, traditional oak or larch wood casks may be used. The vat is partially filled with the prepared juice (desulfited if made from sulfur dioxide–preserved juice) to which will have been added part of any required fermentation sugar syrup (usually sucrose). Often the acidity of the juice will have been adjusted prior to use, either by the addition of food-grade acid or, more often, by the neutralization of excessive acidity using calcium carbonate. This is particularly important in the case of fruits or other plant extracts containing oxalic acid, which would otherwise cause difficulties in clarification of the wine at a later stage.

Since most fruit juices are relatively deficient in nitrogenous components, nutrients such as diammonium phosphate, ammonium chloride, ammonium sulfate, or ammonium carbonate are added at a level of 0.1–0.3 g/L juice. Yang and Wiegand (1949) recommend the addition of urea at a level of 0.5 g/L to juices of berry fruits, such as raspberry and blackberry, to increase the fermentation rate. Yeast hydrolysate may also be added to provide a source of vitamins and sterols. Fermentation bases prepared from fresh (i.e., nonsulfited juice and nonconcentrated juice) should be treated with some 30–150 mg/L sulfur dioxide to prevent growth of wild yeast strains.

The yeast inoculum will have been prepared in a pasteurized fruit juice base containing approximately 15% (w/v) sugar at 22–25°C. The prepared yeast culture will be added to the juice base to give about 10^6 viable cells per milliliter in a volume equivalent to 1–10% of the total volume and at a yeast dry matter of 12.5 g/L. Normally the juice will have been prepared at a temperature of 12–15°C (never in excess of 20°C) prior to inoculation in order that the initial exothermic aerobic growth of the inoculum does not lead to excessively high temperatures with consequential autosterilization of the base. Over the first few days, the temperature will rise gradually to a level of 20–25°C and must, in any case, not be allowed to exceed 28°C. Additional sugar syrup will be added on two or three further occasions prior to the fermentation's going to dryness (i.e., with no fermentable carbohydrate remaining). Dependent upon the level of alcohol produced and the fermentation conditions, the process may continue for some 6–8 weeks.

In some processes, fermentation may be carried out at a temperature of 5–10°C using a psychrophilic yeast strain (this is not dissimilar to the conditions used in the fermentation of champagne). Understandably, such a process takes much longer to ferment to dryness but preserves the fruit aroma rather better than does fermentation at higher temperatures. Cold fermentation also enables generation of a higher alcohol content and slows the rate of yeast autolysis.

Maturation and Mellowing

At the end of fermentation, the wine is racked off the lees to remove both fruit solids and autolyzing yeasts. The young wine will normally not have developed a definitive bouquet and requires maturation before packaging. Maturation of fruit wines is normally undertaken at a temperature of 7–15°C; further racking will be carried out after periods of 1 and 2 months and then at 2-month intervals over a period of about 1 year. If it is intended to produce very high-quality products, the wine would normally be racked every 6 months until the product is deemed to be suitably mature. During the maturation process, a secondary malolactic fermentation will normally occur, but only at the higher maturation temperatures. This secondary fermentation converts malic and citric acids into lactic and citromalic acids, respectively.

Subsequent mellowing of the wine results from the chemical and enzymatic changes that take place throughout the maturation process and provides a diversity of esters, alcohols, aldehydes, ketones, and acetals, which combine to provide the bouquet and flavor of the wine. Important in this process is the small amount of O_2 incorporated into the wine each time that racking is done and the further diffusion of O_2 through the staves of the barrel.

Final Processing

The final stage of the process is the blending, sweetening and flavoring (if required), and stabilization of the wines. The blending process is done both to ensure consistency of product character and to reduce the strong aroma and flavor characters of certain wines. Although there is some preference for single-fruit wines, many are blends, especially with apple wine, which is relatively low in flavor.

Wines can be sweetened using either sugar or fruit juice, the latter also serving to increase the "natural" fruit content and flavor of the wine. In some cases, it may be necessary also to adjust the acidity of the wine by the addition of an approved food-grade acid, such as citric or tartaric acid.

Clarification of wines prior to bottling involves the treatment with gelatin, albumen, isinglass, bentonite, potassium ferrocyanide, or salts of phytic acid; the two last treatments are intended to reduce the level of soluble iron complexes that would otherwise cause blackening of the wine (qv), but with fruit wines they are frequently inadequate. After fining, the wine would normally be filtered through a layer of kieselguhr or through cellulose pulp. Traditional filtration processes through asbestos are nowadays considered unsafe procedures. Alternative clarifying procedures include chilling the wine prior to or after fining and the use of microfiltration systems.

The wine, after clarification, will normally be either flash pasteurized and hot filled into the bottle or will be treated to give a residual SO_2 content. In certain countries, bright wine may be treated with up to 200 mg of dimethyl pyrocarbonate (DMPC) per liter. This antimicrobial breaks down on contact with water to give reactive products that destroy any low level of microorganisms and leaves only minimal residues of methanol and CO_2; however, as with diethyl pyrocarbonate, which is now banned, there are some concerns about the extended safety-in-use of DMPC because of the production of methyl carbamate in DMPC-treated wines (Sen, Seaman, and Weber, 1992; Sen et al., 1993).

Defects of Fruit Wines

Reports of "food poisoning" following consumption of homemade elderberry wines were believed to have been caused by high histamine levels. Pogorzelski (1992) investigated the potential for production of histamine in Polish elderberry juices and wines. He found higher levels of histamine in depectinized and pulp-fermented juices than in juices from hot macerated fruits. However, these levels were low by comparison with those from the elderberry fruit wines themselves, in which concentrations up to 12.36 mg/L were found, dependent upon the method of fruit treatment and the yeast strain used for fermentation. The elderberry contains high levels of the amino acid histidine, which serves as substrate for the enzyme histidine decarboxylase. This enzyme is present in the fruit itself and also in certain preparations of pectolytic enzymes and can be produced by many fermentation yeasts (Pogorzelski, 1992).

Flavor defects include taints caused by the presence of hydrogen sulfide, acetic acid, mouse, diacetyl, bitter compounds, and yeasty and moldy taints. Microbial spoilage per se results generally from growth of film yeasts such as species of *Hansenula*, *Brettanomyces*, *Pichia*, and *Torulopsis* within the bottled wine, growth of acetifying bacteria, and (rarely) growth of anaerobic bacteria. The last cause bittering, but this is usually restricted to low-acid wines.

Turbidity can result from the growth of fermentative yeasts, which will also increase the CO_2 level and may cause bottles to burst. Alternately, turbidity may be caused by chemical precipitation of residual tannins, pectins, or proteins (especially in the presence of traces of copper). Blackening is caused by the precipitation of iron-tannin complexes, browning is caused by oxidation of tannins and anthocyanin pigments, and a bluish-white cloud is caused by precipitation of ferric phosphate. The development of browning as a result of polymerization of anthocyanin pigments in blackberry and raspberry juice wines can be avoided by fermentation of flash-pasteurized depectinized juice in the presence of 50 mg of sulfur dioxide per liter (Rommel et al., 1992; Withy et al., 1993).

6.4.4 Fruit Pulp Fermentations

Fermentation of pulp permits a higher level of extraction of anthocyanins and other pigments from fruits such as bilberries, cherries, strawberries, and elderberries and facilitates the extraction of juice from most fruits, particularly from plums. A consequence is that the final clarification of the wine is simplified.

Because of the high level of yeasts on the fruit, it is necessary to ensure the addition of a vigorous pure yeast inoculum to compete effectively against the wild yeast population. Since the pulp is a richer source of nutrients than the juice, a vigorous fermentation occurs and nutrient supplementation is not necessary. Natural fermentation can be effective but often produces unpleasant or unsatisfactory flavors. Ideally, the fermenting pulp is held beneath the surface of the fermenting liquid in special tanks in order to exclude the pulp from contact with air. If such tanks are not used, the development of CO_2 in the fermenting pulp will raise the pulp to the surface, where rapid oxidation of pigments will occur together with growth of oxidative yeasts. At the completion of fermentation, the wine is racked off and the fermented pulp is then pressed to give an alcohol-rich must that is immediately sweetened and refermented, thus increasing the yield.

6.4.5 Alcohol-Fortified Wines

Fortification of fruit wines with rectified alcohol, of agricultural origin, is practiced in a number of countries to increase the overall alcohol level generally up to 15% ABV but sometimes as high as 22% ABV. Usually this process is undertaken toward the end of, or immediately following, the fermentation period and before maturation. In most instances, the fruit wine is fermented without chaptalization (i.e., addition of fermentation sugar) to an alcohol level of 5–6% ABV, and the alcohol is added before maturation in order to permit the harmonization of the added alcohol and the wine. The final product is sweetened by addition of sugar following clarification.

6.4.6 Sparkling (Carbonated) Fruit Wines

As with the production of sparkling ciders, fruit wines may be carbonated either by injection of CO_2 (to a pressure of about 6 kg/cm^2) or by a *cuvée close* secondary fermentation in a pressurized closed tank. The wine is sweetened by the addition of a small quantity of sugar syrup and inoculated with a suitable yeast culture. The process retains all the developed CO_2 in the wine, which will have a pressure of some 45 kg/cm^2. The wine will be filtered and then bottled using an isobaric bottling machine. Critical to the process is the selection of the yeast to be used, which should not generate any specific flavor characteristics. Strains of yeast used for secondary fermentation of champagne are very suitable for this purpose.

6.5 FRUIT SPIRITS AND LIQUEURS

The distillation of wines, fruit wines, cider, and perry to produce fruit spirits is practiced widely. While the best known are brandies such as cognac and armagnac derived from grape wines, traditional spirits such as cider brandy from cider, poiré from perry, slivovitz from plum wine, and kirsch from black cherries have been produced for many years.

Also widely practiced is the alcoholic infusion or decoction of whole or macerated fruits, sometimes with the addition of herbs, spices, or other plant material to produce apéritifs, liqueurs, and other strong alcoholic beverages with fruit flavor and aroma.

6.5.1 Fruit Spirits (Fruit Brandies)

Wine fermented from fruits, or in some instances fermented fruit pulp, is double-distilled to produce a spiritous beverage containing up to 70% ABV, although in many cases only a single distillation is used, giving a product with an alcohol level of 25–55% ABV. The methods of production vary from area to area but share many common features. Traditionally, distillation takes place soon after completion of fermentation (i.e., without any significant maturation period) in a copper pot still heated over an open flame or in a Charentais-type double-distilling still. Generally, the wine will not have been filtered and will contain a small amount of yeast and other sediments. In a two-stage process, the first distillate will contain about 28–30% ABV. During the second distillation, the "heads" and "tails" containing high concentrations of aldehydes and fusel oils, respectively, are separated. The middle fraction will have an alcohol content of about 70% ABV.

The distilled product may be stored in used oak or larch barrels for several years in order to mature and develop a desirable bouquet. However, some fruit brandies (e.g., kirsch) are not stored in wood in order to maintain the clear colorless appearance. For retail sale, the distilled products are generally diluted to an alcohol content of 35–40% ABV or are blended with fruit juices to yield fruit liqueurs (qv).

The quality of the product will be influenced by the quality and variety of the fruit used for the initial fermentation, the pH of the juice or the pulp, the choice of fermentation yeast, the extent to which depectinization was undertaken, and the level of use of sulfur dioxide in the primary fermentation.

A major problem with some fruit spirits, such as plum brandy, relates to the level of methanol liberated into the fruit pulp by depectinizing enzymes and subsequently distilled into the final product. Paunovic (1991) reported methanol levels up to 15 mg/L in pectinase-treated plum pulp, with residual levels of methanol in the distilled alcohol of up to 12.4 mg/L. He also showed that the level of methanol in the pulp could be reduced significantly by high-temperature treatment of the pulp and careful selection of the fermentation yeast strain: products of "spontaneous" fermentation tended to have high methanol levels.

Examples of some fruit spirits and brandies are presented in Table 6–11.

6.5.2 Apéritifs and Liqueurs

Many well-known brands and styles of fruit-flavored liqueur are marketed worldwide; such products are often produced by "secret" recipes. Additionally, a number of traditional "country" recipes are used widely. Some examples are presented in Table 6–12. Normally the process involves the extraction of macerated fresh, dried, or fermented fruit pulp with either a neutral distilled agricultural alcohol or with a spirit distillate obtained from the specific fruit wine. However, some products are compounded by blending fruit spirit or neutral spirit with fruit juice, often with added sweetening. Most products have alcohol contents in the range 20–28% ABV.

Gorinstein et al. (1993) investigated the effects of processing variables on the characteristics of a liqueur prepared from persimmon. They observed that the organoleptic qualities (aroma and flavor) of the liqueur could be influenced significantly by the condition of the fruit extracted; the relative ratios of fruit and alcohol; the concentration of alcohol; the extent to which the fruit had been fermented before extraction; and the quality of the distilled spirit. The method of choice was to extract macerated, freeze-dried fruits with a middle distillation fraction containing 52–86% ethanol, which was obtained from a 21-day 20–24°C *S. cerevisiae* var. *ellipsoides*-fermented pectinase-treated fresh fruit pulp. The final product contained 27% ABV, 30% total sugar, and 30% persimmon extract.

An example of a traditional UK product is sloe gin, which is produced by steeping chopped, destoned sloe (wild plum; *Prunus spinosa*) fruits, gathered from the hedgerows, in gin for a period of 3–4 months. A typical domestic recipe would require 5 kg

Table 6–11 Some Fruit Spirits and Brandies

Type	Synonym	Country of Origin	Fruit	Special Characteristics
Alise	Sorbier	France	Rowanberries	
Aprikosengeist	Apricot brandy	Germany	Apricots	
Calvados	Cider brandy	France	Apple	Double distillation from fermented pulp; aged in wood
Eau-de-vie-de-Cidre	Cider brandy	France	Apple	Distilled from cider (cf. Calvados); not usually aged
Kirsch		Germany	Black cherry	Distilled from fermented black cherry pulp
Mirabelle	Plum brandy	France	Yellow plum	Distilled from fermented plum pulp
Poire Williams		Switzerland, Germany, France, Italy	Williams pear	Distilled from fermented pear pulp; not aged in wood
Slivovitz	Plum brandy	Former Czechoslovakia, Hungary, Bulgaria	Plum	Distilled from fermented plum pulp; aged in wood

of fruits, 1.2 kg of sugar, 25 mL of almond essence, and 12 L of gin. After steeping, the concoction is filtered and bottled, usually after standardization of the alcohol content to a level of 22–25% ABV. The liquor has a medium to dark red–purple coloration and is a favorite winter drink in the United Kingdom, widely associated with field sports. If preferred, the sloes may just be pricked with a large needle; this then permits the extracted fruit to be stored under a layer of fresh gin and consumed as a liqueur gin. Variations on the recipe include the substitution of vodka for gin and use of damsons or other bitter stone fruits.

6.6 MISCELLANY

The production of fruit-based alcoholic drinks provides a broad range of product types with variations in alcohol content from <1.2% to 55% ABV or more. It is not surprising that local specialities prevail in different countries and regions; it is also not surprising that many diverse products are marketed in the international drinks market.

Table 6–12 Some Fruit Liqueurs and Aperitifs

Type	Country of Origin	Fruit	Alcohol Type	Alcohol Content (% ABV)	Special Characteristics
Amaretto	Italy	Apricot (*Armeniaca vulgaris*) and kernels	Eau-de-vie	24	Distillation of fermented fruit pulp, then infusion of macerated fruits, kernels, and sugar
Cerasella	Italy	Cherry (*Prunus cerasus*)	Neutral spirit or brandy	28	Infusion of macerated fruits; clear, white
Karpi	Finland	Cranberry (*Vaccinium macrocarpon*)	Neutral spirit	30	Infusion of macerated fruits
Lakka, Suomuurain	Finland	Cloudberry (*Rubus chamaemonas*)	Neutral spirit	30	Infusion of macerated fruits and sugar
Rabinowka	Eastern Europe	Rowanberry (*Sorbus aucuparia*)	Neutral spirit	25	Infusion of macerated fruits and sugar
Sloe gin	England	Sloe berry (*Prunus spinosa*)	Gin	22–25	Infusion of whole or macerated fruits and sugar
Cider Royal	England	Cider apple (*Pirus malus*)	Cider brandy + cider	20	Blending of cider brandy and cider or apple juice
Reishu	Japan	Melon (*Cucumis melo*)	Neutral spirit	15	Infusion of macerated fruits
Wisniak, Wisnioka	Poland	Cherry (*Prunus cerasus*)	Vodka	25–30	Distillation percolation of alcohol; sweetened
Jerzynowka	Poland	Blackberry (*Rubus fructicosus*)	Neutral spirit or eau-de-vie	25–30	Infusion of macerated fruits
Cassis	France	Black currant (*Ribes nigrum*)	Eau-de-vie	15–20	Distillation of fermented pulp before infusion of macerated fruits; sweetened
Zolataya Osen	Mount Caucasus	Damson (*Prunus spinosa*), apple (*Pirus malus*), and quince (*Pirus cydonia* Lin)	Neutral spirit	20–25	Distillation percolation of alcohol; sweetened

The composition of fruit makes it eminently suitable for the preparation of alcoholic products.

Worthy of consideration, however, are the opportunities that the new era of biotechnological processes present in this area—opportunities, for instance, to develop special strains of fermentation yeasts by genetic manipulation: yeasts with the capacity to carry out fermentations at extremes of temperatures, to metabolize fruit polysaccharides (starch) and oligosaccharides, or to produce specific metabolites in order to generate selected flavor profiles. Other opportunities include the use of immobilized yeasts and malolactic bacteria to undertake continuous fermentations of fruit juice substrates.

In tropical countries, attention is already drawn to the development of new alcoholic products from indigenous fruits (see section 6.5.1). In addition to their use as beverages, such developments include the generation of locally produced fuel alcohol (gasohol) based on fermentation of fruit residues, fruit-processing effluents, and other fruit wastes. Work in such areas has been reported from Mexico (López-Baca and Gómez, 1992), Japan (Tanemura et al., 1994) and other countries. Fermentation of wastes can also lead to the establishment of local industries for the production of enzymes, food colors, flavors, antimicrobials, etc. (see, for instance, Cook, 1994). All such developments will benefit local and national economies.

An interesting approach that is worthy of further development is the production of fermented *non*alcoholic beverages enriched in vitamins and other nutrients. Lebrun (1991) has patented a process to produce low-alcohol (<1% ABV) petillant fruit-flavored drinks by fermentation of a mixture of fruits, plant extracts, and sugar.

Of greater interest is the patent of Chavant and Rollan (1992). These workers have developed a process to produce nonalcoholic fruit-based acid drinks enriched with microbial polysaccharides, folic and orotic acids, and a diverse range of B-group vitamins. The novelty arises from the sequential fermentation by selected strains of microorganisms, including the yeasts *Pichia membranaefaciens* and *Candida pseudotropicalis*, and bacteria including *Lactococcus lactis*, *Lactobacillus fermentum*, *Lactobacillus raffinolactis*, *Bifidobacterium* spp., *Leuconostoc* spp., and *Acetobacter* spp. Essentially, they have taken the technology used for many years in the production of the fermented dairy product "kefir" and applied it to develop a new concept in food beverages.

Such studies suggest potential for the development of new food and nonfood products from the fermentation of fruit and their juices.

REFERENCES

Adesina, A.A., Oguntimein, G.B., and Obisanya, M.O. (1993). Kinetic analysis of the fermentation of mango juice. *Lebensmitel-Wissenschaft and Technologie*, **26**, 79–82.
AICV. (1998). Code of Practice. Association des Industries des cidres et vins de fruits de l'UE, Brussels.
Anon. (1994a). European Parliament and Council Directive 94/35/EC of 30 June 1994 on sweeteners for use in foodstuffs. *Official Journal of the European Communities*, **37**, L237, 3–12.
Anon. (1994b). European Parliament and Council Directive 94/36/EC of 30 June 1994 on colours for use in foodstuffs. *Official Journal of the European Communities*, **37**, L237, 13–29.
Anon. (1995). European Parliament and Council Directive 95/2/EC of 25 March 1995 on additives other than colours and sweeteners for use in foodstuffs. *Official Journal of the European Communities*, **38**, L61, 1–40.

Beech, F.W. (1993). Yeasts in cider-making. In *The Yeasts*, Vol. 5, *Yeast Technology*, ed. A.H. Rose and J.S. Harrison. Academic Press, London, pp. 169–214.

Beech, F.W., and Carr, J.G. (1977). Cider and perry. In *Economic Biology*, Vol. 1, *Alcoholic Beverages* ed. A.H. Rose. Academic Press, London, pp. 139–313.

Beech, F.W., and Davenport, R.R. (1983). New prospects and problems in the beverage industry. In *Food Microbiology: Advances and Prospects*, ed. T.A. Roberts and F.A. Skinner. Academic Press, London, pp. 241–256.

Broussous, P., and Ferrari, G. (1990). Procédé de fabrication d'une boisson pétillante faiblement alcoolisée à base de kiwi. *Brevet Europeen* No. 0 395 822.

Chavant, L., and Rollan, S. (1992). Procédé de fabrication d'un nouveau levain glucidique, nouveau levain obtenu, fabrication de boisson fruitée à partir de ce levain et produit obtenu. *Brevet d'Invention Francaise* No. 2 672 614.

Cook, P.E. (1994). Fermented foods as biotechnological resources. *Food Research International*, **27**, 309–316.

di Baggi, V., Ghommidh, C., Navarro, J.-M., and Crouzet, J. (1986). Fermentation alcoolique de la pulpe de mangue. *Sciences des Aliments*, **6**, 407–416.

Dinsdale, M.G., Lloyd, D., and Jarvis, B. (1995). Yeast vitality during cider fermentation: Two approaches to the measurement of membrane potential. *Journal of the Institute of Brewing*, **101**, 453–458.

Dinsdale, M.G., Lloyd, D., McIntyre, P., and Jarvis, B. (1999). Yeast vitality during cider fermentation: Assessment by energy metabolism. *Yeast*, **15**, 285–293.

European Council Regulation on the Organic Production of Agricultural Products and indications referring thereto on agricultural products and foodstuffs. (1991). Regulation 2092/91 as amended subsequently.

Gorinstein, S., Moshe, R., Weisz, M., Hilevitz, J., Tilis, K., Feintuch, D., Bavli, D., and Amram, D. (1993). Effect of processing variables on the characteristics of persimmon liqueur. *Food Chemistry*, **46**, 183–188.

Jarczyk, A., and Wzorek, W. (1977). Fruit and honey wines. In *Economic Microbiology*, Vol. 1, *Alcoholic Beverages*, ed. A.H. Rose. Academic Press, London, pp. 372–421.

Jarvis, B. (1993). Cider (hard cider): The product and its manufacture. In *Encyclopaedia of Food Science, Food Technology and Nutrition*, ed. R. Macrae, R.K. Robinson, and M.J. Sadler. Academic Press, London, pp. 979–983.

Jarvis, B., Forster, M.J., and Kinsella, W. (1995). Factors affecting the development of cider flavour. In *Microbial Fermentations: Beverages, Foods and Feeds*, ed. R.G. Board, D. Jones, and B. Jarvis. *Journal of Applied Bacteriology Symposium Supplement 1995*, **79**, 55–185.

Jarvis, B., and Lea, A.G.H. (2000). Sulphite binding in ciders. *International Journal Food Science and Technology*, **35**, 113–127.

Joshi, V.K., Bhutani, V.P., Lal, B.B., and Sharma, R. (1990). A method for preparation of wild apricot (chulli) wine. *Indian Food Packer*, September–October, 50–55.

Lebrun, F.-A.-L. (1991). Boisson naturellement pétillante et son procédé de fabrication. *Brevet d'Invention Francaise*, No. 2 651 240.

Lloyd, D., Moran, C.A., Suller, M.T.E., Hayes, A.J., and Dinsdale, M.G. (1996). Flow cytometric monitoring of rhodamine 123 and a cyanine dye uptake by yeast during cider fermentation. *Journal of the Institute of Brewing*, **102**, 251–259.

López-Baca A., and Gómez, J. (1992). Fermentation patterns of whole banana waste liquor with four inocula. *Journal of the Science of Food and Agriculture*, **60**, 85–89.

Maloney, S. (1993). Effects of sulphite in yeasts with special reference to intracellular buffering capacity. PhD thesis, University of Bath, Bath, UK.

Millard, P.S., Gensheimer, K.F., Addiss, D.G., Sosin, D.M., Beckett, G.A., Houch-Jankoski, A., and Hudson, A. (1994). An outbreak of cryptosporidiosis from fresh-pressed apple cider. *Journal of the American Medical Association*, **272**, 1592–1596.

Mitchell, L.A. (1999). Phenols and tannins of apples and ciders. PhD thesis, University of Surrey, Surrey, UK.

Mosso, K., Aboua, F., Angbo, S., Konan, K.E., Nyamien, M.N., and Koissy-Kpein, L.M. (1990). Evolution des sucres totaux, de l'extrait see réfractométrique et de l'alcool, au cours de la fermentation vineuse à température alternées de la banane poyo, de la banane plantain et de la papaye. *Actualités des Industries Alimentaires et Agro-alimentaires*, September, 767–772.

Nagy, S., Chen, C.S., and Shaw, P.E., eds. (1993). *Fruit Juice Processing Technology.* Agscience, Auburndale, FL.

Paillard, N.M.M. (1990). The flavour of apples, pears and quinces. In *Food Flavours*, Part C, *The Flavour of Fruits*, ed. I.D. Morton and A.J. Macleod. Elsevier Science, Amsterdam, pp. 1–41.

Paunovic, R. (1991). Uticaj izazivaca i uslova izvodenja alkoholne fermentacije covnog kljuka na sastav vocnih rakija. *Archiv za Polloprivre Nauke*, **52**, 186, 171–183.

Pogorzelski E. (1992). Studies on the formation of histamine in must and wines from elderberry fruit. *Journal of the Science of Food and Agriculture*, **60**, 239–244.

Pollard, A., and Beech, F.W. (1963). The principles and practice of perry making. In *Perry Pears*, ed. L.C. Luckwill and A. Pollard. University of Bristol, Bristol, UK, pp. 195–203.

Poullet, H. (1994). Procédé de fermentation de moût de Carambole. *Brevet d'Invention Francaise*, No. 2 695 136.

Rommel, A., Wrolstad, R.E., and Heatherbell, D.A. (1992). Blackberry juice and wine: Processing and storage effects on anthocyanin composition, color and appearance. *Journal of Food Science*, **57**(2), 385–391, 410.

Rose, R. (1991) Boisson alcoolique pétillante à base d'orange. *Brevet d'Invention Francaise*, No. 2 657 878.

Sen, N.P., Seaman, S.W., and Weber, D. (1992). A method for the determination of methyl carbamate and ethyl carbamate in wines. *Food Additives and Contaminants*, **9**, 149–160.

Sen, N.P., Seaman, S.W., Boyle, M., and Weber, D. (1993). Methyl carbarnate and ethyl carbamate in alcoholic beverages and other fermented foods. *Food Chemistry*, **48**, 359–366.

Senanchek, J., and Golden, D.A. (1996). Survival of *E. coli O157:H7* during fermentation of apple cider. *Journal of Food Protection*, **59**(12), 1256–1259.

Seward, R., Willetts, J.C., Dinsdale, M.G., and Lloyd, D. (1996). The effects of ethanol, hexan-1-ol and 2-phenylethanol on cider yeast growth, viability and energy states: synergistic inhibition. *Journal of the Institute of Brewing* **102**, 439–443.

Swaffield, C.S., Scott, J.A., and Jarvis, B. (1994). Influence of selected microbial isolates on development of mature cider flavours. Proceedings of the Society for Applied Bacteriology Summer Conference 1994. *Journal of Applied Bacteriology*, **77** (Suppl.), xi.

Tanemura, K., Kida, K., Ikbal, Matsumoto, J., and Sonada, Y. (1994). Anaerobic treatment of wastewater with high salt content from a pickled plum manufacturing process. *Journal of Fermentation and Bioengineering*, **77**, 188–193.

Teotia, M.S., Manan, J.K., Berry, S.K., and Sehgal, R.C. (1991). Beverage development from fermented (*S. cerevisiae*) muskmelon (*C. melo*) juice. *Indian Food Packer*, July–August, 49–55.

Tjomb, P. (1991). Boissons an kiwi: c'est l'effervescence, *Revue de l'Industrie Agro-Alimentaire No. 468*, October, 42, 44–45.

Willets, J.C., Seward, R., Dinsdale, M.G., and Lloyd, D. (1997). Vitality of cider yeast grown micro-aerobically with added ethanol, butan-1-ol or iso-butanol. *Journal of the Institute of Brewing*, **103**, 79–84.

Williams, R.R., Faulkner, G., and Burroughs, L.F. (1963). Descriptions of the principal varieties of perry pears. In *Perry Pears*, ed. L.C. Luckwill and A. Pollard. University of Bristol, Bristol, UK, pp. 61–172.

Withy, L.M., Heatherbell, D.A., and Fisher, B.M. (1993). Red raspberry wine—effect of processing and storage on colour and stability. *Fruit Processing*, **3**(8), 303–307.

Yang, H.Y., and Wiegand, E.H. (1949). Production of fruit wines in the Pacific Northwest. *Fruit Products Journal*, **29**, 8–12, 27, 29.

Chapter 7

Production of Thermally Processed and Frozen Fruit

Gerry Burrows

7.1 INTRODUCTION

Most fruits have a definite harvesting season, usually of very short duration, meaning that although there is an abundant supply during this period it is not available at other times of the year. The fruit itself has a limited keeping time. Modern forms of transport make available fruits that are not in season locally (e.g., strawberries to Britain in February), but they are expensive.

Fruit quickly deteriorates, the two main causes being the multiplication of microorganisms and biochemical activity. The fruit tissue is living and respiring when it is harvested, with the biochemical processes catalyzed by a number of enzymes. In order to preserve the fruit, the enzymes must be inactivated or the fruit must be preserved in some other way. However, even when biochemical activity has been stopped, microbial infection can still occur and cause further deterioration.

So that fruit can be easily available all year round, different methods of preservation have to be applied. These methods often alter the characteristics of the fruit to a greater or lesser degree. Some methods may extend the shelf life by a few weeks while others can give a shelf life of 2 years or more. The methods commonly used for long-term preservation are canning, bottling, and freezing, while aseptic preservation is used only in a limited way for particulate fruits. Each of these methods has both advantages and disadvantages, and these are discussed in this chapter. The choice of method of preservation may well be determined by the raw material. Varieties that are suitable for canning with the addition of artificial color may not be suitable for freezing, where no added color can be used.

7.2 RAW MATERIALS

The raw material must be sound, ripe, and free from blemishes and disease. If the raw material is unsuitable for heat processing because it is of the wrong variety or it is underripe, the final product may have a poor color, texture, and flavor. Victoria plums that are not fully ripe and still have a green color will not give a satisfactory product however much coloring matter is used because the natural chlorophyll turns brown during the heat process required for canning. Some varieties of pears acquire a pink coloration after heating, particularly if they are not thoroughly cooled. The chemistry of this reaction is not fully understood, but it is believed to be caused by the natural pigments in the fruit combining with any metal ions present to form complexes.

Some fruits, such as apples and pears, tend to go brown when peeled and exposed to the air, thus causing a poor color prior to being processed. The browning effect can be prevented by placing the fruit in dilute brine or a solution of ascorbic acid.

Plums and gages sometimes turn brown when the fruit is exposed to the headspace between the surface of the liquor and the lid of the can, but if this part of the process is properly controlled by eliminating as much air as possible from the headspace by use of steam-flow closure, discoloration can be prevented.

Certain soft fruits, such as strawberries and some varieties of raspberries, lose their color during heat processing and may produce brown hues. Artificial colors can be used in these instances, but new varieties have been developed that will retain their color. This is particularly important where fruit is canned in fruit juice with no other additives.

7.3 CANNING OF FRUIT

7.3.1 Cannery Hygiene

High standards of hygiene are required to avoid losses through the product's being spoiled or contaminated by debris on the line or by pests attracted by the debris. The schedules for cleaning and housekeeping must be documented and operated by properly trained staff. Equipment should be selected and sited with ease of cleaning in mind. In order to demonstrate due diligence there should be a record to indicate that cleaning has been carried out and that it has been checked. A simple tick sheet should suffice.

Cleaning should be monitored both visually and microbiologically. The former can be carried out as part of a preproduction check by the line supervisor, while there are rapid luminescence methods available to assess quickly the level of microbial contamination.

Conveyors, pipelines, and any other surfaces that come into contact with food should be constructed from smooth, durable materials and should be as free as possible from crevices and inaccessible corners.

Cooling water should be chlorinated or brominated, and all postprocess handling equipment should receive special attention to avoid the possibility of postprocess spoilage.

All storage and production areas must be properly screened to prevent ingress by pests such as rodents, birds, and insects, which can cause damage to raw materials. It is advisable to engage the services of a specialist pest control company to ensure that all protective measures have been employed.

7.3.2 Factory Reception

Fruit is delivered to the factory by lorry or, if in small amounts, by trailer, usually in crates, boxes, or small punnets, depending on its susceptibility to damage.

Each load is weighed on arrival to determine the payment to be made to the supplier. Samples are taken from each delivery and measured against the raw material specification to ensure that the quality is suitable for processing. Some factories operate a sliding scale of reduction of payment if the fruit quality is not quite right to cover the cost of extra inspection and trimming needed to bring it back into specification. The sampling may reveal that the fruit is too ripe to be stored and that it must be canned immediately before further deterioration takes place. If foreign material, such as glass, is found, the load may be rejected.

Once accepted, the fruit must be canned with the minimum of delay, particularly if it has been picked during the heat of the day. Deterioration and mold growth will occur quickly in warm fruit. Ideally, fruit should be canned on the day of picking, and many factories during the fruit season work a shift pattern that goes on into the night to ensure that this happens.

If fruit does have to be stored for any time, it should be placed in a cool, dry environment. Care must be taken to prevent the fruit from being attacked by birds, insects, or other pests. Chilled stores operating at 5–8°C may be used.

7.3.3 Peeling

Methods used for peeling fruit vary from hand peeling to a variety of mechanical methods; in all cases there will be a need for visual inspection and some hand trimming to remove any remaining skin or blemishes. It is possible to peel these away mechanically, but the process is very wasteful.

Abrasion peeling, on its own, tends to be wasteful as the whole fruit has to be rubbed down to remove the skin and blemishes. It is more simple to control peeling losses using lye (caustic soda). The fruit is passed through a hot solution of caustic soda (sodium hydroxide, NaOH), and the degree of peeling can be adjusted by varying the strength and the temperature of the solution and the residence time. In practice the temperature is usually maintained at, or close to, boiling point. The contact time may vary from 1 to 2 minutes in 2% to 10% solution. The loosened skin is removed by jets of water, which also remove all traces of caustic soda. Great care must be taken to avoid accidents with boiling caustic soda solutions. Staff must wear suitable protective clothing to prevent the caustic soda's coming into contact with the skin. It is important that all traces of caustic soda are removed from the fruit so that it is safe to consume.

Some fruits, particularly apples and pears, may be knife peeled by machines on which the fruit is impaled and rotated against a knife that follows the contours of the fruit. The core may be removed at the same time. These machines tend to be wasteful, particularly if the fruit has not been carefully size graded.

Flame peeling has been used for apples, with the loosened skin being removed by mild abrasion under jets of water. Peeling losses are low, but great care is needed to ensure that all the charred skin is removed, as this is very unsightly in the finished pack.

7.3.4 Blanching

Some fruits may require blanching prior to filling, particularly if they are to be solid packed, since the softening and shrinkage enables them to be more readily filled into the cans. The hot filling also reduces the processing time where heat penetration may be slow, but conversely, cooling of the contents of large cans may be slow, with consequent loss of quality.

The advantages of blanching may be slightly offset by the loss of nutritional values that occurs during the operation. Therefore, blanching times should be kept as short as possible. The nutritional-value loss can be minimized by blanching in steam rather than in water, as the leaching losses are lower.

7.3.5 Choice of Cans

It is preferable to use plain-bodied cans for some fruits, such as apples, as these help to keep the color and flavor bright and fresh. This is the result of a chemical reaction between the fruit and the tinplate. The presence of tin gives the fruit a brighter color. However, other fruits, such as rhubarb and plums, must be filled into lacquered cans to prevent the acid in the fruit from reacting with the tinplate. Table 7–4 lists the can sizes used for fruits in the United Kingdom.

7.3.6 Filling

Irrespective of whether the cans are filled by machine or by hand, regular checks must be made on the temperature of the product at the time of closing the can, as this may affect the subsequent exhausting and pasteurizing processes. The filled weights must also be controlled. This may need to include the weights of the constituent parts of the pack (e.g., fruit[s] and syrup) in order to ensure compliance with recipe requirements or legal standards. Correct filling not only is desirable for economic reasons but it is also technically important.

All weight-control data and temperatures should be recorded on charts kept near to the filling area so that trends can be observed quickly and any necessary corrective action taken.

7.3.7 Syrup

Apart from some solid-pack products, fruits are usually canned in a sugar syrup, although there is an increasing use of fruit juice in these more health-conscious days. The syrup is usually prepared from granulated sugar obtained from beet or cane, but it is possible to use other sugars, such as dextrose, corn syrup, glucose syrup, or invert sugar.

Sugar strengths are normally expressed in degrees Brix (°B), which is a measure of the percentage of sugar by weight in an aqueous solution at 20°C.

The UK Canners Code of Practice defines light syrup, syrup, and heavy syrup. These vary according to the type of fruit, a distinction being made between the following three classes:

Class A: Apples (other than purée or solid pack), bilberries, blackberries, black currants, cherries, damsons, gooseberries, greengages, loganberries, plums, raspberries, red currants, rhubarb, and strawberries

Class B: Apricots, peaches, pears, pineapple, fruit salad, and fruit cocktail

Class C: Prunes

Table 7–1 details the different syrup strengths for the different classes of fruit.

Sugar density can be checked by using a direct Brix reading hydrometer. If the temperature is other than 20°C, then a corrective factor has to be applied. It is often more convenient to use a refractometer to measure the strength of the syrup. These are available as bench models or as hand-held instruments, which are particularly useful when an instant reading is required.

7.3.8 Cut Out

When examining cans of fruit to determine their quality, measurements are made of the drained weights of the fruit and the density of the covering liquor. This examination is known as the "cut out" and is best carried out after canning and when the contents have reached equilibrium (after 48 hours or more).

Syrup may be added at 45°B, but when the can is examined it will probably cut out at 25 to 30°B. This is because it has been diluted by the water naturally present in the fruit. The amount of dilution will depend on the type of fruit, its variety and state of ripeness, and the ratio of fruit to syrup in the can.

If the weight of the fruit packed in the can is known, it is possible to calculate fairly accurately the strength of the syrup used. The weight of drained solids in any particular can is not constantly proportional to the filled weight, as it can be influenced by a number of factors, such as the exhaust time and temperature, the condition of the fruit, and the strength of the syrup.

Canned fruits packed in 40 to 45°B syrup generally give the approximate drained weights expressed as a percentage of the filled weight as shown in Table 7–2. The variations are caused by the differences in texture. If 30°B syrup is used, the drained weights will be approximately 2% less than those shown.

7.3.9 Exhausting

Exhausting is the term given to the process of removing air and entrapped gases from the can before closing. It can be achieved by several methods, depending on the type of product.

A product containing a thin liquid will only rarely occlude gases below its surface and, therefore, only requires the removal of air in the headspace. Viscous or semisolid

Table 7–1 Syrup Strengths for Different Classes of Fruit

	Class A (% w/w)	Class B (% w/w)	Class C (% w/w)
Light syrup	15	15	10
Syrup	30	22	15
Heavy syrup	40	30	20

Table 7–2 Drained Weights as Percentage of Filled Weights for Fruits Canned in 40–50°B Syrup

Fruit	Drained Weight as Percentage of Filled Weight
Blackberries	65–90
Black currants	75–95
Cherries	85–100
Damsons	85–95
Gooseberries	85–100
Loganberries	70–90
Raspberries	65–95
Strawberries	55–80

products may contain a considerable amount of entrapped air when filled into the can. Fruit tissues also may contain CO_2 evolved from the respiratory process.

The syrup used for covering fruits is normally filled as hot as possible (ideally greater than 80°C) so that the air in the headspace is partially displaced by the steam from the hot liquid.

Can closing is often preceded by steam-flow closure, in which steam is injected into the headspace from jets on the seamer. This has the effect of sweeping the air from the headspace immediately before the lid is seamed on, thus creating a partial vacuum as the steam condenses.

7.3.10 Closing

Cans are closed by placing the lid on a filled can and sealing it to the body by the formation of a double seam (Figure 7–1). As the term implies, the seam is formed by two operations in the seamer. During the first operation the can end seaming panel is rolled together and interlocked with the can body flange. The second operation completes the seam by pressing it to the required tightness. Details of double-seam technology and

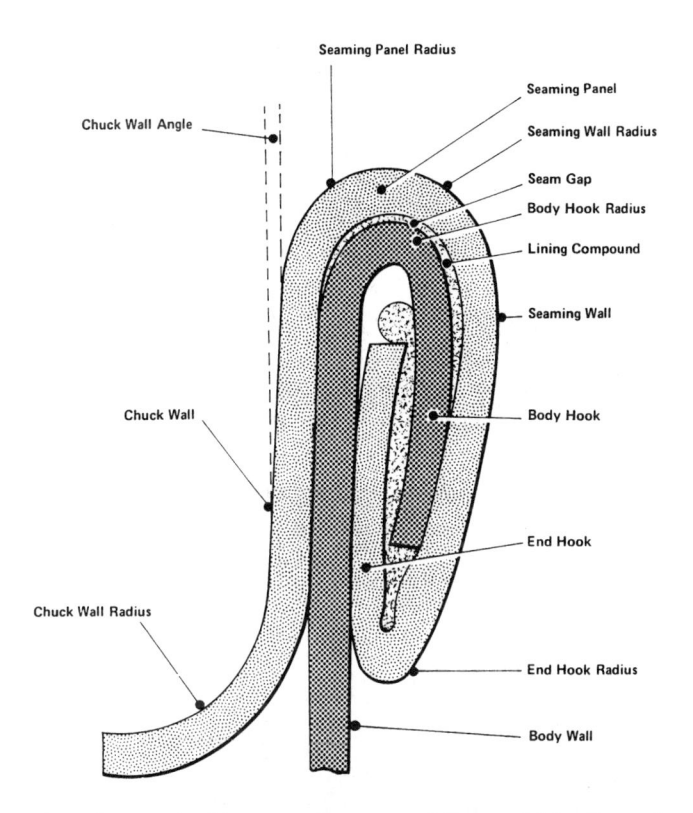

Figure 7–1 Cross section of a can double seam. Courtesy of Carnaud Metalbox plc.

the methods of measuring the seam dimensions can be found in the can manufacturers' seam manuals.

For a number of reasons, it is important to control the temperature of the can contents at the time of closure. The air enclosed in the can will affect the final vacuum, which, in turn, will influence the shelf life of the product by controlling internal corrosion.

The choice of the closing temperature depends on the type of product and the techniques used for preparation, which may preclude hot filling. In these cases, alternative methods of obtaining a vacuum must be employed.

7.3.11 Processing

The most important part of the canning process is the destruction of bacteria and spores by heat. Bacteria in their active state are not particularly heat resistant, but spores often present considerable heat resistance and, as a consequence, temperatures up to 130°C are required to destroy them.

In order to achieve microbiological stability of canned fruit, it is necessary to submit the sealed can to a heat process that will destroy, or render inactive, all microorganisms capable of causing spoilage. Absolute sterility is seldom achieved and, therefore, sterilization is not a strictly accurate term. The term *processing* is generally used to describe the heat treatment.

However, pH plays a part in the heat preservation of fruit. Under acid conditions (pH values less than 4.0) bacteria will not multiply and consequently only a pasteurization process is necessary. This process is normally carried out by immersing the sealed can in boiling water or steam at atmospheric pressure for relatively short times.

It is not possible to give actual processes for the different products and can sizes, but an indication can be obtained from Technical Bulletin No 4 produced by the Campden and Chorleywood Food Research Association in the United Kingdom.

For products with a pH value between 4.0 and 4.5, there are a few bacteria that are able to multiply in this range, and these products require a longer process or the acidification of the product. Acidification is usually achieved by the addition of citric acid.

Some fruits, such as prunes, are processed for much longer than is needed to inactivate the bacteria, the extra treatment being required to achieve the correct texture of the fruit.

Fruit is normally processed in boiling water or in steam at atmospheric pressure. This can be achieved in batch retorts, continuous retorts, or agitating cookers. The basic processing vessel is the static retort, which may be vertical or horizontal. It is equipped to receive steam and water and has suitable drain valves and vents. The cans are loaded into cages, which are placed into the retort and completely covered with the heating medium. If steam is used, care must be taken to vent off all the air to avoid the possibility of cold spots in the retort, which may cause some cans to be incorrectly processed.

Some batch retorts have a rotating cage in the heating medium, which has the effect of reducing the processing time because the agitation of the contents causes better heat transference.

Continuous-pressure cookers may also be used. These are equipped with specially designed ports or valves that allow the containers into and out of the processing chamber while allowing a constant steam pressure to be maintained. There are considerable advantages in using such equipment, as there is less variability in the treatment of the cans and much less manual handling. Some continuous cookers cause the cans to rotate against the outer retort casing, thus causing agitation with its attendant advantages (Figure 7–2).

Hydrostatic cookers are sometimes used for fruit, but they require such large quantities of cans to fill the infeed and outfeed legs as well as the heating chamber that they are not always the most cost-effective way of processing fruit.

After processing it is important that the cans are correctly and swiftly cooled and properly dried and that the filled cans are stored under the correct conditions. They should be kept in a cool, dry, and clean area, preferably away from direct sunlight. Sudden changes in ambient temperature should be avoided, as the can contents may be

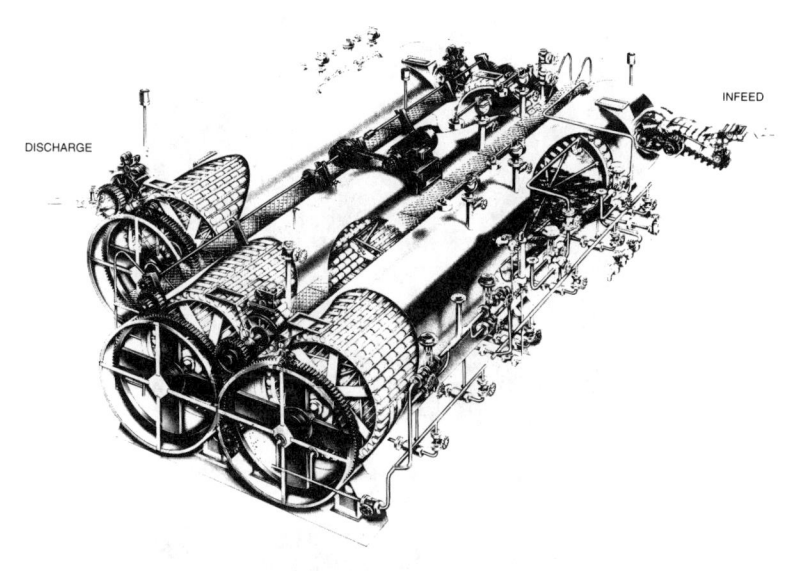

Figure 7–2 A continuous rotary cooker and cooler (Sterilmatic) suitable for processing cans of fruit. Courtesy of the FMC Corporation Food Machinery.

below the dew point of the atmosphere. This may result in condensation, which will cause rusting of the exterior of the can.

7.3.12 Can Vacuum

In general, the presence of an adequate vacuum in a can after processing is a sign of good canning practice. The exceptions are carbonated drinks, which have a positive internal pressure, and milk products, which have little or no vacuum.

The presence of a vacuum ensures that the can ends are concave, thus allowing a visual means of detecting cans that have an internal pressure resulting from gas formation caused by spoilage.

Fruits, being of a highly corrosive nature because of their acidity, require a high vacuum of 25 mm Hg or above, so that any hydrogen formed by corrosion will dissipate into this vacuum. The high vacuum also ensures the presence of a low amount of O_2, thus ensuring that corrosion is slow.

The vacuum in a can is normally measured with a Bourdon-type gauge, which has a sharp, tapering hypodermic needle projecting through a thick rubber washer. When pushed with the hand on the can lid, the needle penetrates the tinplate while the washer acts as a seal. The reading on the dial is taken before the gauge is removed from the can and the seal broken (Figure 7–3).

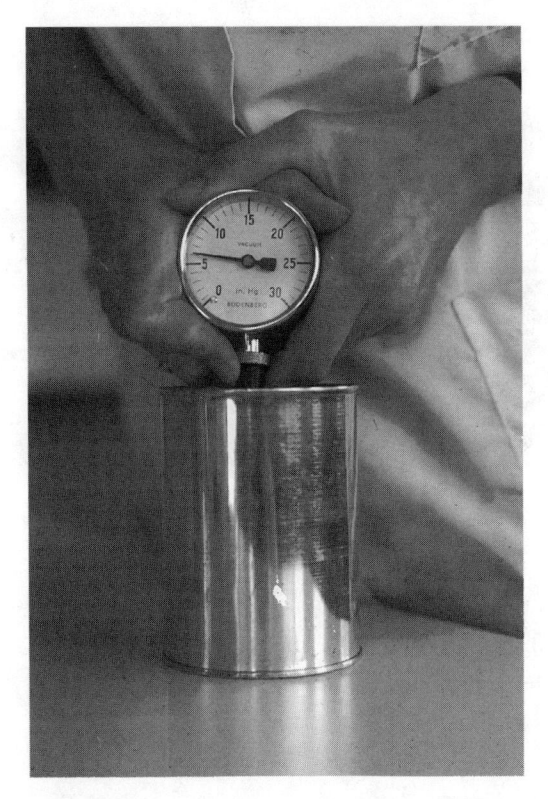

Figure 7–3 Taking the reading of a can vacuum using a Bourdon-type gauge.

7.3.13 Finished-Pack pH Values

The pH of most canned fruits lies between 2.7 and 4.3. The list shown in Table 7–3 is not exhaustive and is laid out in ascending order of average pH value.

Table 7–3 The pH Values of Finished Packs of Some Fruits

Fruit	pH	Fruit	pH
Rhubarb	3.2–3.6	Damsons	2.9–3.4
Purple plums	2.7–3.3	Blackberries	2.8–3.5
Loganberries	2.7–3.1	Greengages	3.0–3.5
Golden plums	2.9–3.2	Acid cherries	3.1–3.4
Victoria plums	2.8–3.3	Sub-acid cherries	3.2–3.4
Gooseberries	2.7–3.3	Strawberries	3.2–3.8
Apples (solid)	2.8–3.3	Sweet cherries	3.7–4.4

7.4 VARIETIES OF FRUIT

7.4.1 Apples

Apples for canning must be of good size and shape, evenly graded, and of good quality. The best English apple for processing is the Bramley Seedling, which is probably the most well-known cooking apple. Newton Wonder has also been canned with some success. Golden Delicious is grown more on the European mainland and will give a product similar to the Bramley Seedling, particularly if the acidity of the covering liquor is chemically adjusted with citric acid.

Cox's Orange Pippin has been used for sweet packs, such as sliced apples in syrup. All the apples used must be of a good, evenly graded size and of a good shape, so that the resulting peeled segments are approximately the same size. It is not possible to use windfalls or other damaged fruit, as the removal of the bruised or discolored sections will impair the quality of the finished pack.

Ideally, when the apples arrive at the factory, the quality should be such that washing and grading are unnecessary and the fruit can move to a machine that will core and peel in one operation. After peeling, the apples are dropped directly into a tank of dilute brine (1.5%) to prevent their coming into contact with air, which will cause browning by oxidation or enzyme activity.

The peeled apples are moved quickly through a machine to remove the seed cells and then hand trimmed to remove any remaining blemishes. Following this they are cut into sections, usually sixes or eights, depending on the size of the apple. The sections are kept in dilute brine for a number of hours, sometimes overnight. This improves the texture, color, and keeping quality of the canned apples and, by removing all the air entrapped in the fruit tissues, renders the can less likely to internal corrosion. If this is not effective, then the apples must be gently blanched in a stainless steel blancher.

Immediately prior to use, the apple segments are spread onto a white conveyor belt, where they are sprayed with fresh water to remove the brine and visually inspected to remove any remaining discolored pieces.

There are a variety of ways of canning apples, some of which are outlined as follows.

Solid Pack

This is by far the most common way of canning apples, often in A10 cans for the catering trade. (See Table 7-4 for a list of major can sizes used in the United Kingdom.) After the inspection procedure, the prepared apple pieces are blanched by passing the sections through a steam tunnel or by immersion in hot water for a sufficient time to render the slices pliable. Filling should be carried out immediately to prevent any appreciable drop in temperature. Ideally the temperature should be above 70°C. The cans are then topped up with boiling water, taking care to maintain the headspace at between 8 mm and 10 mm, as any excess air will give rise to internal headspace rusting. After filling, the cans should be thoroughly exhausted to ensure that all air is removed. When the cans have been seamed, they are heat treated in a retort for sufficient time to raise the can center temperature to above 70°C. This time will be reduced considerably if the cans are agitated during processing. If the contents are hot filled and there is no

Table 7–4 Can Sizes Used for Fruits in the United Kingdom

Common Name	Metric Size (mm)	Imperial Size	Capacity* (mL)
3-Piece cans			
Picnic	65 × 78	211 × 301	235
A1	65 × 102	211 × 400	315
U8	73 × 62	300 × 207	230
14Z	73 × 105	300 × 402	400
ET	73 × 110	—	425
UT	73 × 115	300 × 408¾	445
A2	83 × 114	307 × 408	580
A2½	99 × 119	401 × 411	850
A6	153 × 152	603 × 600	2630
A10	153 × 178	603 × 700	3110
2-Piece cans			
ET (round)	73 × 100	—	425
Pie can (taper)	153 × 32	603 × 104	450
Bowl	83 × 60	307 × 206	240
Bowl	105 × 71	404 × 213	495

* Capacity to nearest 5 mL.

delay before closing, the cans may just be immediately inverted. Following the heat process, the cans are thoroughly cooled in cold chlorinated water.

Apples in Syrup

After blanching, the apple segments or, in some cases, the dice are filled into cans and hot syrup is added. The cans are then seamed using steam-flow closure and are pasteurized in boiling water for 10 to 25 minutes, according to can size, following which they are properly water cooled.

Apple Sauce (Purée)

Good-quality applesauce cannot be prepared from waste produced during canning of apple segments. It is possible, however, to use undersize or oversize fruit that is in good condition. The apples are peeled and sliced or chopped into pieces about 6 mm thick, which are cooked in water or steam until they are soft. The apple is then passed through a pulper with screens of 3 mm or less to remove any seeds or remaining core. Dry sugar is added to the pulp, which is then mixed and reheated. It is possible to use jacketed pans or a screw heat exchanger, but the best results are obtained using a flash pasteurizer.

If the apple purée has been heated in a continuous-screw cooker, it can be filled directly into the cans, the lids seamed on, and no further processing is necessary. If the temperature has fallen because the purée has been tipped from the jacketed pan in which it has been prepared, the cans will require further heat processing after seaming. This is carried out in a retort.

Apple Pie

This is produced in a 603-mm-diameter tapered pie can. As it requires pressure processing, there are two main problems to overcome. The quality of the filling must be retained while moisture absorption must be prevented to avoid soggy pastry.

A filling will contain apple segments or dice combined with starch and, if so desired, a clove-based spice mix. The starch is necessary to hold the apple in suspension so that there is a homogeneous mixture within the can. Formulation of the pastry is critical and will require much product development to suit a particular process. The pH of the filling must be controlled to keep it below 4.0 by the addition of citric acid.

A similar product, known as apple pudding, may be prepared by placing a slab of pastry into a can, filling with sliced apple and a little syrup, and placing a pastry lid on top before exhausting, seaming, and processing.

Plain-bodied cans and lacquered ends are used for apple products. Owing to the reaction of the fruit with the tinplate, the plain body gives the apple a bright color. Fully lacquered cans may impart a dull color.

It is very important that apples do not come into contact with any iron or copper as this will cause a chemical discoloration of the fruit. All knives, preparation, and filling equipment must be made from stainless steel.

7.4.2 Apricots

Most apricots are canned in their country of origin, and the most popular variety is Bulidon. They should be delivered to the cannery fresh and must not be overripe.

The fruit is size graded and the stone is removed by splitting the fruit into two halves. The apricot halves are then washed by water sprays and allowed to drain before being filled into the can.

Apricots are rather acid, and a heavy syrup (50°B) is recommended. After seaming and pasteurizing, the cans are properly cooled before being packed and stored.

7.4.3 Bilberries

These are sometimes referred to as blueberries or blaeberries. They are not processed to a great extent. In some areas this product is in demand by confectioners and in this case the fruits are usually canned in A10 cans.

The pack is very corrosive, and it is important that good canning practice is meticulously observed and that fully lacquered cans are used.

7.4.4 Blackberries

Cultivated blackberries are used for processing. The most commonly used varieties are Black Diamond and Himalayan Giant. They should be processed when fully ripe, as the flavor is best at this stage and the seeds are less noticeable. The fruit should be picked into shallow baskets or trays to prevent crushing and it should be processed as quickly as possible after picking, as any delay will cause deterioration.

The blackberries are tipped onto an inspection belt where any leaves, calices, and other foreign material are removed. They are then washed in a flood washer and sprayed with clean water as they emerge on an inclined belt. Following this they are filled into cans, hot syrup is added, and the cans are seamed and pasteurized.

7.4.5 Black Currants

Probably the best variety for processing is Baldwin. Only firm, ripe fruit should be used. Underripe currants should be avoided as they have tough skins and a poor color and flavor. Removal of the stalks by hand is both costly and time consuming, so, on a commercial basis, strigging is carried out mechanically. It can only be done on fresh currants if they are very firm and dry, and it is best carried out on hard frozen fruit. Only small amounts are removed from the cold store at any one time to prevent thawing. After being strigged the fruit is visually inspected to remove any foreign material.

When the frozen fruit is canned, extra exhausting time must be allowed to give the fruit a chance to reach normal closing temperature. Heat must be applied fairly gently to prevent the currants from bursting.

7.4.6 Cherries

There are three types of preserved cherries: the sweet white type; the red sour fruit, which is most often pitted; and the characteristically flavored maraschino.

The sweet cherry varieties that are most suitable for canning are Napoleon Bigarreau, Kentish Bigarreau, and Elton Bigarreau, all being "white" varieties. These usually require color to make them red. As these cherries are packed into nonlacquered cans, erythrosine is generally used as a colorant because it is relatively stable with regard to dissolved tin. If the cherries are to be used as part of a fruit salad, the amount of color must be very carefully controlled to prevent subsequent "bleeding" of the dye, which could be picked up by the other constituents of the pack. Current legislation should be checked to ensure that the maximum permitted amount of erythrosine is not exceeded.

None of the black-colored sweet cherries has been found to be totally satisfactory for canning, although Hertfordshire Black is occasionally used.

The most suitable varieties of sour cherries are Morello (by far the most popular), May Duke, Kentish Red, and Flemish Red.

The maraschino derives its name from a Dalmatian sweet-sour liqueur originally prepared from amarasco bitter cherries. These are usually preserved as glacé cherries or as bottled cocktail cherries, and the best variety is Royal Anne. The cherries are normally received preserved with sulfur dioxide in barrels. If they are to be canned, all of the sulfur dioxide must be leached out or the cans will be severely corroded.

When canned, the cherries should be fairly ripe to give the maximum amount of flesh in relation to stone.

When the fruit arrives at the factory, the stalks are removed by a machine consisting of a series of contrarotating rollers. The cherries are then inspected and graded prior to being filled into cans.

Some packs are produced from pitted fruits, which have had the stones removed by a pin-like probe that passes through the center of each cherry, after which the fruit is visually inspected for any remaining stones. After being filled, the cans are exhausted, processed, and cooled.

7.4.7 Gooseberries

Some fruits do not retain the best flavor when canned, but gooseberries most definitely improve in flavor. The most important varieties for processing are Careless and Keepsake. They should be picked as they reach full size but before they have become soft or have shown a color change.

On arrival at the factory, the fruit is passed through a cleaning and grading machine in which the leaves and other light vegetable material are removed by air currents supplied by fans. The stalks and blossom ends are removed in a carborundum peeler or snibber. Care must be taken to ensure that just the stalks and blossom ends are removed and to avoid cutting deeply into the fruit. The fruit is washed with water sprays to remove the debris. The snibbing operation also causes the skin of the gooseberry to be punctured or scarified, which will allow the added syrup to penetrate into the fruit and thus prevent it's becoming wrinkled.

After the gooseberries are filled into the can, syrup is added, and the can is seamed using steam-flow closure. The cans are then pasteurized for 5 to 30 minutes depending on their size.

This is one of the fruits that benefits from being packed in plain-bodied cans. Use of lacquered cans gives a dull-looking pack, whereas the plain cans result in a bright, attractive pack.

7.4.8 Grapefruit

Grapefruit for canning must be ripe, sound, and size graded, ideally between 7.5-cm and 10.2-cm diameter. Any smaller or larger fruit are diverted to produce juice. It is uneconomic to process small fruit owing to the large percentage of peeling losses and other waste. Large fruit give too few segments in a can.

The most difficult aspect of the preparation of the fruit for canning is the separation of the segments from the peel and the white surrounding membrane. Most of the bitter principle, naringin, is found in the white membrane, which must be completely removed.

Grapefruits are peeled by one of two methods. In the cold-peel method, the peel is stripped off manually to remove both albedo and the outer membrane of the segments at the same time. This method results in a low yield and gives segments, which occasionally show signs of overtrimming. The more common method is hot peeling, in which the grapefruits are scalded by steam or in water at about 95°C for about 5 minutes to make the peel plump and puffy and easy to remove. After slight water cooling, the peel is removed by hand using a short stainless steel knife. The stem end of the fruit is slit or scored, and the grapefruit is popped out of the peel. Some of the albedo or inner white portion of the peel adheres to the fruit and this is stripped away.

There are mechanical peelers that successfully remove the skin from the fruit. The grapefruit are treated for about 12 seconds with 0.5% to 2.5% caustic soda solution at 95°C to dissolve any adhering albedo. The caustic soda is washed off with water sprays.

The major disadvantage of the high-temperature method is that the heat causes the fruit sacs to expand, and these may eventually burst.

The peeled grapefruit is sectioned either mechanically or by hand, the seeds are removed, and the sections are filled into cans. Syrup is added and the can end is seamed on.

7.4.9 Fruit Salad

Fruit salad is the largest fruit pack canned out of the normal fruit season. Because it is produced as an out-of-season product and also because the fruits used originate from a number of sources, it consists largely of recanned fruits. Special packs of the individual fruits are canned into A10 or larger cans to be reused later. The raw material has to be carefully selected to withstand the double amount of heat processing it will receive. This is particularly important for pears and apricots, which must be properly ripe but not so soft that they will break down during the process.

The British Fruit and Vegetable Canners Association (BFVCA) Code of Practice states that canned fruit salad should contain fruits in the following proportion of the filled weight of fruit: peaches 23–46%, apricots 15–30%, pears 19–38%, pineapple 8–16%, cherries or grapes 5–15%.

The syrup drained from the individual canned fruits may be used in the preparation of the syrup to be used in the fruit salad after it has been filtered and adjusted to achieve the specified strength.

The fruits are usually prepared to give approximately the same count of each in the can. In order to achieve this, it is necessary to include peach and pear quarters, apricot halves, pineapple slices, and whole cherries.

The cherries used in this pack must be carefully dyed with erythrosine. It is advisable to check the current legislation regarding the maximum permitted amount of erythrosine as this may change. Care must be taken not to use too much coloring as this may bleed in the finished pack, thus passing on a pink tinge to the other fruits, particularly the white-fleshed pears and the pale yellow pineapple. As an added precaution, it is advisable to fill the pears first into the can, followed by the pineapple, apricots, and peaches, with the cherries being added last. This will prevent the pears picking up the color.

7.4.10 Fruit Cocktail

Fruit cocktail is similar to fruit salad but is prepared with diced fruit. The BFVCA Code of Practice states that fruit cocktail should consist of diced peaches 30–50%, diced pears 25–45%, pieces or diced pineapple 6–25%, cherries 2–15%, and, sometimes, seedless grapes 6–20%.

Both fruit salad and fruit cocktail are best packed into plain-bodied cans with lacquered ends, which will result in a final pack with a bright color and a true, sharp flavor.

7.4.11 Fruit Pie Fillings

This is a range of products that has become quite popular in recent years. It can be used hot as a filling in a pie or it can be used cold as a dessert topping or accompaniment. The fruit, either single or mixed, is prepared in the normal manner. For convenience, frozen fruit is often used. It is blended with a syrup that contains 3–4% starch and about 40% sugar, together with added color if appropriate. When fruit is defrosted, it has high enzyme activity. This can be minimized by accelerating the thawing process, either in a tank or by a continuous process, and pasteurizing the fruit shortly after thawing is completed. The use of a thermoscrew heat exchanger is ideal for this purpose.

Owing to the solid nature of the can contents, which results in a slow rate of heat penetration, it is necessary to give pie fillings a longer process than particulate fruit in syrup, juice, or water. Many of the pie fillings are hot filled, and this will enable a reduction of the processing time.

7.4.12 Loganberries

Loganberries originated as a cross between a wild blackberry and a raspberry and are reasonably successful as a processed pack. They have to be handled in a manner similar to that for raspberries to prevent them from bursting, with a resultant loss of juice.

The fruit should be packed ripe but not black. It must not be underripe, otherwise the core of the fruit, which is tough, will be difficult to cook. Because of its greater resistance to heat transfer, an underripe berry in a can may retain gas-producing microorganisms in the core that will survive the normal heat treatment. These microorganisms could then cause the can to swell.

Loganberries should be canned immediately on arrival at the factory, as they tend to go moldy very quickly. They should be inspected carefully for maggots, and any suspect fruit must be removed.

When loganberries are filled into the can, they must not be pushed down tightly, as they easily crush.

7.4.13 Oranges

This pack is canned from fresh, ripe, sound fruit. The washed oranges are passed through a steam or hot water bath for about 1 minute to aid peeling, and are then cooled in cold water. They are peeled by hand and allowed to dry before being segmented by hand. The individual segments are immersed in cold 2.5% hydrochloric acid for about 2 hours to remove the stringy fibers adhering to them. The acid is subsequently removed by running water, and the washed segments are immersed for about 20 minutes in a 1% solution of caustic soda before being washed again.

The separated orange segments are very fragile and must be handled gently. Usually they are conveyed and washed in large volumes of gently flowing water.

The cans are filled with the segments and inverted to drain off any excess water. Syrup is added (approximately 17°B) and the can is vacuum seamed and pasteurized.

7.4.14 Peaches

Peaches for canning should be of a uniform size and symmetrical shape with a good yellow color and a close tender, but not too soft, texture. In general, the yellow clingstone varieties are most suitable, the most popular being Tuscan, Phillips, and the Midsummer type.

The fruit should be picked when it has achieved its greatest size and is still at the firm stage of maturity. Care must be taken when handling the fruit to avoid bruising. Often the fruit is cooled prior to transportation to minimize spoilage. When received at the factory, the fruit is graded by variety and maturity to prevent mixing on the production lines.

The first operation in the canning process is halving and pitting. This can be done by hand or machine. The peaches are cut in half and the pits (or stones) are removed and discarded. Any damaged halves or undersized fruits are separated for use in a sliced pack, or if unsuitable for this, for use in pie filling.

The peach halves are peeled by immersion in a hot caustic solution (1–2.5%) for 30 to 60 seconds. The caustic soda and loosened skin are removed by passing the peach halves under powerful water sprays. Steam peeling may be used, which loosens the skin, which is then removed by rotating rubber or abrasive rollers.

The peeled peaches are blanched for 1 to 2 minutes in steam or hot water at 80°C to remove the final traces of caustic soda and to inactivate the oxidase enzyme that causes the exposed surfaces of the fruit to turn brown in contact with air.

The peaches are visually inspected, to remove any damaged or blemished halves, and they are graded. The largest and most perfect halves are the Fancy grade, and fruit of smaller sizes and less perfect are known in diminishing order as Choice grade, Standard grade, Second grade, and Pie grade. Sliced peaches are prepared from any damaged halves or smaller-sized fruit. These are cut into segments and visually inspected to remove any damaged or blemished pieces.

Peach halves are carefully filled into the cans in such a way as to prevent air being trapped in the hollow left by the pit. Hot 30°B syrup is added, and the can is closed using steam-flow closure.

The cans are pasteurized in hot water or atmospheric steam and are then cooled quickly before being labeled and packed.

7.4.15 Pears

The most popular variety of pear, which lends itself particularly to canning, is the Bartlett (William bon Chretien). It has good texture and flavor with a bright color and is of a uniform size and shape. The Conference pear is also canned, but because of its lower natural acidity the liquor has to be acidified with citric acid.

Pears for canning should be allowed to grow to full size, but they should be picked while still green and hard. Ripening is carried out in lug boxes or crates in a well-ventilated store. The pears are inspected daily to determine the correct ripeness for processing. A penetrometer may be used to measure the degree of ripeness.

The fruit is mechanically size graded before being peeled, cored, and halved using curved knives. After trimming, the halves must be immersed in 1–1.5% brine solution to prevent enzymatic browning, which will occur if they are left in contact with the air. Care must be taken to wash all of the salt from the pears before canning.

The pears are filled into the cans, hot syrup is added, and the end is seamed on using steam-flow closure. The cans are pasteurized at 100°C for 12 to 30 minutes, depending on the can size. Occasionally, to soften their hard texture sufficiently, the pears need a longer pasteurization time than is required to achieve commercial sterilization.

7.4.16 Pineapple

Pineapple product is not canned in Europe, but it is used as a principal ingredient in fruit salad and fruit cocktail.

The fruit is harvested when it has reached full maturity but before it has turned soft. On arrival at the factory, it is size graded to reduce peeling losses. Peeling is carried out on a machine that removes the core and the outer shell and cuts off the shell portion at the ends of a cylinder of flesh. The detached shell is treated by another machine that removes any remaining flesh for use in crushed pineapple or juice.

The pineapple cylinders are visually inspected and trimmed to remove any remaining shell or foreign material. The cylinders are then sliced to give rings, and a further inspection is carried out to remove any misshapen rings and blemishes. Some of the rings are segmented for use as pineapple pieces or chunks.

The fruit is filled into the can, which is vacuum treated to remove any air from within the tissues. Hot syrup is added, and the can is seamed and pasteurized prior to being cooled and packed into cases.

7.4.17 Plums and Damsons

The best canning varieties are Yellow Egg, Pershore, Victoria, and damsons. Some varieties of plums show a tendency to form deposits of gum close to the stone. Usually this cannot be detected on the surface of the fruits. The phenomenon is generally attributed to physiological changes that take place during growth and the enzyme activity associated with the stone. During heat processing, the spots of gum absorb water and swell to form large lumps of jelly that, in extreme cases, may cause the plum to burst and spoil the appearance of the pack. The only way to minimize the problem is by careful selection of raw material.

When the fruit arrives at the factory, the stalks are mechanically removed and the plums are washed. The fruit may be size graded before being inspected and filled into cans. Plain-bodied cans are used for green or yellow varieties, while fully lacquered cans are used for the red or purple types.

After filling, the syrup is added. It may be possible to use artificial coloring in the syrup, but it is advisable to consult current legislation before doing so. The exhausting must be thorough to remove any air, which may cause corrosion of the internal lining of the can, and the heat process must be sufficient to inactivate the stone enzymes, which could otherwise accelerate the corrosion process. The pasteurization time will vary from 10 minutes to 30 minutes, depending on the can size and the method of preparation.

7.4.18 Prunes

Canned prunes are an excellent product and, like fruit salad, may be canned outside their normal harvesting season. Dried prunes are produced in the United States and some European countries. Those selected for canning are usually 80 to 90 grade, which is an indication of fruit size, being the number of fruit per 454 g.

Prunes are inspected for broken or defective fruit and for foreign material, after which they are washed and may be blanched. It is possible to soak prunes overnight, but this will cause much of the natural flavor to be lost in the discarded soak water. The prunes are filled into the cans, allowing space for them to swell as the syrup is absorbed during processing and storage. A UT can (see Table 7–4) will require about 20 prunes, depending on the actual size grade. A low Brix syrup or fruit juice is added, followed by a thorough exhaust or a vacuum filler to ensure that all the air is removed. This reduces the risk of internal corrosion within the can. The heat process given is longer than is required for sterilization, as the fruit requires softening. The can used should be fully lacquered, usually with a white acrylic lacquer.

7.4.19 Raspberries

Raspberry is one of the best-flavored British fruits and it is popular as a canned pack as it retains its flavor well. Two of the main varieties are Malling Jewel and Glen Clova. They should be picked when ripe but firm, and certainly they should never be canned after they have become soft, as they will break down completely in the container. Raspberries should be picked into small shallow punnets to avoid unnecessary damage to the fruit.

The raspberries should be canned as quickly as possible after arrival at the factory. It is not advisable to carry fruit overnight as mold grows very quickly, but if the practice has to be followed, then the fruit should be chilled to 1–5°C.

The fruit is so fragile that it should not be washed, as the action of the water will cause it to break down. The punnets should be shaken gently onto a wide white inspection belt, where inspectors will remove any foreign bodies such as calices, insects, leaf, twig, and stones. Unfortunately, some insects, such as earwigs or beetles, will crawl into the dark recesses of the hole left when the plug has been removed, where they may escape detection.

The raspberries are gently filled into fully lacquered cans, a hot syrup or fruit juice is added, and the can is seamed using steam-flow closure. The legislation in some countries may permit the use of artificial coloring matter in this product.

Processing consists of pasteurization at 100°C to achieve a can center temperature of 85°C. The processed cans are cooled thoroughly and are gently handled during labeling and casing to prevent damage to the contents.

7.4.20 Rhubarb

The petiole of this plant is regarded as a fruit for food purposes, as it is canned in syrup and consumed as a dessert. It forms one of the largest fruit packs in the United Kingdom and has the added advantage of maturing very early in the season. Probably the best variety for canning is Champagne, but there are others nearly as good, such as Sutton, Victoria, Albert, and Timperley. When the rhubarb is harvested, the leaves are stripped by hand and the bottom end of the stalk is not cut. It is broken from the plant to avoid drying out and the subsequent splitting of the stalk.

When the rhubarb arrives at the cannery, the first operation is to cut off the top and the base and split any extra-thick sticks lengthwise to avoid over-large pieces. The sticks are cut into pieces approximately 25 mm in length, usually by machine. The pieces are then washed in an agitating washer and are inspected. Rhubarb may be lightly blanched to aid filling.

After filling, hot syrup is added to the rhubarb. Artificial coloring matter may be used in the syrup subject to current legislation. Steam-flow closure may be used when seaming to ensure that all O_2 is excluded from the can.

Because of the high acid content of the rhubarb, it may quickly corrode the internal surface of the can. Fully lacquered cans must be used, but even so the shelf life of canned rhubarb is unlikely to be more than a year, more usually 6 to 9 months.

Rhubarb can also be canned as a solid pack. After slicing, washing, and inspection, the fruit is blanched at 70°C for 2 to 3 minutes. Artificial color may be added to the blancher to aid the color of the finished pack. When blanched, the rhubarb is hot filled into the cans, seamed, and pasteurized.

7.4.21 Strawberries

Unfortunately, the color and flavor of most strawberries deteriorates considerably as the fruit is processed. Both growers and research establishments are continually searching for new varieties that will be more resistant to heat processing. However, in the United Kingdom the variety Cambridge Favourite is used almost exclusively for canning where the addition of artificial color is permitted. There are other red-fleshed varieties that are better for freezing and in canning packs where no artificial color is added.

In order to obtain a good-quality product, strawberries should be handled quickly and carefully. Ideally, the fruit should be delivered in shallow trays or punnets. If the fruit arrives in top-class condition, it may be possible to avoid washing it. However, experience has shown that washing is usually necessary to remove any adhering soil, straw, or leaves, particularly following rain. Washing also helps to remove any residues resulting from spray treatment for fungus or pests.

The fruit should be decalyxed as it is picked, but care must be taken in the factory to remove any residual calices, as these are most unpleasant in the finished pack.

Strawberries may be size-graded before packing so that the small fruit are packed into smaller cans, thus avoiding the situation of two or three large berries in a small can. Fully lacquered cans should be used, and the fruit should be pasteurized for about 6 to 10 minutes at 100°C. The pasteurizing time is determined by the time taken for the temperature at the center of the can to reach 85°C.

7.5 BOTTLING

Bottling is not a large industry, although a jar of good-quality fruit looks really appetizing on the supermarket shelf. The process is quite simple and is readily carried out in the home.

The choice and preparation of the fruit is similar to that for canning. As the container is transparent, good visual quality is essential. The pack should be free from blemishes and the fruit should be well size-graded. As fruit shrinks on cooking, it should be packed as tightly as possible without damaging it, to avoid giving the impression of an underfilled jar. Filling is assisted if the surface of the jar is wet so that the fruit slips easily. Consequently, the jars are used directly after the jar washer, where they are inverted and a jet of water is used to remove any possible foreign bodies. After the fruit has been added, the jars are usually completely filled with water, fruit juice, or syrup. A small amount of covering liquor is tipped from the jar to give a controlled headspace and then steam is injected immediately before the cap is added to evacuate any remaining air.

The heat treatment may be carried out either in water in a tank heated by steam coils or in a steam/air mixture. In either case, the temperature must be brought up slowly to the processing temperature to avoid thermal shock to the jar, which could result in shattering. Water cooling is not usually employed. If it is used, the hot water must be slowly replaced by cold and the water level must not cover the caps in case any air is drawn into the jar by the induced partial vacuum.

The filled jars should be handled as little as possible for 24 hours after processing, as the fruit is very soft and any excessive movement may cause it to break down. Bright light, such as sunlight, should be excluded from the store rooms as it will cause the color of the fruit (particularly red varieties) to bleach or turn brown.

7.6 FREEZING

Apart from being a major processing method in its own right, freezing has largely supplanted sulfiting as a means of preserving berries intended for further processing, such as jams, conserves, and pie fillings. If properly frozen, stored, and defrosted, frozen fruit is a high-quality raw material having properties very close to those of fresh fruit.

The freezing process does not destroy microorganisms as does heat sterilization or pasteurization, but it essentially retards their growth. Spores may survive the storage process, and any damaged microbiological cells may be viable after a period, thus causing consequent spoilage. This means that the fresh fruit must be as free as possible from microorganisms and that scrupulous hygiene is required within the factory to prevent cross-contamination. It is important to handle the material quickly, as bacteria can grow if there is a delay between the stages of processing.

Prior to freezing, the fresh fruit is normally washed to improve the microbiological standard and physical appearance. A fast, continuous wash will minimize the leaching of color and flavor. A close inspection should be carried out to remove any foreign material that might still be adhering to the fruit, together with any blemished pieces.

Fruits in general do not require blanching before freezing. They can be individually quick frozen (IQF), they can be packed in sugar or syrup, or they can be puréed before freezing.

A rapid freezing process is essential to preserve the quality of the fruit. At the freezing point, the ice crystals formed will rupture the cell walls and disrupt the intercellular structure, thus releasing enzymes and substrates that will cause a considerable increase in the rate of deterioration of flavor and color at this stage of the freezing process. It is important that the fruit is taken below the freezing point as quickly as possible. Rapid freezing yields small ice crystals, whereas a slow-freezing process will cause the formation of larger ice crystals that will, in turn, cause greater physical damage to the cell walls. Product damaged in this way will be soft and disintegrated when thawed.

7.6.1 Freezing Methods

There are a number of ways of freezing fruit.

Room Freezing

Room freezing entails placing the product on metal or plastic trays in a cold room at −25°C or colder and allowing it to freeze. If the product is required to be free flowing, it is passed through a breaker while still frozen. This process is often employed for black currants before strigging to prevent excessive damage to the fruit.

Blast Freezing

In this process, the product is packed into its final packaging, which may be retail packs or 10-kg boxes, and it is frozen in tunnels with a rapid air velocity at a temperature of −30 to −40°C. This method is not practicable if free-flowing product is desired.

Spiral Belt Freezing

This method is similar to blast freezing in that the product can be in its final packaging. The fruit can also be in pieces, particularly where the pieces are quite large, such as melon balls. The fruit passes along a slatted belt, which spirals upward in a chamber where refrigerated air is blown downward (Figure 7–4).

Fluidized Bed Freezing

The fruit is fed onto a vibrating perforated bed through which refrigerated air at −30°C is blown from below. The vibrating action of the bed and the turbulence of the air prevents the fruit from sticking together and gives rise to a free-flowing product. Care must be taken if using this method for soft fruit. Damage to the fruit can be minimized by careful adjustment of the loading of the vibrating bed, the rate of vibration, and the velocity of the air (Figure 7–5).

172   FRUIT PROCESSING

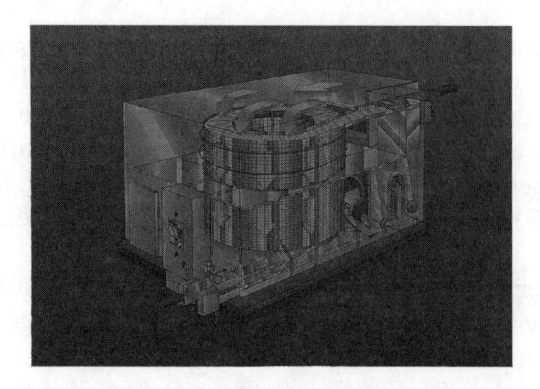

Figure 7–4 A spiral belt air-blast freezer, GYRoCompact. Courtesy of Frigoscandia Equipment AB, Helsingborg, Sweden.

Liquid Nitrogen Freezing

The fruit is fed onto a mesh belt that passes through a tunnel, where liquid nitrogen is sprayed onto it. This method is only suitable for small tonnages and it is expensive to operate. It should, therefore, be used only for fruit for which a premium price can be obtained or where other methods would cause excessive damage.

7.6.2 Storage

During cold storage, deterioration still takes place, albeit very slowly. It can be shown that for any given storage temperature there is a finite time before there is a detectable deterioration of flavor.

As has been said earlier, a major factor in the success of commercial quick freezing is the small size of the ice crystals formed during the process. However, if the storage temperature is not correct or there are fluctuations of temperature, the ice crystals will

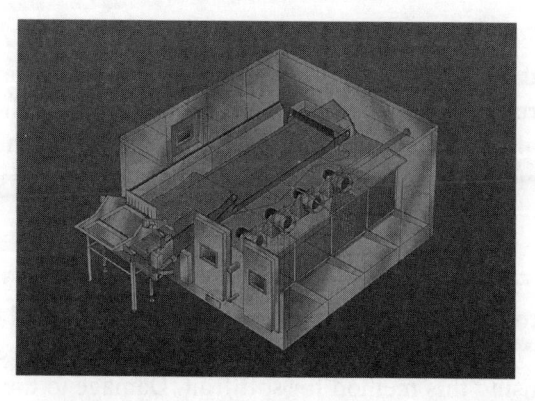

Figure 7–5 A fluidized bed air-blast freezer—FLoFREEZE MX. Courtesy of Frigoscandia Equipment AB, Helsingborg, Sweden.

combine during storage to form large crystals. This obviously defeats the object of fast freezing. Therefore, the control of temperature during storage and distribution is as important to the quality of the food as is the original process.

It is generally considered that below –12°C microbial growth ceases or is, at worst, extremely slow. Again, it is essential to maintain the correct storage temperature; the product must not be allowed to thaw, as this will allow any surviving microbes to grow. Thawing followed by refreezing will cause the ice crystals to grow larger, causing rupture of the cell structure and a product with a poor texture.

Immediately after freezing, the products are placed into commercial cold stores where the temperature is maintained at between –23°C and –30°C. The products are then distributed in insulated containers that are equipped with their own refrigeration units. Where refrigeration units are not available, the low temperature can be maintained by using dry ice, i.e., solid CO_2, on racks in the container. This is only satisfactory for journeys of relatively short duration. In order to ensure that the product temperature does not rise on transfer from the cold store, it is advisable to load the containers from the cold store using port doors, where the container is backed flush to the door to form a seal, thus preventing warm ambient air from coming into contact with the product.

On arrival at the retailer, the products should be stored in a small cold store or be placed directly into the display cabinet, where the temperature is maintained at –18°C. Care must be taken not to overload these cabinets, as thawing can occur if the product is above the marked load line.

7.6.3 Packaging

In addition to the requirements common for all food containers, such as attractive design, consumer convenience, construction from nontoxic materials, and economic costs, frozen food packaging must be able to withstand the low temperatures required by the process and it must prevent the product's suffering from freezer burn during storage. The pack must stay in one piece even when it becomes damp as the product is thawing. Strength is also required, as the product is handled in the retail display cabinet.

Tests show that the rate of weight loss from a pack during frozen storage is governed by the permeability characteristics of the packaging material. A pack with a good moisture barrier but not completely filled with product is at risk of ice separation caused by sublimation from the product if it is subjected to fluctuating temperatures.

Another requirement of the packaging is that it should be capable of being handled by high-speed machinery. Much use is made in the frozen food industry of form, fill, and seal packaging.

Packaging can take the form of waxed cartons, direct film wraps and bags, or overwrapped trays. All of these can be used for fruit. Cartons can be used for fragile fruits such as raspberries, bags can be used for more robust products such as rhubarb or apple pieces, and the overwrapped trays may be used for precooked fruit pies.

7.7 ASEPTIC PACKAGING

Aseptic packaging is another form of packaging heat-processed fruit products. In contrast with canning and bottling, where the product is filled into containers, sealed, sterilized, and cooled, aseptic technology requires the product to be sterilized or pasteurized using some form of heating and then filled under sterile conditions into sterilized packaging and finally sealed. The process is not used very extensively for particulate fruits. It is reserved more for juices and purées (see Chapter 5).

The chemical and physical changes are not so severe with fruit products as with others. Many of them are acid and, therefore, relatively low temperatures are required for sterilization. Traditionally many of the fruit products have been hot packed.

7.7.1 Sterilization/Pasteurization

This is carried out either by direct heating with steam or by indirect heating using some form of heat exchanger (e.g., tubular, plate, or scraped surface).

The advantage of the aseptic process is that it is necessary to use a high temperature (approximately 150°C) for only a few seconds, which ensures that more of the flavor and the nutrients are retained compared with the lower temperature and longer sterilization time demanded by the canning process. The short time is made possible because the product is heated without a container to restrict the heat transfer by holding the contents in bulk. In aseptic processing, the product is heated quickly in a thin film. Certain changes do occur when processing these products at elevated temperatures. Oxygen is removed during the process, which helps in the color retention of certain products. However, in other products, such as bananas, the browning or discoloration caused by oxidation is extremely serious. In these cases, the processing system should be arranged to include efficient deaeration as soon as possible after peeling.

Over the years, the processing of particulate fruits has been achieved using a number of methods. Particles up to 5 mm in size can be processed in a tubular heat exchanger, whereas larger particles, up to 15 mm, can be processed in a scraped-surface heat exchanger with a 15- to 18-mm gap between the rotor and the heating surface.

There are special types of heat exchangers that incorporate product flow separation, thus permitting different residence times for particles and liquids. A typical device is Rotahold, which consists of a cylindrical vessel fitted with infeed and discharge ports. Inside, a set of fork blades is mounted on a central shaft that can be rotated at an adjustable speed to suit the residence time. The fork blades are separated so that the particles are retained and the liquid passes through the particles. A set of stationary blades between the infeed and outlet prevents the particles from short-cutting the system.

Particles and liquids can be processed separately, the particles being batch processed in a jacketed vessel that rotates to ensure even heating. The hot liquid is added from another heating vessel and the mix is then filled into containers.

Aseptic processing is also used for purées, such as fruit pie fillings, in a number of ways:

- steam injection, where live steam is injected directly into the purée
- a tubular heat exchanger, where the product passes through a stainless steel tube that is heated on the outside by steam (the tube may be coiled for greater efficiency)
- a plate heat exchanger, which allows the product to flow through alternate channels in a pack of embossed plates with the heating medium flowing through the other channels
- a scraped-surface heat exchanger, in which the product is moved along a heated tube by a rotating scroll or blades.

Good hygiene is of paramount importance in aseptic processing. The whole technology is based on achieving sterility before the product is filled into the packaging. This means that there must be extensive hygiene schedules and cleaning instructions, and these must be carried out assiduously.

7.7.2 Packaging

There is a wide variety of packaging available for aseptic products, including

- rigid containers, such as metal cans and glass bottles and jars
- semirigid containers, such as plastic cans, bottles and tubs, aluminium trays, and paper-laminated cartons
- nonrigid containers, such as plastic pouches and bags

The choice of packaging is determined by the food product, its expected shelf life, marketing appeal, and cost. There are other criteria affecting the choice, including tamper evident to prevent product contamination, physical protection of the product, and also the need to provide the consumer with information about the product.

Aseptic products have the advantage shared by canned and bottled products: they can be stored at ambient temperatures. In these days of high energy costs, this must be an advantage.

7.8 CONCLUSION

In order that all methods of thermal processing and freezing of fruit are successful, great care must be taken with the selection of the raw material and the hygiene standards within the factory. Each process should be subjected to an HACCP (Hazard Analysis and Critical Control Point) study to ensure that all hazards are identified and suitable controls are implemented (see Chapter 11).

SUGGESTED READING

Anon. (1984). *Handbook for the Fruit Processing Industry*. A/S Kobenhavns Pektinfavrik, Denmark.
British Fruit and Vegetable Canners Association. *Code of Practice on Canned Fruit and Vegetables*.

Campden and Chorleywood Food Research Association. (1980). *Revised Technical Bulletin*, No. 4, Chipping Campden, Gloucestershire, UK.

Carnaud Metalbox plc. *Double Seam Manual.*

Cherry Burrell Application Engineering Department. (1982). *Aseptic Processing*. Cherry Burrell Corporation, Cedar Rapids, IA.

Cruess, W.V. (1958). *Commercial Fruit and Vegetable Products*, 4th ed. McGraw-Hill Book Company Inc., New York.

Hanson, L.P. (1976). *Commercial Processing of Fruits*. Noyes Data Corporation, Park Ridge, NJ.

Holdsworth, S.D. (1992). *Aseptic Processing and Packing of Food Products*. Elsevier, London.

Holdsworth, S.D. (1983). *The Preservation of Fruit and Vegetable Food Products*. Macmillan Press, London.

Nagy, S., Shaw, P.E., and Veldius, M.K. (1977). *Citrus Science and Technology*, AVI, Westport, CT.

Ranken, M.D., ed. (1988). *Food Industries Manual*. Blackie Academic and Professional, Glasgow, Scotland.

Rees, J.A.G., and Bettison, J., eds. (1991). *Processing and Packaging of Heat Preserved Foods*. Blackie Academic and Professional, Glasgow, Scotland.

Shapton, D.A., and Shapton, N.F. (1991). *Principles and Practice for the Safe Processing of Foods*. Butterworth-Heinemann, Oxford.

Tressler, D.K., van Ardsel, W.B., and Copley, M.J. (1968). *The Freezing Preservation of Foods*. Vol. 1—Refrigeration and Equipment. AVI, Westport, CT.

Various. (1982). Papers presented at *A Seminar on Aseptic Processing and Packaging*, 18–19 October 1982. Centre for Continuing Education, University of Auckland, New Zealand.

Willhoft, E.M.A. (1993). *Aseptic Processing and Packaging of Particulate Foods*. Blackie Academic and Professional, Glasgow, Scotland.

Chapter 8

The Manufacture of Preserves, Flavorings, and Dried Fruits

Roger W. Broomfield

8.1 PRESERVES

The brief for this chapter was that it should be "state of the art." For those involved in the manufacture of preserves, it often appears that there is more art in the job than science. There is a mystique in "stirring the pot," which at the time often seems difficult to explain: experience is a very valuable commodity. Nevertheless, when one, in the relative peace away from the manufacturing floor, tries to seek an explanation to a problem recently encountered (and problems there will always be), a scientific reason always comes to the fore. Unfortunately, however, there are two, three, or more possible explanations for the same problem, and the sorting out of the interlinked relationships of these can cause sleepless nights.

It is now 50 years since the book that became the standard work on commercial jam making first appeared; the first edition of *Jam Manufacture* by G.H. Rausch was published in 1950. Much of the information it contained is most relevant today, although many of the suggestions are dated when compared with the requirements of today, particularly those relating to food hygiene. Governing jams, jelly jams, marmalade, lemon curd, and mincemeat at that time was the 1953 *Preserves Order* (Statutory Regulation and Order 1953 No. 691, as amended), which had replaced earlier (1944) legislation. This has, in turn, been replaced in the United Kingdom by the *Jam and Similar Products Regulations* (1981) (Statutory Instrument 1981 No. 1063), which enacted the EC Directive *79/693/EEC*, and which has itself been amended by the EC Directive *88/593/EEC*.

Apart from differences in prescribed minimum fruit contents (and the introduction of several additional categories of product), a major difference between the 1953 and the 1981 regulations is the prescribed content for total soluble solids. For jam, the

1981 regulations prescribe a minimum figure of 60% soluble solids (as measured by refractometer) compared with 65% (for a hermetically sealed container) and 68.5% (nonhermetically sealed) in the 1953 order.

What is jam? A summary description from the 1981 regulations defines it as "a mixture of fruit . . . and sweetening agents brought to a suitable gelled consistency, with or without other permitted ingredients" (or words to that effect).

The traditional understanding of jam was that of a self-preserved cooked mixture of fruit and sugar (honey often qualified as a sugar). The higher figure of 68.5% soluble solids did give a degree of self-preservation, which is not achieved by the 1981 limit of 60%. The degree of preservation is related to the water activity of the product, but there are other factors affecting spoilage. In the light of experience, two jams can be prepared and stored similarly and have identical analytical characteristics in relation to soluble solids content, pH, and titratable acidity, and yet the one made from apricots will spoil more quickly than the one made from black currants. Research into this phenomenon could prove useful.

The 1981 UK legislation prescribes other products such as reduced sugar preserves. These, by themselves, are not self-preserving but need the presence of a permitted preservative to prevent spoilage.

For the labeling requirements of preserve products, one should refer to the appropriate legislation.

8.1.1 Ingredients

The ingredients in preserve manufacture are fruit, sweetening agents, and any other permitted ingredient that may be required. The last group includes gelling agents, acids, buffer salts, and colors, although other substances such as spiritous liquor may be used, particularly in specialty products. In fruit curds, a source of fat (normally either butter or margarine) is a mandatory ingredient; fruit curds also contain egg. That traditional British commodity, mincemeat, comprises, at its most basic, sugar, apples, vine fruits, citrus peel, a source of fat, spices, and acetic acid.

8.1.2 Fruit for Jam Manufacture

"You cannot make good jam without good fruit" is a maxim often overlooked, for it has frequently been the case that overripe or underripe fruit has been used for processing. Underripe fruit rarely has the flavor characteristics and the developed color of fully ripe fruit, and the natural pectin present at that stage is often unusable to the jam maker; pectin becomes more soluble, and so more available, as the ripeness of the fruit increases. Overripe fruit often has poor flavor and is subject to microbiological spoilage. In addition, enzymatic action will have broken down the pectin, leading to disintegration of the structure of the fruit.

Very few large commercial concerns now produce jams and marmalades from fresh fruit during the harvesting season; most use fruit that has been preserved one way or another. Three types of preservation are used: freezing, canning, and sulfiting. The first two of these are dealt with in the previous chapter and are discussed very briefly here.

Freezing is a good method for maintaining fruit quality, although, to the jam maker, the structural changes that occur in the fruit through ice crystal formation may adversely affect the texture of the finished preserve.

In the United Kingdom, canned fruits tend to be limited to imported fruits such as apricots, peaches, and pineapples. These fruits are often packed to the purchaser's requirements. A development of this is the use of aseptic packing, either truly aseptic, where the fruit is sterilized by heat and then cooled and filled aseptically into sterile containers, or hot filling into semibulk containers, such as internally lacquered drums of 200-L capacity, that are sealed and crash cooled by immersion in cold water.

Chemical preservation of fruit intended for jam manufacture using sulfur dioxide is widely practiced and is easy and cheap to carry out. A sulfur dioxide concentration of 2000 to 3000 mg/kg of fruit is achieved by using a 6% aqueous solution, or the gas may be added directly to the fruit. The use of salts of sulfurous acid is also common but can result in problems later on in the manufacture of the jam because of the presence of metallic (e.g., sodium) ions, which affects the pH of the finished jam. Storage of sulfite-preserved fruit may be in polythene-lined drums or, where whole fruits are not present, in bulk-storage tanks. The presence of whole fruits in bulk-storage tanks can lead to poor fruit piece distribution as a result of floating of the particles.

There are legal constraints on the use of fruit preserved with sulfur dioxide: it may not be used, for example, in Extra Jams or Extra Jellies. The limitations on its use are mainly concerned with its removal during subsequent processing. For example, it is easier to boil off sulfur dioxide using open pans than when processing under vacuum. Atmospheric pollution must also be considered because sulfur dioxide is blamed for acid rain. A factory producing 1000 tonnes of jam from sulfited fruit pulp by open pan boiling, with discharge of the evaporated steam with its entrained sulfur dioxide into the atmosphere, can result in the release of up to 1 tonne of the gas.

Before many fruits can be used in jam manufacture, a preprocessing operation is often necessary, for example, peeling, slicing, dicing, and/or cooking. Many of these operations can be carried out at the jam maker's factory; more often these days, the fruit is bought in as a ready-to-use ingredient. Some examples of fruit processing are discussed later in this chapter.

Types and Varieties of Fruit

Almost any type and variety of fruit can be used in the manufacture of preserves, but, inevitably, some fruits will find more favor than others. The manufacture of jam has been in decline for many years, and, in consequence, many of the fruits used in earlier times are no longer commercially available, although they may sometimes still be obtained by the maker of specialty products. A typical example is the strawberry, where simple economies have resulted in an almost total loss of British-grown fruit for use by the larger British manufacturer.

Apples are used in the manufacture of many types of preserves, especially bakery preserves and in mincemeat. For use in the former, the fruit is cooked and sieved to produce a purée, while in the latter it may be used as precooked purée or in an uncooked macerated or diced form. The most popular apple in use today is the Bramley Seedling.

Black currants can be used as sulfited black currant pulp or in whole fruit form (rather than sieved). The latter should be precooked before inclusion in the jam-making process to prevent shriveling of the berries and a consequent toughening of texture.

Many fruits have lost popularity over the years, or may not ever have been as popular as the mainline varieties of today, for example, the gooseberry, damson, and plum. Earlier writers have promoted the use of underripe plums, and it is felt that this may have had a deleterious effect on the sales of plum jam. While the use of sulfited fruit enables manufacture from a basically cheap raw material to be carried out all year round, there are differences in flavor from similar jams made from sulfited or fresh/frozen fruit. These comments apply to all fruit varieties, and the manufacturer has to choose between the cost and flavor constraints when bringing a product onto the market. The fickle nature of the consumer must also be considered.

Citrus fruits and, in particular, the bitter orange are used in the manufacture of marmalade; indeed the term *marmalade* is reserved for products made with citrus fruits. The oranges (or other fruit) must be preprocessed before the marmalade can be made.

8.1.3 Other Ingredients

Sweetening Agents

The most commonly used sweetening agent is white sugar (sucrose), either as the dry ingredient or as an aqueous solution or syrup. Glucose syrup is also in general use in preserves, although the compositional requirements of the end-product must be considered because it often contains very small quantities of sulfur dioxide. Other commercially available syrups (e.g., high fructose syrup or invert syrup) can be used, as can honey. These syrups must be used carefully because the invert sugar content of the finished product can affect the quality of the set as well as the potential for crystallization, particularly in jams of higher total soluble solids content (such as some bakery jams).

Acids and Buffers

"Fruit" acids, commonly citric or malic acids, are often added to adjust the flavor profile and to achieve the optimum pH for setting. Buffer salts such as trisodium citrate are also used to modify pH. Lemon juice may be substituted for citric acid.

Fats (for Curds and Mincemeat)

Butter or margarine is normally used in fruit curds, while beef suet or its vegetable equivalent, normally in shredded form, may be used in mincemeat.

Vine Fruits (for Mincemeat)

Details are to be found later in the chapter.

Citrus Peel (for Mincemeat)

The peel is separated from the center of the citrus fruit and often held in brine prior to dicing, tenderizing, and soaking in sugar syrup. This is drained prior to use. Normally, lemon and sweet orange peels are used, although it is not unknown to use bitter orange peel in mincemeat.

Gelling Agents

Pectin is the gelling agent that occurs naturally in fruit, and it governs the setting characteristics of the product. The pectic substances that occur naturally in the middle lamella of plant tissues change with the changing maturity of the fruit. They can be regarded as the glue that holds the plant cells together. As the fruit ripens, an insoluble pro-pectin is converted into soluble pectin. The pectin molecule is composed of long chains of partially methoxylated polygalacturonic acid (Figure 8–1). In riper fruit, enzyme action will break the chains into shorter lengths, producing pectin with inferior setting characteristics. The quantity and the effectiveness of pectin also vary with the variety of fruit. Citrus fruits and apples contain more pectin than do cherries, for example. It is prudent, therefore, for the jam maker to estimate the useful pectin content of the fruit before considering large-scale production.

Because of the need to produce a consistent product in the face of natural variations, pectin often has to be added to the product. This is essential when depectined fruit juices are used in the production of jelly jams. Commercially, pectin is extracted from both apples and citrus fruit, or rather, more normally, from apple residue after the extraction of the juice, and from citrus peel. From UK production, apple pectin is sold as a liquid while citrus pectin is sold as a powder. Both sources will normally have been standardized before sale.

Naturally occurring pectin is termed *high methoxyl*, with a sufficiently high percentage of its hydroxyl groups methylated for it to fall into the "fast-set" category. This will form a gel with high sugar content solutions (refractometer solids between 60 and 70%) and with pH values between 2.8 and 3.5. The manufacturers of pectin chemically modify the pectin molecule in order to affect the speed of setting of the gel. A "slow-set" high-methoxyl pectin will have between 60 and 68% of its groups methylated, while a "fast-set" pectin has between 68 and 75%.

When a high-methoxyl pectin is dissolved in water, it forms a slightly acidic solution, and there is a tendency for the carboxylic acid groups to dissociate; this is controlled by adjusting the pH. The addition of sugar has a dehydrating effect on the pectin and will reduce its solubility. To form the gel, cross-linkages are formed when these two effects are combined.

The set, its quality, and speed of production are governed by the degree of methylation and the quantity of the pectin and also by the sugar content (and the types of sugar present), the pH, and the temperature of the solution. A fast-set high-methoxyl pectin will have a higher setting temperature than a similar gel made using a slow-set high-methoxyl pectin. For the jam maker, it is sometimes desirable to use a slow-setting pectin, for example, where the flotation of particles of fruit to the surface of the jam

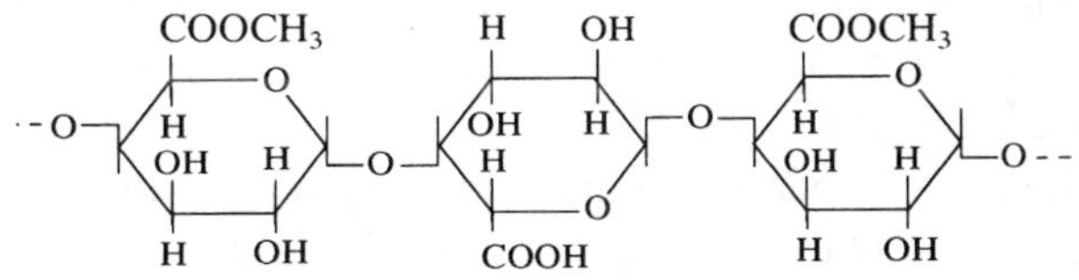

Figure 8–1 The structure of part of the pectin molecule.

is not a problem. Normally, however, a compromise is needed, and a balance between the variables achieved.

Modifications to the pectin molecule to reduce the number of methoxyl groups while maintaining the designation "high methoxyl" have been mentioned. If further modification takes place with the removal of more methoxyl groups, the pectin becomes "low methoxyl" and will produce gels at lower sugar contents. Further chemical modifications can produce "amidated" pectins, where up to 20% of the methoxyl groups have been converted to amide groups. Both ordinary low-methoxyl and amidated pectins behave similarly and are dependent on the presence of calcium for gel formation, with the gel strength being less dependent on sugar content. Indeed, some of these pectins can form gels with sugar contents of below 20%. These are important in the manufacture of "Reduced Sugar Jams," where the low sugar content prevents gel formation using the pectin naturally present in the fruit.

Other gelling agents, such as alginates or others prescribed in the regulations, may be used, often in collaboration with other stabilizers in the manufacture of reduced-sugar jams. It is recommended that the intending user experiments with various combinations until a satisfactory mix is arrived at.

8.1.4 Product Types and Recipes

Compositional definitions from the 1981 UK Regulations include Extra Jam, Jam, Reduced Sugar Jam, Extra Jelly, Jelly, UK Standard Jelly, Marmalade, Reduced Sugar Marmalade, Mincemeat, Fruit Curds and Fruit Flavour Curds, and Sweetened Chestnut Purée. There are two other categories that are deserving of mention: No-Added-Sugar Fruit Spreads and Diabetic Jam. Mention will also be made of products prepared specially for manufacturers: Bakery Jams. The regulations also include many more definitions, including those of the permitted sweetening agents and other authorized ingredients.

Sweetened Chestnut Purée is a mixture of sweetening agents and puréed chestnuts (minimum 38 g/100 g) with minimum refractometer solids of 60%.

Jams and Jelly Jams are similar except that the former products are made with whole fruit, fruit pulp, or fruit purée while the jelly jams are made using fruit juice. There are two qualities of jam: Jam and Extra Jam. Three qualities of Jelly Jam are specified: Jelly, Extra Jelly, and UK Standard Jelly. Each will be dealt with separately.

The general fruit content for Jam is 35 g of edible fruit per 100 g of product, and the product can be made from fruit, fruit pulp, or fruit purée (or a mixture of these). There are some exceptions to this general limit, most notably black currant (25 g/100 g) and ginger (15 g/100 g). The fruit used for this category of product may have been preserved with sulfur dioxide, and in recipe calculations, the added preservative (and any other ingredients such as water) used in the manufacture of the pulp must be considered. The term *pulp* here is a jam maker's term for a ready-to-use ingredient and should not be confused with the legal definition.

For Extra Jam, the general minimum fruit content is 45 g of edible fruit per 100 g, with exceptions similar to those for Jam: black currant at 35 g/100 g and ginger at 25 g/100 g. Extra Jams cannot be made from purée, and the fruit must not have been preserved with sulfur dioxide. In addition, Extra Jams cannot be made with mixtures that include apples, pears, clingstone plums, and several other fruits. It must be pointed out that this does not prevent the manufacturer from making products using mixtures where the fruit content is above the legal minimum for Extra Jam, but the product cannot be called an Extra Jam.

The recipe for a product is a very personal thing, and its makeup is often jealously guarded, for commercial viability often depends on its contents being unknown to others. Below, therefore, are the basic ideas of formulation, which may prove useful to the reader.

When formulating recipes, it is normal to specify a standard theoretical output and relate ingredient quantities to this. It is necessary to know the actual fruit contents (both "total," for the requirements of the labeling legislation and "edible," for the recipe itself). Certain analytical details of the ingredients also need to be known.

It may be necessary to add a quantity of a gelling agent to the product: pectin is the most common in Jams and Extra Jams, and should be added as a solution.

Sugar (sucrose) is the predominant sweetening agent in most Jams and Extra Jams. During the boiling operation, some of the sucrose will be inverted to form glucose and fructose (invert sugar). One molecule of sucrose ($C_{12}H_{22}O_{11}$) combines with one molecule of water (H_2O) to form one molecule of glucose and one of fructose (both $C_6H_{12}O_6$). Thus 1.000 kg of sucrose will produce 1.052 kg of invert sugar. For the purpose of the illustrations that follow, it is assumed that 25% of the sucrose is converted into invert sugar during processing; in which case, 0.987 kg of dry sugar will produce 1.000 kg of the mixture of sucrose and invert sugar. This supposition is not atypical of normal processing.

An Example of Formulation of a Recipe

To prepare a raspberry jam that is to contain the UK minimum fruit content of 35 g edible fruit per 100 g of product would use that amount of whole raspberries. For a seedless raspberry jam, the contribution of the raspberry seeds (approximately 4% by weight) would need to be considered, and 36.4 g of raspberries per 100 g of product would be needed. The exact figure is determined empirically.

The label declarations on the raspberry jam would be "prepared with 35 g of fruit per 100 g," and it would have "total sugar content of 66 g per 100 g." If the particular jam is to be made from sulfited pulp, which is specified as having a total (edible) fruit content of 92% by weight, the 100 g of product would contain 38.04 g (35 ÷ 92%) of the sulfited pulp. The total soluble solids of the pulp, as measured by refractometer, is found to be 8% and so the 38.04 g of pulp contains 38.04 × 8% = 3.4 g of sugar solids.

By experiment, the pectin requirement is set at 10.0 g of pectin solution per 100 g of product. This pectin solution contains 10.0% refractometer solids and, therefore, contributes 1.00 g of solids. The final refractometer solids of the jam are to be 66%; so the sugars to be added are 66 minus 3.04 (from the fruit) minus 1.00 (from the pectin), i.e., 61.96 g per 100 g of product. This is added as dry sugar (sucrose), and it is assumed that 25% of the sugar is inverted during processing. The amount of dry sugar that has to be added is, therefore, 61.96 × 0.987 = 61.16 g per 100 g (i.e., the "sugar equivalent" is 101.32% of the sucrose content and reflects the refractometer readings).

Finally, by experiment, it is found necessary to add buffer salts (citric acid and, perhaps, sodium citrate) in small quantities to produce satisfactory conditions for setting (see also section 10.5 on pectin in Chapter 10). In addition, the use of added color may be needed, especially where sulfited red fruits are used, as the whole of the color does not return during the boiling that drives off the sulfur dioxide.

These calculations may be (perhaps more simply) represented using a *pro forma* recipe sheet as shown in Table 8–1.

The sugar could be added as sucrose syrup (or "sugar solution," as defined in the UK's Specified Sugar Products Regulations 1976 [S.I. 1976 No. 509]), in which case, due account would have to be taken of its concentration. Such syrups are often supplied at 67% sucrose (67°B by refractometer), and if this were used, the recipe sheet would appear as in Table 8–2. It is also quite normal, technically acceptable, and often commercially beneficial for up to 30% of the added sugar to be replaced by commercial glucose syrup. Such syrups may be slightly less sweet than sugar or invert syrup, and it is often thought that a decrease in sweetness is desirable in today's preserves.

If 25% of the originally calculated added sugar were to be replaced with a glucose syrup having 80% total soluble solids, the recipe sheet would now appear as in Table 8–3.

This is only one example; the proportion of sugar solution and glucose syrup can be varied as desired.

Small errors will be apparent in these calculations; no account of the contribution of acid and buffer to the refractometer solids has been taken and the degree of inversion has been assumed to be 25%; in practice, this may not be the case.

Once a formulation has been decided upon, it is advisable to carry out experimental boils to ascertain the exact needs in terms of added pectin, acid, and buffer.

At first sight, the first of these recipes may appear to be the most economic (neglecting any differences there might be in the prices of sugar and glucose syrup), as it exhibits the least amount of required evaporation. This is not necessarily the case because of the possibility of the precipitation of pectin from solution if insufficient water is present. Experience has shown that a boil size of 120 g minimum per 100 g of output is required (for normal jams) if this phenomenon is to be avoided, for once

Table 8–1 A *Pro Forma* Recipe Sheet for Jam

Sugar Equivalent (kg)	Ingredient	Quantity (kg)
3.04	Raspberry pulp (8.0% TSS*)	38.04
61.96	Sugar	61.16
1.0	Pectin solution (10.0% TSS*)	10.00
	Color	
	Citric acid	} If required
	Sodium citrate	
Refractometric solids, total 66.0		Boil to output 100.00

*TSS represents total soluble solids (as measured by refractometer), or refractometric solids.

Table 8–2 *Pro Forma* Recipe Sheet Using Sucrose Syrup

Sugar Equivalent (kg)	Ingredient	Quantity (kg)
3.04	Raspberry pulp (8.0% TSS*)	38.04
61.96	Sugar syrup (67°B)	91.28
1.0	Pectin solution (10.0% TSS*)	10.00
	Color	
	Citric acid } If required	
	Sodium citrate	
Refractometric solids, total 66.0		Boil to output 100.00

*TSS represents total soluble solids (as measured by refractometer), or refractometric solids.

the pectin has lost its solubility, the boil seldom can be satisfactorily reclaimed and properly set jam produced.

The UK (and EU) requirements for an Extra Jam are prescribed, in that the minimum fruit content and the type of fruit to be used are defined. The formulation for a Raspberry Extra Jam is calculated similarly to the previous recipes. To make a product at the statutory minimum fruit content, and with a sugar content of 66 g/100 g, the recipe might read as in Table 8–4.

The basic difference between Jams and Extra Jams is the additional fruit content. This is an oversimplification, because of the differences relating to residual sulfur dioxide and the use of added colors, for example. To produce a product that is acceptable legally, one must refer to and comply with the legislation of the country in question.

Jelly jams, in all their guises, are similar to normal jams, but the source of fruit is juice rather than whole fruit, fruit pulp, or fruit purée. The recipes can be worked out in the same manner as those outlined previously. UK and EU legislation provides for three categories of jelly jams: Jelly, Extra Jelly, and UK Standard Jelly. A Blackberry Jelly

Table 8–3 *Pro Forma* Recipe Sheet Using Glucose Syrup

Sugar Equivalent (kg)	Ingredient	Quantity (kg)
3.04	Raspberry pulp (8.0% TSS*)	38.04
46.48	Sugar solution (67°B)	68.46
15.48	Glucose syrup (80°B)	19.35
1.0	Pectin solution (10.0% TSS*)	10.00
	Color	
	Citric acid } If required	
	Sodium citrate	
Refractometric solids, total 66.0		Boil to output 100.00

*TSS represents total soluble solids (as measured by refractometer), or refractometric solids.

Table 8–4 Recipe for Raspberry Extra Jam

Sugar Equivalent (kg)	Ingredient	Quantity (kg)
3.60	Raspberries, fresh or frozen (8.0% TSS*)	45.0
62.00	Sugar	61.19
0.4	Pectin solution (10.0% TSS*)	4.00
	Citric acid } If required	
	Sodium citrate	
Refractometric solids, total 66.0		Boil to output 100.00

*TSS represents total soluble solids (as measured by refractometer), or refractometric solids.

must contain a minimum of 35 g of blackberry juice per 100 g of product, a Blackberry Extra Jelly a minimum of 45 g, while the UK Standard Blackberry Jelly harks back to older legislation and is required to contain the juice of at least 35 g of blackberries per 100 g of product.

The juices used for jelly manufacture are normally clarified and are often treated with a pectinase enzyme to destroy the natural pectin. In this case, it will obviously be necessary to add pectin in order for the product to set!

Marmalade is also compositionally prescribed in UK and EU law (as well as in that of many other countries). It is a jam that is manufactured exclusively from citrus fruit, and the normal fruit to be used is the Bitter (or Seville) Orange. The legislation requires marmalade to contain a minimum of 20 g of citrus fruit per 100 g of product. The fruit is further defined as pulp, purée juice, aqueous extract, or any combination thereof with the proviso that at least 7.5 g is derived from the center of the fruit. Although the bitter orange is the fruit normally used in the UK product, other citrus fruits such as grapefruits and sweet oranges are finding favor.

Jelly marmalades are among the most popular of marmalades and are made with juice extracted from the fruit, which is normally clarified before use. It is often concentrated in the country of origin and exported in this condition; the degree of concentration must be known exactly if the fruit content of the finished product is to be correctly assessed. The marmalade may or may not contain thin-cut shreds of peel, which are usually about 25 mm long, 1.5 mm wide, and 1.5 mm thick (i.e., without albedo). They are almost invariably tenderized and soaked in syrup before use in order to obtain an even distribution of shreds throughout the marmalade.

Marmalades containing medium or fine-cut peel are normally made with a purée of the center of the fruit and shreds of peel. The mechanical cutters that separate the peel from the center are set to provide peel with the maximum amount of albedo. The shreds have a maximum length and a width between 1.5 and 3.0 mm. Such marmalades normally have the peel pieces tenderized (by cooking in water or steam) and then mixed back with the precooked and sieved center purée before the mixture is used to make marmalade. Pulps made up of peel and center purée are available commercially, having been prepared in the country of origin and packed either aseptically or with the preservative sulfur dioxide. Whether the pulp is made in the factory where the marmalade is to be prepared or elsewhere, it is essential to ascertain its exact fruit content in order to establish the fruit content of the finished marmalade.

Marmalades with thick or coarse-cut peel contain much larger pieces of peel—sometimes up to 8 mm wide and 50 mm long—which are, after tenderizing and prior to marmalade manufacture, sometimes impregnated with some of the recipe requirement of sugar to increase their density in a way similar to the peel used for jelly marmalade. This helps to achieve a good distribution of shreds in the finished marmalade. It is generally recognized as producing a marmalade with better flavor (especially when fresh fruit is being used) and helps to achieve a good "clean" set (with a minimum of syneresis). The method is, however, considered to be too time consuming these days, being both inefficient and expensive.

Formulations for marmalades are worked out in an identical manner to the examples shown earlier for raspberry jam, but it cannot be emphasized too much that in order to produce a consistent product, one must know precisely the way in which the raw material is treated and its composition. A more detailed discussion of the preparation of citrus fruit for marmalade manufacture is found later in this chapter.

The basic principles outlined previously apply equally well when applied to Reduced Sugar Jams and Marmalades. It must always be remembered that, in these products, preservatives are permitted (as prescribed in the legislation) because the products themselves are not self-preserving. It must also be remembered that if a low-methoxyl or amidated pectin is used, the setting characteristics of the product are dependent on its calcium content. Some of these reduced-sugar products are sold as "made especially for diabetics" because their formulations include only half the amount of added sugar present in ordinary jam. This follows a requirement in the UK Food Labelling legislation. Traditional Diabetic Jam, however, uses sorbitol as a replacement for the added sugar, as sorbitol, a polyhydric alcohol, does not affect the metabolism of the diabetic. The same manufacturing methods are used, but care must be taken to ensure that the products are not contaminated by sugar. Many commercially available pectin powders are standardized using sugar, and so the maker of Diabetic Jam should ensure that the ingredients are not so treated.

8.1.5 Methods of Manufacture

General

Figure 8–2 is a flow diagram of the basic processes in preserve manufacture. This can be extended to include all the different stages in the process and in such a form could form the basis of a hazard analysis for the product. The addition of control points (which may or may not be critical control points) could be most useful in the production of a full HACCP (hazard analysis critical control points) assessment. Control points have not been added into this basic flow chart.

Fruit Preparation

Almost all fruit will require preprocessing in one way or another before it is in suitable condition for incorporation into the mix to be used for the preserve (Figure

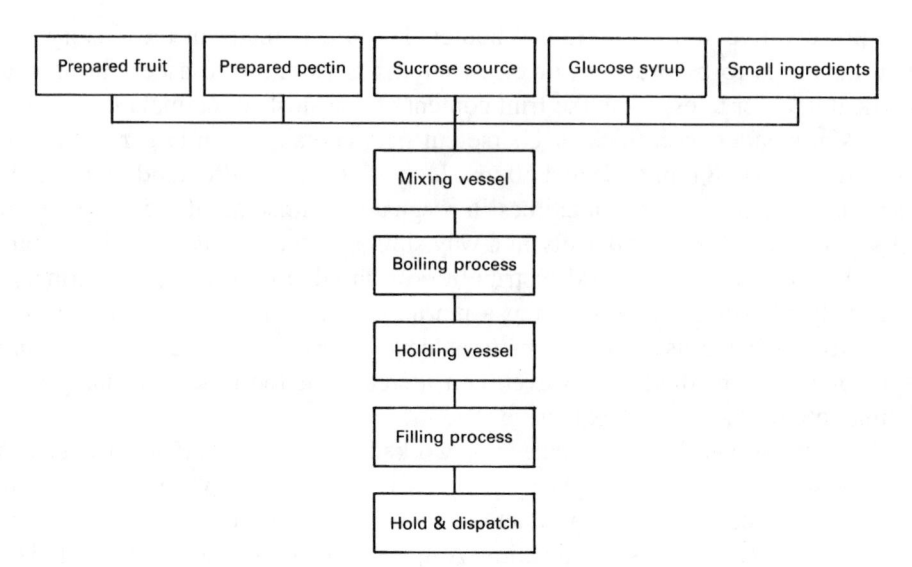

Figure 8–2 Flow diagram of the basic processes in preserve manufacture.

8–3). This particularly applies to citrus fruit. The preprocessing of other fruits will often involve trimming and perhaps tenderizing to facilitate subsequent osmosis and to give a satisfactory mouthfeel to the preserve. Some fruits (e.g., apricots) may be canned as solid pack to extend the period that the fruit is available, and this also achieves the tenderizing of the fruit.

All fruit should be properly sorted before use to exclude foreign bodies or other unacceptable material, including that from the source plant.

The processing of citrus fruit is, of necessity, more complicated. The peel is separated from the center of the fruit, and the two parts are treated separately, often being recombined in natural ratio for use in marmalade manufacture. Sometimes, however, it is necessary for the two parts to be kept separate, only to be recombined when the marmalade is being produced. Sometimes these ingredients may not be recombined in a natural ratio, where, for example, a high shred content is required. Shredless marmalades are also manufactured. Figure 8–4 shows a flow chart for citrus fruit preparation. As with the general chart, this can easily be extended and used in an HACCP presentation.

Boiling and Filling

For normal jams and marmalades, some evaporation is needed to move from the mixing bowl stage to the finished product. Reduced-sugar products, however, can be formulated so that evaporation is not necessary. For some products, a two-stage process may be preferred if certain potential problems (such as the floating of fruit particles or pieces in the finished product) are to be avoided.

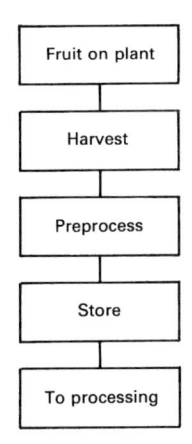

The Manufacture of Preserves, Flavorings, and Dried Fruits 189

Figure 8–3 Preprocessing stages in preserve manufacture.

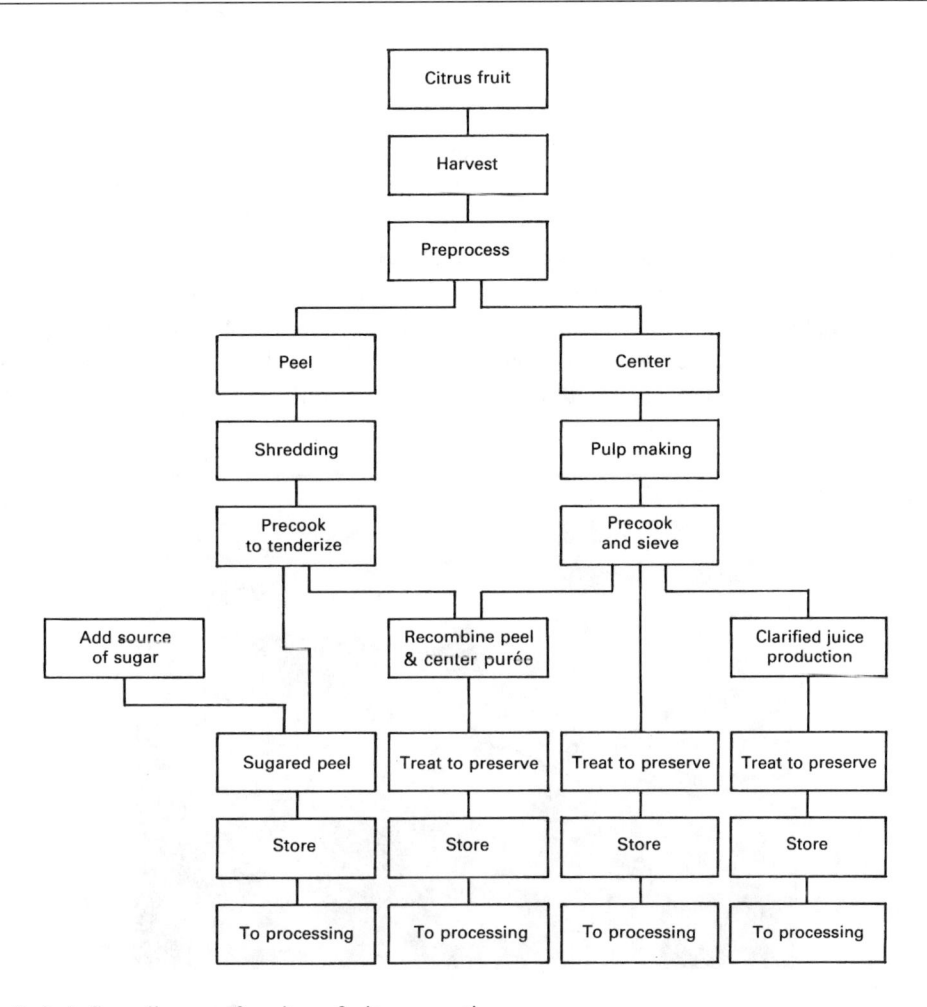

Figure 8–4 A flow diagram for citrus fruit preparation.

The boiling of jams is outlined as follows. Because this process can vary greatly, not only with the method of boiling used but also with the different types of fruit and even with different batches of the same fruit, acquired experience is invaluable.

Traditionally, jam boiling was carried out using open (atmospheric) copper boiling pans, but these days pans are now usually stainless steel (Figure 8–5). The pans are hemispherical, with a large lip extension to assist in the prevention of boilover. The hemispherical part of the pan is steam jacketed, with high-pressure steam providing the heat source. Extra heating may be provided by the use of internal steam coils. The capacity of the pans may range from a few kilograms up to 100 kg or more, but calculations should be carried out to balance the input and output; the input should not be so large as to allow boiling over and consequent wastage, while the output should not be so small as to allow burning of the jam onto the heated surface of the pan. Frothing of the jam, especially at the beginning of the boil, can present a problem—indeed, it can result in boiling over—but the use of a very small amount of vegetable oil often reduces this. For jelly marmalade, an alternative antifoam such as dimethyl polysiloxane may be used, as the use of vegetable oil can cause cloudiness of the marmalade. The tipping of boiling jam (a somewhat dangerous practice!) from the pan into troughs for subsequent pumping to the filling line is rarely carried out, as pans with bottom-opening valves are now largely used.

Preserves may also be boiled under vacuum using either batch (Figure 8–6) or continuous methods as well as by a combination of atmospheric pressure and vacuum. The capacity of batch vacuum cookers also varies from a few kilograms up to several tonnes. Purée-type jams may be manufactured on a continuous basis using a plate evaporator, while for products with defined fruit pieces, scraped-surface heat exchangers operating both at atmospheric pressure and under vacuum are used as evaporators.

After boiling, but prior to filling, the jam is normally held in a reservoir so that quality testing can be carried out to ensure that it is of the right consistency when filled.

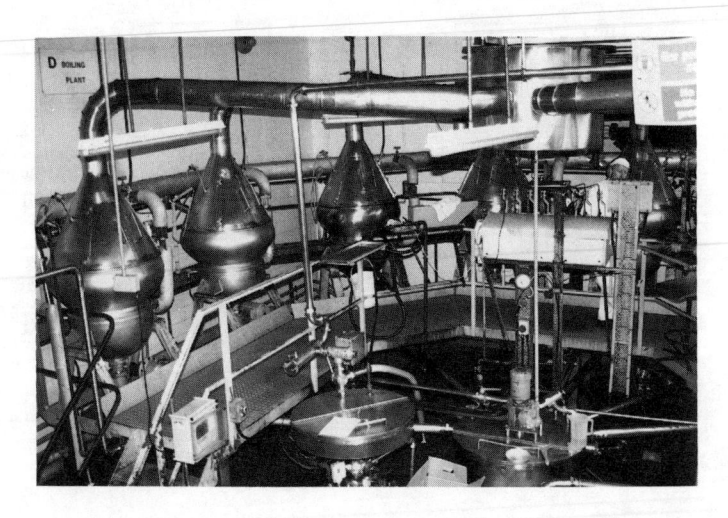

Figure 8–5 Atmospheric boiling pans. Courtesy of RHM Ledbury, UK.

Figure 8–6 Batch vacuum boiling pans. Courtesy of RHM Ledbury, UK.

Total soluble solids, pH, and temperature all need to be monitored to ensure compliance with specification, and the consistency of the product at the time of filling will in no small way affect the appearance and texture of the filled and cooled jam.

From the holding reservoir (which is generally gently agitated to ensure an even distribution of the fruit pieces and is often heated to maintain a proper filling temperature), the jam is filled into an appropriate container. The size of these will vary from the 20-g portion pack to the 20-tonne bulk tanker for the large bakery user.

Portion packing is an art in itself (Figure 8–7), for the packaging material is normally a plastic that is thermally formed into the receptacle prior to the deposition of the jam, which must not be so hot as to deform the container but sufficiently hot to ensure a good gel formation as well as commercial sterility. These packs are normally heat sealed with foil.

Preserves may be filled into glass jars (capacity from 1 oz up to 2 lb) using volumetric piston filters (Figure 8–8). These are often of the rotary multihead piston type with filling speeds of 300 or more units per minute. Each head should be considered as a separate filling machine, and machines are available where every head is adjustable. Here, great accuracy can be achieved with an overall average filling weight very close to the target weight (which will have been established by statistical analysis). It is very necessary to ensure that the material being filled is of the proper quality to ensure a good distribution of fruit particles as well as an appropriate gel strength.

Preserves for bakery use are packed in large containers, normally at much lower temperatures to prevent caramelization of the product. Formulation here is very important so that the gel of the jam in the finished application is of an appropriate texture and strength. It must also be acceptable to the user in the state in which it is delivered, and often these two conditions seem almost irreconcilable. Packs vary from plastic pails (Figure 8–9), sometimes with a heat-sealed foil lid, to polythene bag-lined fiberboard cartons to polylined 200-L drums to bulk tankers. In the last case, the jam may sometimes be transported in an unfinished condition where the user will, for

192   FRUIT PROCESSING

Figure 8–7 Portion pack filling. Courtesy of CPC (UK) Ltd.

example, add citric acid in a prescribed amount to achieve a pH in the finished product that will enable the gel to be formed at the required speed.

8.2   FRUITS PRESERVED BY SUGAR: GLACÉ FRUITS

In one sense, this method of preservation of fruit is an extension of the jam-making principle, for in many jams, the producer tries to exhibit large fruit pieces within the gel. With glacé fruits, the object is to maintain the original shape of the fruit while raising the total soluble solids of that fruit to a level that is self-preserving, normally 72%, or

Figure 8–8 Volumetric jar filler. Courtesy of RHM Ledbury, UK.

Figure 8–9 Plastic pail filling. Courtesy of Broadheath Foods Ltd.

above (as measured by refractometer). In order to maintain the original fruit shape, it is sometimes necessary to pretreat the fruit to modify its structure. In the case of cherries (the most popular glacé fruit), this is achieved using calcium, which, in effect, toughens the fruit by changing the pectin structure.

Cherries provide a good example of the process. After picking, the fruit is stored in a sulfur dioxide solution of sufficient strength to prevent deterioration of the fruit. This solution also contains a calcium salt—some writers have advocated calcium hydroxide while others recommend calcium chloride. After a minimum period of storage and when the texture of the sulfited cherries is considered satisfactory, the processing begins. The first stage is the partial removal of the preserving sulfur dioxide, which is achieved by successive leachings with hot water. This process has the additional effect of softening the internal structure (this is of benefit not only in the final texture of the fruit but also in facilitating the osmosis by which the sugar enters the fruit).

The cherries are strigged and pitted after they have been size graded. The pitted cherries are then boiled in water to complete the removal of the sulfur dioxide and the softening operation. If the cherries are to be colored by the synthetic dye erythrosine (E127), the coloring takes place at this stage. If natural colors are to be used, the color is added later in the sugaring process. After boiling, the cherries are transferred to a vessel containing a sugar solution at approximately 20%. Over a period of at least 7 days, the strength of the sugar solution is increased in stages on a daily basis until a final concentration of between 70 and 74% is achieved. As osmosis takes place, the concentration of the syrup in the vessel decreases. The vessels used for the sugaring are interconnected and syrup of the highest concentration is placed in the final-stage vessel. As this decreases in strength, it is passed into the penultimate stage, and so on. As the process proceeds, the cherries increase in total solids content and the syrup strength decreases (i.e., the glacé process is a countercurrent exchange). The process is outlined in Figure 8–10.

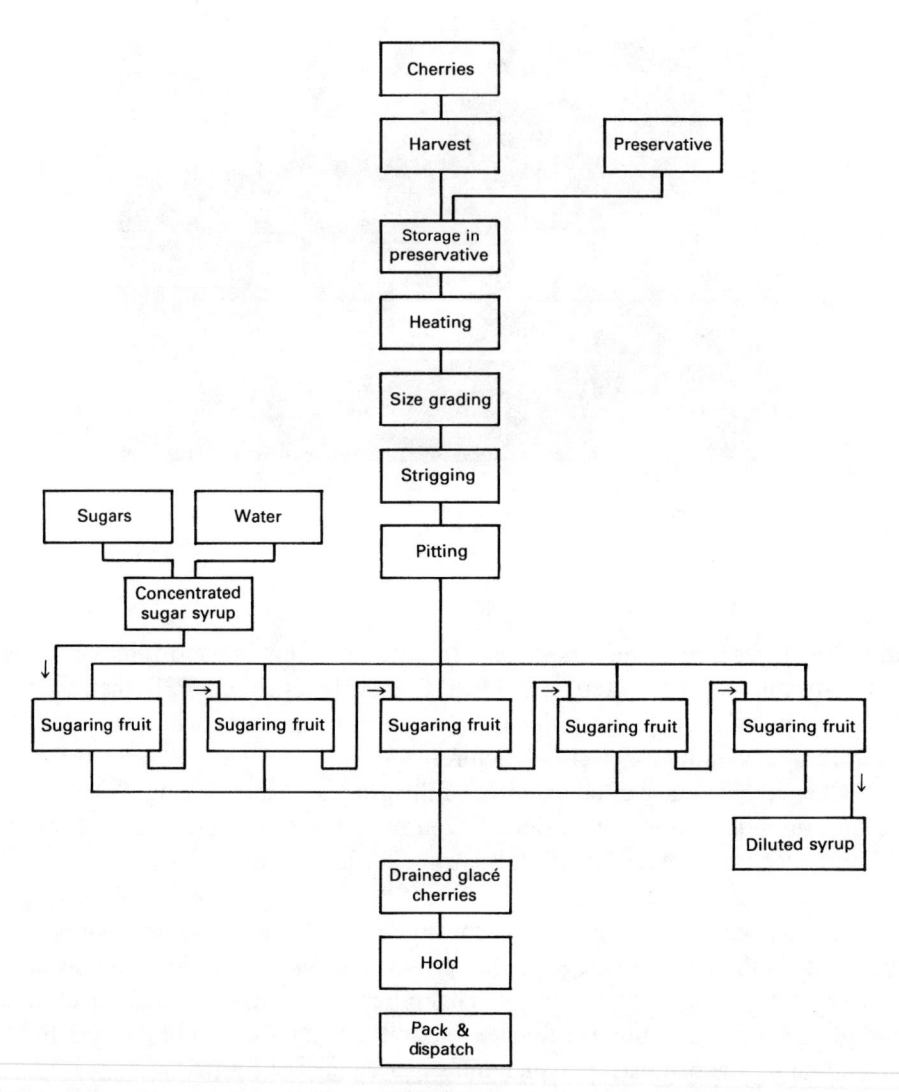

Figure 8–10 A flow diagram for preparing glacé cherries.

Other soft fruits may be treated similarly except that, in most cases, the initial treatment with sulfur dioxide is not used. Fruits such as apricots, peaches (both of which are normally glacéd "on stone"), and the trimmed halves of pears or slices of pineapple are sometimes used fresh, although they may be canned in a heavy syrup for glacéing at a later date.

A commodity that has been traded for several centuries is sometimes referred to as "Ginger Sweetmeat" and originated in China. It is, in fact, the rhizome of the ginger plant (*Zingiber officinale*), which has been harvested, trimmed, and impregnated with sugar by a method similar to that used for cherries.

The rhizome becomes more fibrous with age, and so the time of harvest is important in determining the texture of the finished product. That harvested early in the season is used for sugaring, while the later-harvested (and so tougher) rhizomes are dried for flavor extraction or grinding. The harvested rhizomes are stored in vats until required for use, when they are trimmed, cut, and sorted into the required shapes. The preservative is removed by leaching, and the sugaring process continues by immersion of the pieces in successively stronger sugar solutions. The glacé (or sugared) ginger is then packed to the customer's specification prior to shipping.

Another commonly available sugared product is citrus peel, normally available as "cut mixed peel." Formerly, this too was prepared in a manner similar to that used for cherries, using "half caps" (i.e., a hemisphere of the peel of the fruit). The peel was supplied to the manufacturer packed in brine (a solution of common salt) in large barrels. This peel was debrined by successive soakings in clean water and tenderized by cooking. The half caps were steeped in successively stronger sugar solutions, but here, quite often, the syrup was drained from the peel, strengthened by the addition of more sugar, and boiled before being added back to the peel while hot.

Nowadays, the peel is generally treated at source. It is diced using the same machinery as that used for the marmalade shred production, tenderized by pressure cooking, and then sugared by cooking with sugar under vacuum. This is a much quicker process. The peel, after sugaring, is drained and packed for subsequent shipment and is generally a mixture of sweet orange and lemon peels. The use of glucose syrup as a partial replacement for sucrose is technically advantageous as it helps in preventing crystallization and improves the appearance of the product by providing it with a greater sheen than occurs with sucrose alone.

A process sometimes used after the sugar impregnation is that of crystallization. Here, the analysis of the preserved fruit or peel is important, as the amount of inversion of the sugar that has taken place during the processing influences the final texture. Crystallization may be achieved by coating the preserved fruit or peel with a supersaturated sucrose solution and then applying heat to drive off moisture and so induce the formation of sugar crystals. The use of the supersaturated sugar solution is the traditional method for the making of crystallized citrus peel, but its use, like many of the processes so far discussed, is somewhat of an art, probably because of the "by hand" nature of the work.

Although not fruit, there are two further items in addition to ginger that merit brief discussion: crystallized flowers and angelica. The second of these is the young stem of the angelica plant (*Angelica archangelica*), an umbelliferous plant introduced from Europe that has become naturalized in parts of the United Kingdom. It should not be confused with other, inedible, umbels! The stems are cut and trimmed and then tenderized and sugared in a similar manner to cherries. They are often colored green.

Flowers such as violets or rose petals are crystallized in a somewhat different manner. The flowers are harvested and coated in a solution of a gum; often gum arabic is used. The flowers are then dusted in icing sugar and dried: this achieves preservation. Alternatively, the petals are coated with a supersaturated sugar solution and then dried to give rigid crystallized petal-like pieces, which may be used for cake decoration. The syrup used for this process is often colored to impart the traditional color to the product. Care must be taken to ensure that the flowers used are edible!

8.3 FRUITS PRESERVED BY DRYING

The most well-known fruits preserved by drying are grapes. The various types produce currants, sultanas, and raisins. Other fruits are also dried, for example, apricots, pears, bananas, and plums (to yield prunes). While the principle is the same for all the fruits, many of the soft fruits are sulfited prior to drying; the processes are elaborated upon as follows.

8.3.1 Dried Vine Fruit Production

It is an oversimplification to say that currants, raisins, sultanas, and the like are dried grapes, although this is essentially the case (Figure 8–11). The history of raisins (the word is here used as a generic term) goes back over 3000 years; it is possible that the discovery was made accidentally when grapes were dried on the vine. Different

Figure 8–11 Raisins. Courtesy of the California Raisin Advisory Board.

varieties of grape originated from different regions of the Mediterranean area, and each was used to produce a raisin characteristic of that variety. For example, grapes for the muscat raisin were (and still are) grown in Malaga and Valencia, Spain, while grapes for currants were planted in Corinth and spread to the Zante region of Greece. Persia was the birthplace of the seedless grape from which sultanas are obtained. With time, viticulture spread to the New World as well as to the antipodes, and, today, raisin production is a worldwide occupation.

Practices vary slightly from place to place, but one essential requisite is a good climate: adequate sunshine, a suitable ambient temperature, and, especially during the drying period, a low rainfall, for the grapes are dried naturally and in the open air.

There are slight differences in terminology from the various growing areas; currants are universally derived from the small dark seedless Zante-type grape (the so-called Black Corinth of California), although other grapes with similar characteristics (e.g., the Australian Carina) are used. In Australia, sultanas are obtained from the Thompson Seedless grape as well as the sultana. In California, raisins are also obtained from the Thompson Seedless while the sultana grape yields sultanas as it does in Europe. Muscat raisins are known as such in Europe and America, while in Australia they are simply known as raisins. This last category is made from large grapes containing seeds, which may or may not be removed mechanically during the manufacturing process after drying. Muscat raisins have a much sweeter taste than the other types.

The processes after grape harvesting vary slightly from place to place; sometimes the grapes are dried naturally without any further treatment or they are pretreated to speed up the drying process. The "natural" raisins of California are produced without intermediate treatment, being spread on paper sheets laid between the rows of vines to dry (Figure 8–12). In 2 to 3 weeks, this reduces the moisture content of the berries

Figure 8–12 Drying raisins. Courtesy of the California Raisin Advisory Board.

from over 70% to about 15%, at which stage the paper sheets are rolled with the dried grapes inside to produce "bundles." These are left for several more days for further drying and to equilibrate the moisture contents of the grapes within the bundle. The grapes are then stored in large containers (which further assists with this equilibration) at the packing plant. Prior to removal from the bulk bins, samples are taken for quality approval (United States Food and Drug Administration standards are applied here). The raisins now pass through a cleaning procedure to remove leaves and large pieces of stems, with the small capstems being removed by a special machine in which the fruit is cascaded through a spinning, grooved, rubber-coated cone. Further cleaning is achieved by vacuum separator treatments, where the lighter stem material is removed. Following washing and grading, the raisins are packed into the packs from a few grams for a snack pack to bulk packs of around 12.5 kg for use by other food manufacturers. A basic flow diagram for the California process is given in Figure 8–13.

In Turkey, woven polypropylene sheets are used, and these are laid on cement or clay beds, often slightly higher than the surrounding ground to reduce the possibility of contamination of the fruit by stones, etc. In Australia, drying racks are constructed. These consist of 8 to 12 tiers made of wire netting within the rack, which itself is about 2.5 m high. The grapes are spread on the mesh at about one bunch deep for drying, either naturally or by the accelerated method outlined as follows (here the individual layers of grapes on the rack are sprayed with the alkaline oil treatment). There is some mechanical drying of the fruit carried out, often by placing racks of grapes in a tunnel and blowing heated air through. When dried to the correct moisture content, the raisins pass through the cleaning and packing processes as outlined previously (Figure 8–14).

For some products, extra process(es) may be introduced. In order to speed up the drying process, the grapes may be sprayed after harvesting with, or dipped in, a solution of potassium carbonate (between 2.5% [w/w] as used in Australia and 4.5% [w/w] as

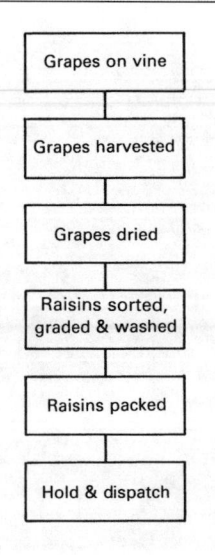

Figure 8–13 A flow diagram for the California process to produce raisins.

Figure 8–14 Bulk packing of raisins. Courtesy of the California Raisin Advisory Board.

used in Turkey) containing "dipping oil," the composition of which varies from place to place. In Australia, it is a mixture of the ethyl esters of fatty acids and free oleic acid used at 1.5% (w/w), while one establishment in Turkey uses 0.5% olive oil. This can reduce the drying period for sultanas from 4 to 5 weeks to 8–14 days. It was originally believed that this process removed the waxy bloom from the surface of the grape, but this is not the case. It is now thought that it modifies the structure of this waxy coat to allow greater moisture permeability and hence allows dehydration to proceed more quickly. It also appears to make the grape more transparent to infrared rays, allowing a better radiant heat uptake. The effect of the treatment is reversed by washing.

Because of their larger size, the muscat grapes are normally alkaline-treated before drying, and the treatment tends to be a little more severe than with the smaller types of grape. A relatively small amount of muscat grape is mechanically dried without the alkali pretreatment and is handled delicately to minimize damage and to preserve the surface bloom. The resulting raisins are called muscatels and are used for high-class outlets, such as health food stores.

Another process that may be applied to some sultanas (and raisins in California) is their treatment with sulfur dioxide. This bleaches the fruit to give a much more golden produce but can leave a residual amount of the preservative in the fruit (up to 2000 mg/kg).

8.3.2 Dried Tree Fruit Production

The drying methods for tree fruits are essentially the same as those for vine fruits (i.e, the sun drying of prepared material). Soft fruits such as apricots, pears, bananas, and plums are dried, as are dates and figs. Some of these fruits have to be treated with sulfur dioxide before drying to prevent spoilage, and this particularly applies to sun-dried cut fruits, such as apricot halves.

Fruit for drying is harvested in the normal way with care to minimize damage to the fruits. Dates and figs are dried in a similar manner to grapes, that is, on prepared cement or clay beds, although the use of trays is quite possible. Stone fruits such as apricots may be dried whole on stone, whole after having the stone removed or as stoned halves. Plums (to produce prunes) are normally dried whole on stone. Fruit for drying should be mature and fully ripe: underripe fruit does not produce such a flavorful commodity.

In Australia, fruit is dried, as with grapes, on trays, but these trays are of a softwood construction (hardwood imparts a staining to the fruit) and are stackable. The fruit is placed in a single layer on the trays for drying. At this stage some of the fruit types are treated with sulfur dioxide. Apart from preserving the fruit, this process softens the fruit tissues, facilitating a faster loss of moisture during drying. It also has an antioxidant effect and thus prevents enzymatic browning of the fruit.

For those fruits that are to be treated with sulfur dioxide, the process is quite simple. Trays of fruit are formed into stacks that are placed in purpose-built fumigation chambers, which are only a little larger than the stacks themselves. Sulfur is then burned in a draught channel under the chamber and the gas enters the chamber and treats the fruit. For legislative reasons (if for no other!), control of the degree of treatment is necessary. After fumigation, the fruit is sun-dried to a moisture content that is adequate for self-preservation when combined with the sulfur dioxide.

Some fruits, such as plums (for prunes), are dried without treatment with sulfur dioxide, and, here, the drying is carried out artificially using a process similar to that mentioned above for some types of grape.

The drying of fruit is a complex subject, and a whole book (rather than a small part of a single chapter) would be needed to do it complete justice.

8.4 FLAVORINGS FROM FRUITS

Many flavorings in present use are called *natural* or *nature-identical* (similar but synonymous terms are also used). Many of these comprise concentrated fruit juice (which in many circumstances is an excellent flavoring but which, in a list of ingredients, need not be referred to as such) or concentrated fruit juices that have been "boosted."

The subject of fruit juice concentration is dealt with in Chapter 5. During the evaporation process, vapors are generated that contain many of the essential flavoring compounds. These vapors are collected and concentrated by fractional distillation before being added back into the concentrate. The recycling may be at the natural ratio (i.e., the amount recycled is the same amount as that collected); sometimes the volatiles are added back at higher than natural ratios to produce a fortified concentrate.

Sophisticated chemical analysis is now able to identify these flavoring compounds, many of which have been synthesized for use in the manufacture of so-called nature-identical materials.

One of the most important sources of flavoring is the essential oil of the citrus fruits. These are extracted from the peels of the fruits (see Chapter 5) and are widely used, not only in food flavors.

8.5 TOMATO PURÉE

Is it a fruit or is it a vegetable? The humble tomato seems to fall between two stools, for while botanically it is undoubtedly a fruit, it is specifically excluded from some fruit juice legislation (for example the UK Fruit Juices and Fruit Nectars Regulations of 1977). It is often used more as a vegetable than a fruit. Tomatoes can be conveniently divided into two categories: the round plum type and the long Italian type. This latter type is more susceptible to spoilage by *Bacillus thermoacidurans*, and, for its treatment, higher processing temperatures are required. The term *tomato juice* is taken here as being the material obtained by sieving tomatoes after they have been chopped and preliminarily treated (see following). Tomato purée is taken as being concentrated tomato juice.

Much of the plant and machinery developed for tomato processing is very sophisticated, and only an outline is given here. Before the fruit is received at the factory, it is checked for quality, particularly for mold growth, an important factor affecting the quality of the end-product.

At the factory, the fruit is checked again and sorted to remove defective fruit and other extraneous material. It is then washed with potable water before being crushed or chopped (this is often termed *breaking* and is linked to one of three types of preliminary treatment). The treatment used depends on the intended use for the end-product.

The cold-break process involves holding the crushed or chopped fruit at ambient temperature for a period in order to destroy the natural pectin by enzyme action. This lowers the viscosity of the tomato juice and allows for higher concentrations of tomato purée to be achieved. In the thermal break, the fruit is chopped cold and the resulting pulp is heated immediately to 90°C before the juice is extracted. The time between chopping and heating governs the amount of pectin breakdown and so governs the viscosity of the end-product. Hot-break juice is achieved by heating the fruit to 90°C before chopping and pulping, the material being recirculated in order to ensure that the enzymes are inactivated before the pectin is broken down. The heating also releases a gummy substance that surrounds the seed to give a high-viscosity end-product, suitable for the ketchup industry.

The heated tomato pulp is then sieved to remove the seeds and skin as well as other fibrous parts of the fruit and any stalk that may be present. The sieving is achieved by passing the pulp through a series of sieves each with smaller apertures than the last, the final one having an opening of 0.4 mm. Figure 8–15 shows a flow diagram from fresh fruit to the juice.

Following the sieving operations, the juice is transferred to the evaporation equipment for concentration. The type of evaporator used will depend on the product required to some extent but is almost always carried out under vacuum. The scraped-surface type is mainly used, often with double or triple effect. The use of a scraped-surface plant allows a concentrate of a much higher viscosity to be prepared. It is not usual to recover

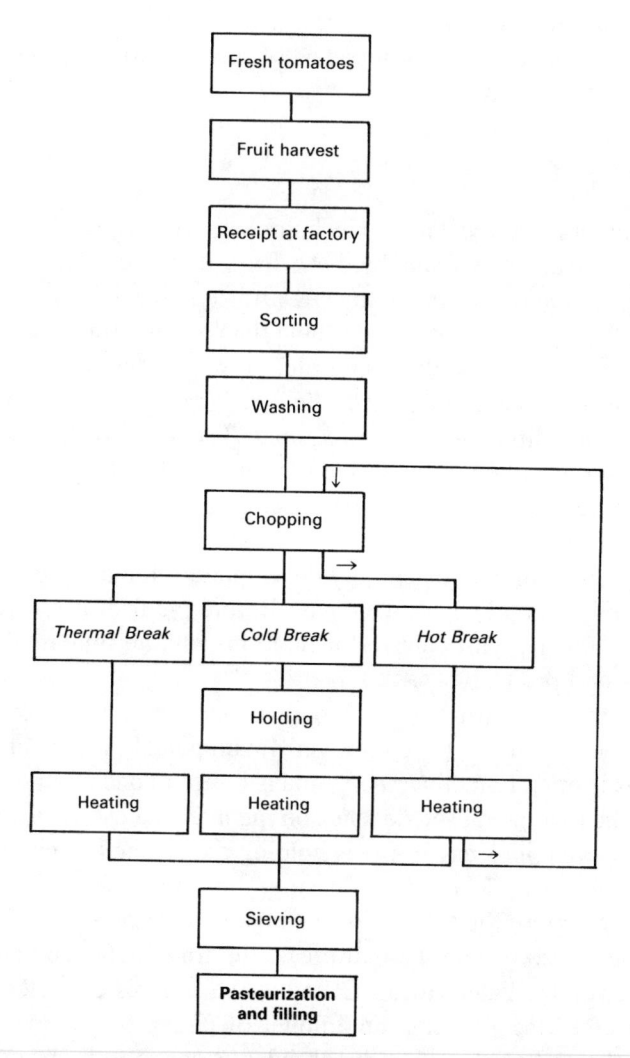

Figure 8-15 A flow diagram for preparation of tomato juice.

volatile aromas during tomato processing. Much concentrate is now homogenized, while a customer specification may require the presence of added salt. Reference to the plant manufacturers' data sheets will provide plant throughputs.

After concentration, the purée may be canned, when it will need to be heated to a suitably high temperature to achieve sterility, followed by rapid cooling to prevent caramelization. Alternatively, the purée may be handled aseptically and bulk filled into containers of 200 kg or more (Figure 8-16). Here the plant used for the evaporation is maintained in a sterile condition and the purée is cooled while still maintaining that sterility for filling into sterile containers. To improve its shelf life, such purée is often chilled at the completion of the processing.

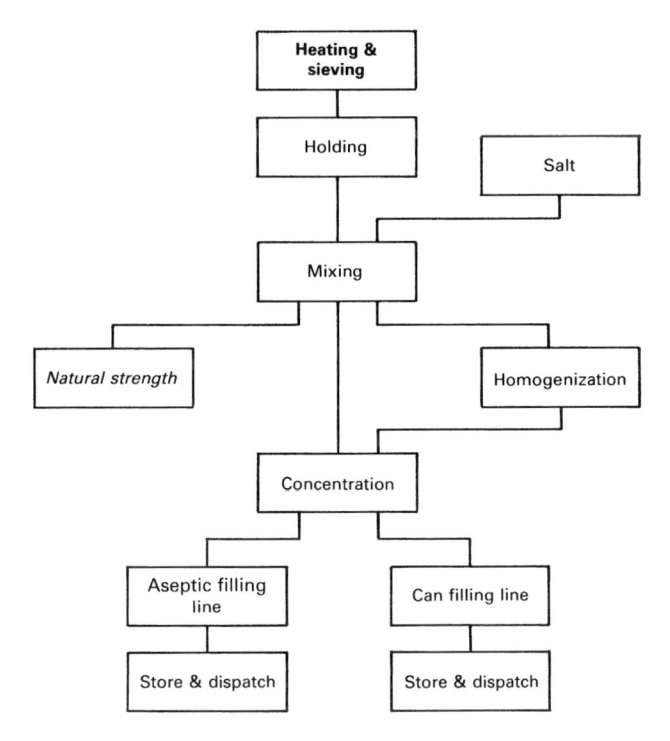

Figure 8-16 A flow diagram for the processing of tomato juice.

The heated sieved juice may be pasteurized as single-strength material for filling into a variety of packs; the evaporation temperatures achieved during concentration must be sufficient to pasteurize the purée.

ACKNOWLEDGMENTS

The author gratefully acknowledges assistance given to him by the following companies and other bodies, together with their staff.

APV Baker Ltd, Crawley, West Sussex RH10 2QB UK.
Broadheath Foods Ltd, Lower Broadheath, Worcester WR2 6RF, UK.
Buderim Ginger (UK) Ltd, Croydon CRO 4NH, UK and Buderim Australian Ginger, Yandina, Queensland, Australia.
CSIRO N. Ryde, N.S.W. 2113, Australia.
California Raisin Advisory Board Food Technology Program, PO Box 281525, San Francisco 94128-1525.
CPC UK Ltd, Esher KT10 9PN, UK.
Pagmat, S.A. Atatürk Cad., No. 150/2 (35210) Izmir, Turkey.
RHM (Ledbury), Ledbury HR8 2JT, UK.
Thomas J. Payne Market Development, 3242 Jones Court, N.W., Washington, DC.
Tesco Stores Ltd, Cheshunt EN8 9SL, UK.
Campden & Chorleywood Food Research Association, especially Members of the Library Staff, Chipping Campden, Gloucestershire GL55 6LD, UK.

SUGGESTED READING

Atkinson, F.C., and Strachan, C.C. (1941). *Candying of Fruit in British Columbia with Special Reference to Cherries*. Contribution no. 573, Horticultural Division, Experimental Farms Service, Experimental Station, Summerland, BC. Canada Department of Agriculture.

Atkinson, F.C., and Strachan, C.C. *Preparation of Candied Fruits and Related Products*, Part 1. Fruit and Vegetable Processing Laboratory, Experimental Station, Summerland, BC. Canada Department of Agriculture.

Campbell, C.H. (1937). *Campbell's Book*. Vance Publishing, New York.

Goose, P.R., and Binstead, R. (1964). *Tomato Paste and other Tomato Products*. Food Trade Press, London.

McBean, D.M. (1976). *Drying and Processing Tree Fruits (Division of Food Research Circular No. 10)*. Commonwealth Scientific and Industrial Research Organisation, NSW, Australia.

Ranken, M.D., ed. (1988). *Food Industries Manual*, 22nd ed. Blackie, Glasgow, Scotland.

Ranken, M.D., and Kill, R.C., ed. (1993). *Food Industries Manual*, 23rd ed. Blackie, Glasgow, Scotland.

Rausch, G.H. (1950 and 1965). *Jam Manufacture*. Leonard Hill (Blackie, Glasgow, Scotland).

Various. (1982). *Grape Drying in Australia*. Dried Fruits Processing Committee, Commonwealth Scientific and Industrial Research Organisation, N. Ryde, NSW Australia.

Various UK Legislation Statutory Instruments, Her Majesty's Stationery Office, London.

Chapter 9

Packaging for Fruit Products

John Bettison

9.1 INTRODUCTION

A wide range of packages containing fruit and fruit products can be seen in the marketplace, and packaging development continues in an endeavor to reduce costs, to improve performance and convenience, and to increase recyclability of the package after completion of its primary function. Demographic studies and forecasts indicate that an increasing proportion of the population will live alone, and this will require more emphasis on single-portion packages and, for the elderly, emphasis on easy-to-open packaging.

The trend toward cost reduction has been led by a reduction in the actual weight of the materials used in the manufacture of containers/closures. We have seen the introduction of thinner steel for cans and closures and lightweight glass bottles, where reductions of up to 30% in glass usage have been achieved.

The recycling of discarded packaging materials is also increasing; some of the plastics materials so collected may be incorporated as reclaim into nonfood contact areas of new plastics containers. Recycled glass containers form the ideal raw material for new glass containers—they require less energy for remelting than raw materials taken from the earth's surface—and new glass food containers now incorporate more than 50% of recycled glass.

The story of packaging therefore appears healthy, but there are always problems on the horizon that require resolution. There are continuous pressures on packaging suppliers from environmentalists, consumers, and legislators to ensure, for example, that no contamination of the product could arise from the extraction of potentially harmful substances out of the packaging material itself. Questions have been raised about substances present in protective lacquers used on metal containers: many of the problems that arise from the use of such materials are simply the nonconformance with irrelevant regulations or delays in achieving approval thereof. Packaging suppliers are well aware of their responsibilities to their customers and to the final consumers, and all such issues are being resolved.

Apart from their involvement in the above activities, packaging manufacturers regard themselves as part of the food industry, and standards of manufacture and hygiene in their plants must be comparable to those in food manufacturing plants. The necessity for packaging manufacturers to gain ISO 9000 Series accreditation (Quality Systems and the Management of Quality Systems) in order to demonstrate to the food manufacturers their approach to quality has, to some extent, been overtaken by the need for packaging suppliers to demonstrate that HACCP (hazard analysis critical control point) principles are also applied in their operations. HACCP principles are now widely recognized as most appropriate to ensure product safety. Their essential aim is the identification, management, and prevention of risk, where risk implies risk to the health of the consumer of the food product. It has become a requirement of food producers under a European Council Directive (the Hygiene Directive 93/43/EEC) to identify and control any activity that is critical to food safety. Suppliers of packaging materials, sealing equipment, etc., to the food industry must likewise understand these principles and apply them during the manufacture of such items.

The packaging user also has a responsibility to ensure that the packaging materials required are obtained from a reputable supplier, to an agreed specification and quality level. Packaging users should take care to ensure that the packaging materials are stored under the conditions specified by the supplier, are used in batch sequence order, and records are kept of usage.

9.2 OBJECTIVES OF PACKAGING

The principal objective of food packaging is to hold the product—ideally in the condition in which it was prepared by the food manufacturer—until it is ready to be consumed. In order to achieve this objective, a combination of properties is required; unfortunately, no one package possesses them all. The most desirable properties demanded from the package, within the declared shelf life period, are the following:

- It must hold the product in such a condition that it is safe to eat.
- It must maintain its integrity and be resistant to normal handling abuse.
- It must protect the contents from the unlawful or deliberate tampering with the closure or body of the package and, if subjected to such abuse, it should be able to indicate that fact.
- It should hold the product, in optimum condition, free from contamination by external odors, flavors, or taints, and the nutritional value of the product must be retained as declared on the label.
- It must be made from nontoxic materials and should not be altered in any way by the product; it should neither take up substances from the product, nor migrate into the product itself.

- It must be capable of withstanding the processing and storage conditions (heat/cold/gas mixtures) to which it will be subjected.
- It should be capable of being handled on high-speed packing lines without suffering failure or abuse, and it should lend itself to automatic in-line testing/control systems designed to remove any failed/abused packs.

Apart from its protective role, the package may also assist in selling the product because of its appeal to the consumer, both aesthetically and from the point of view of economy and convenience in use. In order to achieve these objectives, the following features are to its advantage:

- It should be able to display the product, either visually or by illustration, yet leave sufficient space for nutritional information.
- It should be easy to open, if opening instructions are followed.
- It should be recyclable in some form and it should not present a hazard with regard to its ultimate disposal.
- It may be of a distinctive shape.

Fruits offer almost unlimited possibilities to the imaginative food manufacturer for the preparation of attractive, tasty products. In an attempt to classify this wide range of products so that they may be grouped in a sensible manner to permit their packaging materials/container needs to be discussed, they are divided into four families, as follows:

1. heat-preserved fruit products/fruit-containing beverages (including some that may be carbonated)
2. frozen fruit products
3. dehydrated fruits
4. fresh/chilled fruit and fruit products

9.3 HEAT-PRESERVED FRUIT PRODUCTS/BEVERAGES

The most important are glass jars/bottles, metal cans, semirigid plastics containers, foil-sealed portion packs, and collapsible tubes. A final note will deal with containers used for the aseptic packing of fruit products.

9.3.1 Glass Containers

Properties of Glass

As a container for heat-preserved fruit products, glass is perhaps the perfect choice; it is inert, impervious to odors and gases, and resistant to attack by chemicals originating from fruit products. Glass containers are normally transparent, allowing the consumer the opportunity to examine the product before purchase. The basic disadvantage of glass containers is their fragility, both when physically abused and when subjected to sudden changes of temperature in excess of 50–55°C (thermal shock); however, the glass container itself is eminently suitable for use in microwave ovens, and this adds to its image of convenience.

Manufacture of Glass Containers

Glass containers for use with heat-preserved fruit products are normally made from clear, or flint, glass. A typical mixture of raw materials for glass container manufacture consists of 60% silica sand, 18% soda ash, 17% limestone, and 5% mineral additives, bearing in mind that such raw materials make up 50% or less of the actual batch being fed to the furnace, the remainder being recycled glass. Almost all container glass is melted continuously in regenerative furnaces, in which the ingredients slowly enter at one end and molten glass is slowly withdrawn from the other (Figure 9–1). The operating temperature within the furnace is around 1500°C, and it runs continuously. The molten glass is conveyed from the furnace to the forming machines via a forehearth and feeder, which forms the glass into gobs (beads of molten glass of even weight and uniform temperature around 1100°C), governed by the size of the container being made, and distributes these to the forming machines.

Modern multicavity versions of the IS (independent section) container-forming machines (Figure 9–2) can now produce containers at speeds in excess of 300 per minute and handle over 100 tonnes of glass per day. The gob is loaded into a "parison" mold, which forms the "blank" shape of the container, and is then transferred to a blow mold before being blown to its final shape. There are two processes for making containers, namely "blow-and-blow" and "press-and-blow," and the diagrams show the differences in the two methods. The press-and-blow technique permits more accurate control of the glass and therefore better lightweighting. Narrow-necked bottles, which traditionally were always manufactured by the blow-and-blow process are now switching to the press-and-blow technique.

Containers leave the forming machine at a temperature of around 650°C and, if allowed to cool rapidly, would develop internal stresses and be more susceptible to impact damage. A temperature-controlled tunnel or "annealing lehr" slowly reheats and cools them at a predetermined rate. Lightweighting of glass containers, and the increase in packing line speeds, has necessitated the increasing use of glass-surface treatments in order to permit their smooth flow through such high-speed lines and improve their abrasion resistance. "Hot end" treatments, applied between the forming machines and the annealing lehr, normally consist of spraying the hot containers with a vapor of organic titanium or inorganic tin compounds. "Cold end" treatments are applied to the containers at the exit of the annealing lehr and normally consist of a polyethylene-based dispersion, which increases the lubricity of the containers.

For every type of closure used on glass containers, there is an appropriate glass neck ring (Figure 9–3). The glassmaker will fit the appropriate diameter finish demanded by the closure.

Closures for Glass Fruit Containers

Closures for glass fruit containers are normally made from tinplate, tin-free steel (TFS), or aluminium, lacquered or coated both internally and externally. They are usually provided with a ring of lining compound, which in some cases also extends down the skirt of the closure.

Closure Manufacture. The manufacture of metal closures basically consists of the following operations:

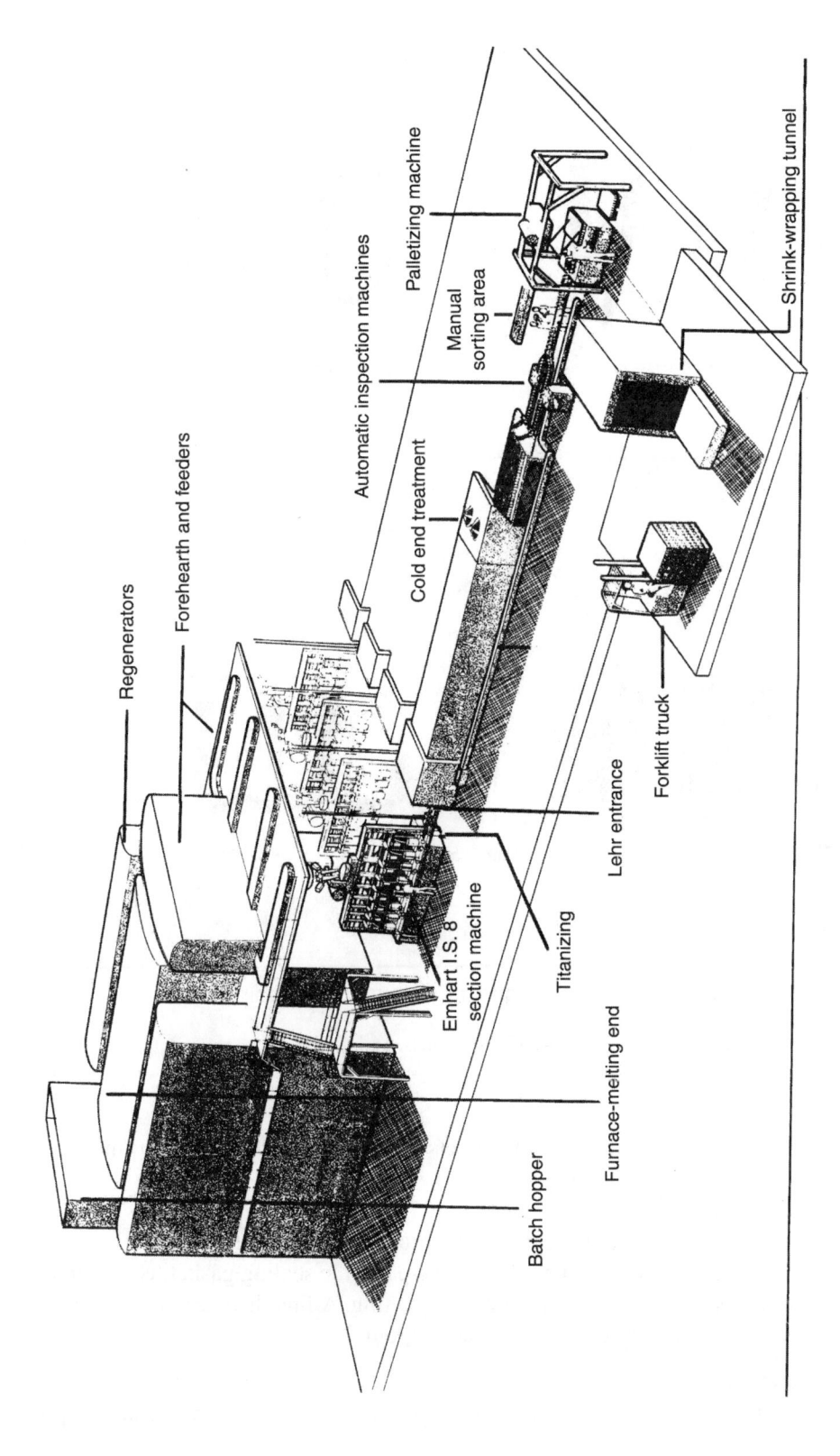

Figure 9–1 A typical furnace. Courtesy United Glass, Ltd.

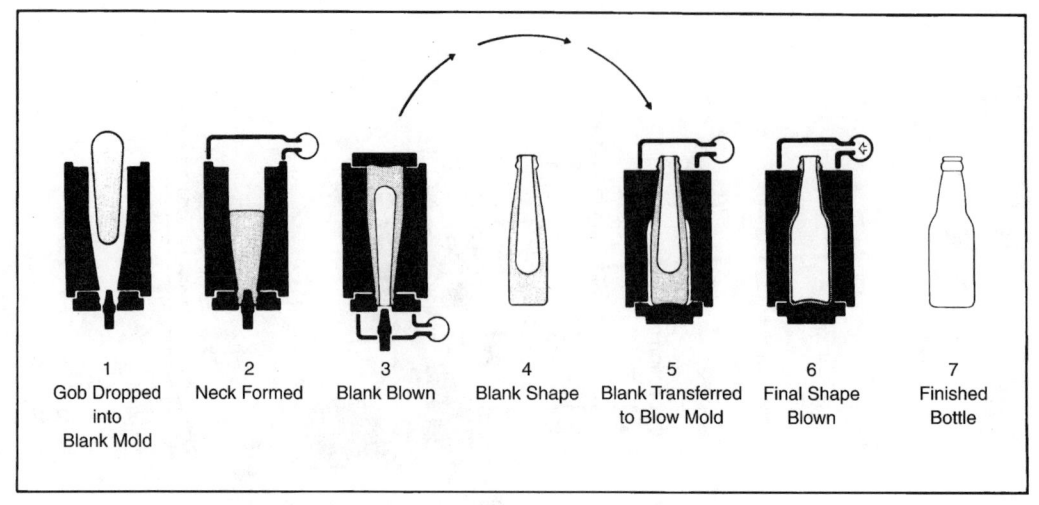

Figure 9–2 Glass container forming. Courtesy United Glass, Ltd.

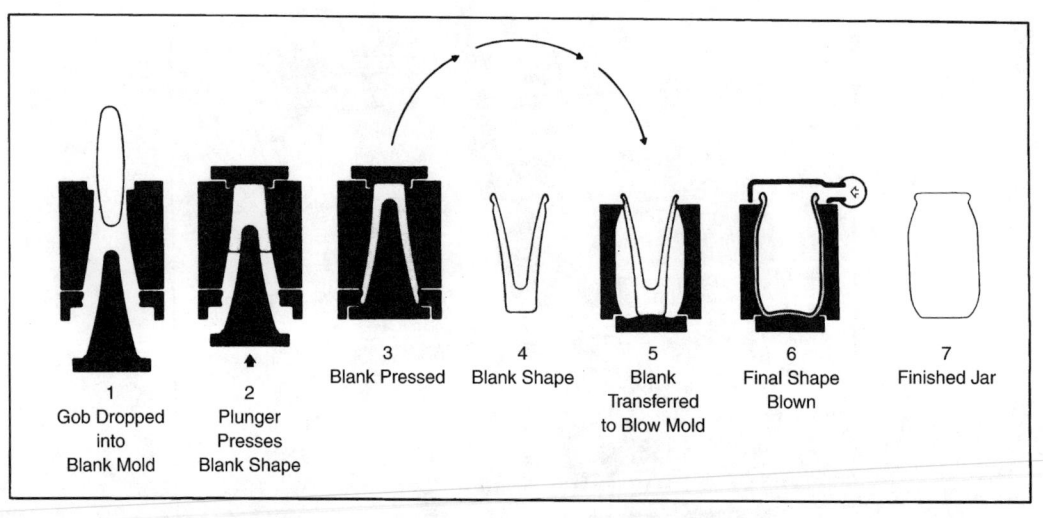

1. The coils of metal are cut into sheets of convenient size for handling through the printing and lacquering units.
2. These sheets are printed and lacquered.
3. The printed, lacquered sheets are then cut into discs, pressed, and curled into shape; liquid lining compound, which creates the sealing gasket, is then injected into the shell of the cap while rapidly rotating. A final heat treatment is given in an oven in order to cure and set the compound.

The vast majority of closures for wide-mouth glass containers for heat-processed fruit products are of the nonventing type. A nonventing closure is normally applied

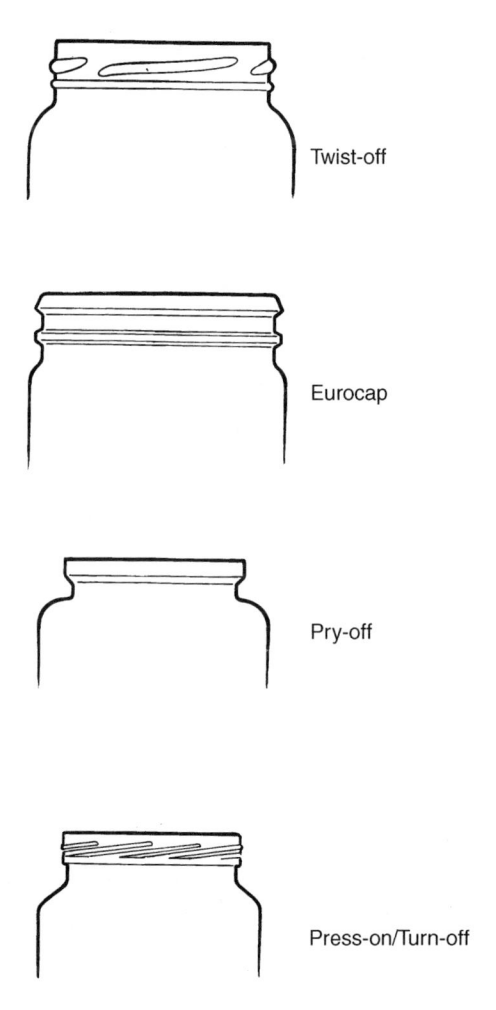

Figure 9–3 Typical glass neck ring finishes.

to a jar under vacuum, thereby giving an immediate hermetic seal, and the closure is maintained in this fully sealed condition throughout subsequent heat treatments by the application, where necessary, of a controlled overpressure to more than balance the internal pressure developed by the product within the jar itself. Calculation charts exist that permit the calculation of the internal pressure developed. When pasteurization is performed in a tunnel open to the atmosphere, it is not possible to impose an overpressure on the containers; under such conditions, the packer has to ensure that no internal pressure will be developed during pasteurization. When filling/packing conditions are such that an internal pressure will be developed, it is necessary to pasteurize the containers in an overpressure retort and apply the relevant overpressure, or change the packing conditions, for example, increase the headspace.

The tinplate/tin-free steel (TFS) twist (TO) style closure is available in diameters between 27 and 110 mm; it is applied to multithread glass finishes and can be removed simply by rotating on opening. They are being applied at speeds of up to 800 jars per minute. For preserves, whole/sliced fruit, and similar products, twist closure diameters of 58, 63, 66, and 70 mm are commonly used for 0.5-kg or smaller capacity jars, whereas 82 mm is normally adopted for the larger, 1-kg jars. For fruit juices, 30-, 38-, 43- and 53-mm diameters are normally used. The 43-mm closure is also popular for minijars of preserves (hotel/airline single portions). The majority of these closures incorporate a button feature in the center panel of the closure itself, which indicates when the closure is satisfactorily sealed (button down) and when it is not (button up).

The tinplate/TFS press-twist (PT) closure differs from the twist closure in that it is applied simply by pressing it down onto the jar without any rotary action; the placement of the lining compound down the skirt of the closure permits it to take up the shape of the threads on the glass neck ring during pasteurization, and it is thus possible to screw it off. Such closures are used in 40-mm diameter for fruit juices, and in 51-mm diameter for fruit-based baby foods; cap application speeds in excess of 1300 per minute are being achieved, and some of these closures are fitted with a plastics tamper-evident band, which ruptures on opening to leave a ring of plastics around the neck of the container. Other types of closure being used on glass containers/bottles for a variety of fruit-based products include the following:

- *Crown:* This is a crimped-on tinplate/TFS pressure-resistant closure, found on some glass bottles of fruit juices and cider. It is cheap, effective, and easy to apply, but it requires an opening tool.
- *Cork:* Traditional cork stoppers, sometimes wired onto the neck of the bottle, are still found on glass bottles used for some strong ciders. More recently, cellular plastics corks have been introduced.
- *Plastics screw closure:* Such prethreaded closures may be made from high-density polyethylene or polypropylene, some of which are fitted with a flexible plastics liner and almost all of which are fitted with a tamper-evident band that ruptures on opening. They are normally of small diameter (26–38 mm) and fitted to bottles containing cider, fruit-based ice-cream toppings, etc.
- *Aluminium roll-on closure:* Such closures, normally of 26- to 28-mm diameter for application to narrow-neck glass bottles of cider and similar beverages, are supplied as a smooth, thread-free shell; rollers on the capping machine rotate to form the screw thread. Many such closures now incorporate a tamper-evident band, when they are referred to as ROPP (roll-on pilferproof) closures.

The Future for Glass Containers/Closures

Because of their inherent properties, glass containers appeal to the producer of high-quality, attractive fruit products and lend themselves to reclosure/reuse with such products as preserves, jams, and sauces, and their recyclability is universally accepted. The continued development of lightweighting will result in long-term cost savings; closure manufacturers are similarly involved in reducing costs, increasing the ease of opening, and creating novel tamper-evident features.

9.3.2 Metal Cans

The most popular materials used for the manufacture of metal cans are tinplate, tin-free steel (TFS), and aluminium.

Tinplate

Tinplate used to manufacture cans and closures is produced by electrolytically depositing a thin coating of tin onto each side of a coil of steel passing through an electroplating bath; this permits the deposition of different amounts of tin on either side of the steel. Tinplate used for the manufacture of fruit cans may have a tincoating weight of 2.8–5.6 g/m^2 both internally and externally, although some unlacquered or partially lacquered cans may carry higher internal tincoating weights up to 11.2 g/m^2. The amount of steel used in can manufacture has also been reduced by an additional cold-rolling process (double-reduced plate). Mechanical methods of strengthening the bodies of metal cans exist, such as the incorporation of circumferential beading or longitudinal panels.

Tin-Free Steel

If tinplate may be regarded as 99% steel and 1% tin, then tin-free steel (TFS) could be regarded as 99.9% steel and 0.1% chromium. In all physical and mechanical respects, therefore, it behaves in a manner similar to tinplate, and it is only in its surface chemical performance that it differs. It is now widely used for the manufacture of caps and closures for fruit products, but its use for the ends seamed onto tinplate fruit cans is still not common.

Aluminium

In comparison to steel, aluminium is about one-third the weight and a much more ductile material. An alloy of pure aluminium with manganese and magnesium is required to impart strength to the material before containers and components suitable for use with, for example, cider beverages, can be prepared.

Lacquers

Although metal cans were manufactured for many years without lacquers, legislative pressures and consumer demands, as well as economic considerations, have led to the wide adoption of fully internally lacquered metal cans for some fruit and fruit products. However, because of the advantages that exposure to tin gives to white fruit (e.g., pears) with regard to color and flavor retention, the use of plain-bodied, or partially lacquered, tinplate cans is still widespread. Lacquers for cans and components are normally applied to the flat sheets of metal on high-speed coating/lacquering units. As an alternative to lacquering, the lamination or direct extrusion of plastics films to either or both sides of tinplate, TFS, or aluminium is now available, with potential uses for can and component manufacture.

Can Manufacture

There are three common methods of making cans suitable for the packaging of heat-processed fruit products and carbonated beverages, namely:

1. the three-piece can with a welded side-seam body
2. the two-piece draw-redraw (DRD) can
3. the two-piece draw wall-ironed (DWI) can

The first, which is still the most widely used technique, is at present applicable to tinplate, whereas the DRD technique is suitable for tinplate, TFS, and aluminium. The DWI technique is used with tinplate or aluminium.

Three-Piece Welded Can. After lacquering/printing of the sheets of tinplate, they are fed into a slitter, which cuts them into rectangular pieces, or body blanks. A bodymaker forms these into cylinders, and the two edges of this cylinder are then drawn together so that they abut with minimum overlap. The edges are cleaned and immediately electrically welded together. A coating of lacquer, known as the side-stripe, is applied internally by roller or spray over the welded seam and cured in a continuous oven. These cylinders are beaded or longitudinally paneled where required, flanged, and conveyed to a double seamer, where one end of the can (known as the makers end) is seamed onto the cylinder. The ability to change the traditional straight-walled shape of the can itself is significant in that (1) it gives the container a distinctive shape, and (2) it permits the necking-in of the body cylinder at one or both ends, allowing the use of a smaller-diameter, and therefore cheaper, end on the can.

Endmaking likewise commences with cutting strips from the lacquered sheets of plate by a scroll-shear unit, stamping out and forming ends from the lacquered circles therein and lining them with a compound in order to create a hermetic seal with the can body cylinder during the double-seaming operation. The manufacture of full-aperture easy-open ends requires closer control of plate specification and lacquering in order to achieve consistent performance. The use of such ends is increasing because of the convenience of opening without the need for a special opening tool. Three-piece welded cans are available in a wide range of diameters and capacities, ranging from 140 to 4150 mL (see also Table 7–4). They serve as containers for whole fruits, sliced fruits, fruit salads, pie fillings, and many other fruit-based products, and they have a long history of reliability.

Two-Piece DRD Can. Here, the body and base of the can are made from a circular disc of metal, coated/lacquered on both sides. The thickness of the body and base of the finished can is identical to the original thickness of the disc itself. The disc is subjected to a series of cupping/drawing operations that reduce the diameter and increase the height, until finally a flanged can complete with bottom end is produced. At the present time, DRD cans are not widely used for fruit products, although the tapered pie can, of dimensions 153×32 mm and formed by a simpler drawing operation, is used for apple pies and similar products.

Two-Piece DWI Can. Here, a shallow cup is formed from a disc of relatively thick plate, and the body of the can is then ironed (stretched and thinned) through concentric metal rings until the final height of the can is achieved. The can body is then trimmed, necked-in, and flanged prior to internal spray lacquering and external decoration/printing. The finished can has a base equal in thickness to the original metal, whereas the body is typically one-third of that thickness. Such cans, made either from tinplate or aluminium of 33 cL or 0.5 L capacity and closed with aluminium easy-open ends, are usually used for the canning of cider, a carbonated product. Similar cans are also used for fruit-flavored carbonated soft drinks, many of which contain a percentage of fruit juice. Because of the relatively weak body wall that results from the DWI manufacturing operation, the use of such cans for foods, even when strengthened by the presence of circumferential beading, tends to be limited to markets where the product is relatively solid and the vacuum level is low. Studies, however, continue regarding the feasibility of using, for example, liquid nitrogen to provide an internal pressure to support the can bodies from collapse. Small-diameter, 15-cL capacity, aluminium DWI cans are also being used successfully for a range of noncarbonated fruit juices.

Corrosion of Tinplate Cans

Internal Corrosion. The relatively low pH/high acidity of fruit products renders them ideal for attack on exposed metal within the container. Apart from the problems that this may cause to the container (pinpoint attack may lead to perforation of the metal and leakage of the container; the creation of hydrogen gas may cause the can to blow or swell), the presence of the by-products of corrosion may cause deterioration of the product—metallic off-flavors, discoloration of fruits containing anthocyanin pigments, and/or the creation of excessively high levels of metallic contamination leading to a (legally) unsalable product. If the fruit product is packed in a plain or partially lacquered can, tin dissolution will initially be high, but this should not normally result in any significant attack on the iron/steel; the can will not therefore perforate, and the product within the can will remain sound. When packing colored fruits, a fully internally lacquered can must be used. Such a can will remain sound as long as the lacquer coverage is perfect; where lacquer coverage overall is good but there are tiny pinpoints of exposed metal, for example, due to scratches, the product will concentrate its attack at these points, causing pitting of the plate that may eventually result in perforation of the can. The corrosion of metal cans in the presence of fruit products is complex and would require a chapter on its own for a full understanding of the issues involved.

External Corrosion. External corrosion of metal cans, in particular the rusting of tinplate containers, can be minimized by controlling the water treatment during cooling and the correct drying of the cans after heat treatment. Cooling water should comply with European Council Directive 80/778/EEC for water used for human consumption and, where this is achieved by the use of chlorine, the total chlorine level should be maintained at 5 ppm or less in order to minimize its corrosive effect on the can. Positive steps should also be taken to dry the external surfaces of the can as quickly as possible, either by air jets positioned in the line and/or by avoiding overcooling of the cans themselves, which will slow down the rate of natural cooling.

The Future for Metal Cans

The continued development of methods of reducing the raw material inputs into can manufacture will ensure that it has an economic future, and easy-open end developments will increase the convenience of the package to the consumer. Their inherent strength and reputation for reliability will continue to serve them well. Shaping of cans will increase, as will the recycling of used cans. The introduction of plastics-laminated steel/aluminium materials may lead to a new lease on life for can makers, and such cans are already in use for processed food products.

9.3.3 Rigid/Semirigid Plastics Containers

Plastics Materials

In order to give satisfactory performance as a container that may be pasteurized after filling, the packaging material must possess both good heat resistance—normally 85–95°C for pasteurized fruit products but lower, 60–65°C, for carbonated beverages—and adequate barrier performance, particularly against moisture and oxygen transmission. Unfortunately, no single plastics resin offers the ideal combination of the required characteristics, and so it is necessary to combine resins, normally by coextrusion, in order to approach this ideal. Among the wide range of plastics materials now available, those that at present find widest appeal in this packing area are as follows:

Polypropylene (PP). This has a high softening point (140–150°C), but its resistance to oxygen transmission is poor. It can be blow molded and injection molded, although it is relatively difficult to thermoform.

Polyethylene Terephthalate (PET). It may be blow molded into high-clarity bottles and tubs. Its gas-barrier properties are good, but its heat resistance is relatively poor unless a heat-setting technique is used.

Polyethylene (PE). This is available as a low-density (LDPE), medium-density (MDPE), or high-density (HDPE) resin. It is widely used as a packaging film but also, in MDPE and HDPE form, for the injection molding of closures and for the blow molding of bottles. It has relatively good resistance to moisture but poor oxygen-barrier properties. As an HDPE bottle, it has sufficient heat resistance to accept hot filling or pasteurization.

Polyvinyl Chloride (PVC). This is available in plasticized or unplasticized (UPVC) form; as UPVC, it has better gas-barrier properties than PE and is easily thermoformed to produce rigid trays.

The gas-barrier performance of a plastics container may be improved by coating it with polyvinylidene chloride (PVdC) or by incorporating PVdC or ethylene vinyl alcohol copolymer (EVOH) within a coextrusion.

Plastics Container Manufacturing Techniques

Three basic techniques are used for the manufacture of rigid/semirigid containers:

Injection Molding. Here, the plastics granules are softened, by pressure and heat, and injected into a cavity mold, where the material sets on cooling. Accurate control of shape and dimensions is achieved. It is commonly used for the manufacture of high-quality pots, tubs, closures, and goblets.

Blow Molding. Here, a thick-walled tube or parison is extruded from the molten plastics material. This parison is transferred to the final-shape mold and blown, while hot, into the shape of the mold—an almost identical technique to the blow-and-blow glass container–making process. It is normally used to make bottles, although some wide-mouth jars are produced in this way.

Thermoforming. Here, a sheet or continuous web of plastics is heated until it softens. It then passes over a platen containing a series of mold cavities, where it is clamped in place and forced into the molds by top pressure, or attracted thereto by vacuum applied beneath the cavities. The molded sheet or web exits from the cavities and the containers therein are cut out. It is normally used to make plastics tubs, trays, pots, and lids. However, it may form part of a complete form-fill-seal operation where, for example, portion-pack containers of preserves/jams are formed, filled with product, and lidded (heat-sealable plastics film or coated aluminium foil) before the containers are cut out from the plastics web.

Plastics Container Closing Techniques

After filling, containers are normally covered with a heat-sealable lidding material. Heat sealing is achieved by a combination of pressure and heat, and, if necessary to achieve increased shelf life, gas flushing with nitrogen or evacuation of the headspace air by vacuum may occur just prior to sealing.

Hot-Filled Barrier Plastics Containers

With these containers, which are common as slab-sided squeeze bottles for products such as tomato ketchup and fruit-flavored dessert/ice-cream toppings, the required barrier/shelf life performance is achieved by the incorporation of a thin layer of EVOH or PVdC within the basic bottle material (PP or HDPE). The bottles are manufactured using a coextrusion blow molding technique, whereby a multilayer parison is produced by the use of two or more extruders coupled to one die. Such containers, for example, may use a six-layer structure, as follows: PP/reclaim/adhesive/EVOH/adhesive/PP. After hot filling, the necks of the bottles are closed with a heat seal–coated aluminum foil disc beneath a plastics dispensing closure, heat sealing normally being performed through the closure itself by induction (heat) sealing. The principal advantages of such bottles are the retention of the essential oil aromas/flavors, the convenience of dispensing the correct amount of product only, and a shelf life of up to 15 months with fruit-based products.

Carbonated Beverage Bottles

These bottles, made from PET by the blow molding technique and used for products such as cider in capacities up to 3 L, are closed with a tamper-evident plastics screw cap or aluminum ROPP closure. In order to achieve the required shelf life, they are normally coated externally (by dipping, spraying, or roller coating) using a water dispersion of PVdC.

Monolayer Plastics Containers

The shelf life performance of such jars and tubs is limited by their gas/moisture barrier properties and their resistance to heat, and they have found widest application with relatively short-shelf-life products outside the heat-processed world, such as fruit-flavored yogurts and similar refrigerated desserts. Increasing in popularity are bicompartmented trays, manufactured from polystyrene (PS) on, for example, Hassia form-fill-seal machines, which hold the fruit component separately from a yogurt/custard component and require similar refrigerated storage.

Some injection-molded, crystal-clear PET tubs, closed with a heat seal–coated aluminum foil lid and plastics snap-on reseal cover cap, are being used for shelf-stable jams/preserves in mainland Europe, and transparent PET wide-mouth jars are also being used for mincemeat and hot-filled jams/preserves with a controlled shelf life, the containers being closed with a metal twist screw cap.

9.3.4 Aluminum Foil Trays

Such semirigid trays are normally manufactured from aluminium foil 0.1 mm or less thick, bearing a PP coating 0.015 mm thick, by a simple pressure-forming technique. They have been used as containers for a limited range of fruit products, including portion-packed jams/preserves, handled on horizontal form-fill-seal units, and closed with a heat seal–coated aluminium foil lid 0.05 mm thick.

9.3.5 Collapsible Tubes

Collapsible tubes are normally manufactured from aluminum, although coextruded barrier plastics laminates are growing in some areas. Typical fruit products packed in this way are tomato paste, fruit sauces, and toppings, the products being in a shelf-stable condition under ambient or chill conditions. Aluminum tubes are manufactured by extrusion, wherein a "slug" or disc of aluminum is forced through a die. The tube resulting therefrom is trimmed to the desired length and internally lacquered. The screw cap or other closing device is applied to the threaded nozzle and the tube is then ready to be filled through its base. After clean filling with product, normally preheated to achieve commercial sterility and cooled, the open end of the tube is folded over and crimped/heat sealed to produce a hermetic seal.

9.3.6 Aseptic Packaging

Aseptic packaging is widely used for a range of fruit products, for example, tomato paste, fruit desserts and puddings, and fruit juices. The increasing ability to fill particulate products is significant and may open new markets for products containing whole soft fruits.

Retail-Size Containers

The most widely used container for fruit juices is the Tetra Brik carton, usually in capacities of 20 cL, 0.5 L, and 1.0 L. The Tetra Pak system (Figure 9–4) uses reel-fed, preprinted, precreased laminated material of the following specification: PE/printing ink/paper/PE/aluminum foil/PE/PE. In a typical Tetra Brik aseptic packaging machine, the web of material is softened and sterilized by hot hydrogen peroxide, and it then enters the top of a sterile chamber. As it descends, it is formed into a cylinder or tube and sealed longitudinally. The product is introduced just prior to this point, and the transverse seals are formed by a set of jaws, which press the tube together while

Machine functions

1. Special trolley with hydraulic lift for handling packaging material.
2. Packaging material.
3. Motor-driven roller for smooth, even feed of the packaging material.
4. Idler. Its action starts and stops the motor-driven roller.
5. Strip applicator which applies a plastic strip to one edge of the packaging material. Later, at the longitudinal sealing stage, this is welded to the opposite edge. The result is a tight and durable seal.
6. Deep bath of heated hydrogen peroxide.
7. Rollers which remove the hydrogen peroxide from the packaging material.
8. Nozzle through which hot, sterile air is blown to dry the packaging material.
9. Product filling pipe.
10. Element for the longitudinal seam which welds together the two edges of the packaging material.
11. "Short-stop element" which completes the longitudinal seam when the machine restarts after any brief halt in production.
12. Photocells which control the machine's automatic design correction system.
13. Transverse sealing is performed by two pairs of sealing jaws, operating continuously.
14. When the filled packages have been cut from the paper tube they fall down into the final folder.
15. In the final folding unit the top and bottom flaps are folded over and sealed onto the package.
16. The finished product is conveyed out of the machine.
17. Pivoting control panel.
18. Easy-access compartment for topping up the oil in the central lubrication and hydraulic systems and the detergent for the automatic cleaning system.
19. Date-stamping unit.
20. Splicing table for packaging material.
21. Bath which fills with water and detergent automatically for external cleaning of the machine.

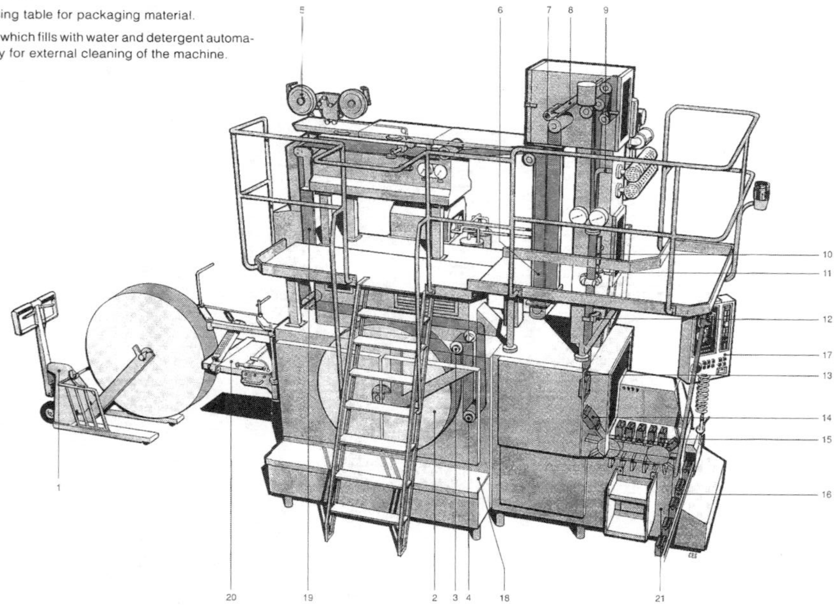

Figure 9–4 Tetra Brik TBA/9 aseptic packaging machine. Courtesy Tetra Pak, Ltd.

squeezing product from the seal area, and then make the seal. The filled pack is cut off and finally formed into its rectangular Brik shape. The filling area, after sterilization, is maintained in a sterile condition by a continuous overpressure of sterile air.

Another popular system is Combibloc, which differs from Tetra Pak by using preformed cartons. Because each carton is sterilized and filled separately, it contains a headspace, which is absent in the Tetra Brik carton.

Other retail-size aseptic packs, the wide range of aseptic plastics container systems available for pots, tubs, trays, and bottles, may be grouped together into four basic types.

1. *Preformed monolayer containers:* The shelf life of such containers is limited by the barrier performance of the material being used, for example, PP or PS. After hydrogen peroxide sterilization, the pots are aseptically filled and finally sealed with a presterilized coated aluminum foil diaphragm.
2. *Preformed multilayer barrier containers:* These are handled on the same machines as above, but offer a shelf life of 12 months or more depending upon the barrier materials being used.
3. *Form-fill-seal multilayer barrier containers:* These use conventional form-fill-seal units with the addition of hydrogen peroxide immersion baths for the base and top webs. The thermoformed containers are filled in a sterile tunnel and then lidded. Typical aseptic form-fill-seal machines are supplied by Erca, Formseal, and Servac; other systems are also available.
4. *Monolayer or multilayer bottles:* Some of these systems handle bottles that are blown with sterile air and sealed across their necks before they leave the mold. Such bottles require only an external sterile wash before removal of the seal/aseptic filling. Multilayer bottles incorporating PVdC give 12 months of shelf life to products such as fruit juices. Machines are available from Remy, Sidel, Serac, and other manufacturers.

Metal cans and glass jars/bottles, although available as aseptic systems, are very rarely used at the present time for fruit products.

Bulk Aseptic Systems

Bulk aseptic systems are of growing importance in the worldwide distribution of products such as tomato paste, fruit purees, and fruit juice concentrates. There are two common bulk systems, one of which employs reusable metal drums; the other uses a plastics/laminate bag within a rigid (wooden or composite) outer. Such units are normally provided with an International Standards Organisation (ISO) pallet base to ease handling.

Drum Systems. Such systems normally use 55-gal (250-L) drums. These are presterilized with wet steam, filled with cooled sterile product, and then closed under aseptic conditions.

Bag Systems. The bulk bag-in-box systems are normally sterilized using gamma radiation (up to 25 kGy) which, after aseptic filling in a sterile atmosphere, may be placed in a box, carton, or drum. Bag capacities range from 6 to 1000+ liters. Laminated materials, including aluminum foil laminates, may be used to give improved barrier properties and longer shelf life, a common material now being a metallized PET/PE laminate.

Systems have been developed that permit the aseptic removal of the sterile product from the bulk pack, thereby permitting its direct use in subsequent aseptic filling operations. Bulk aseptic road/rail tankers with a capacity in excess of 20 tonnes are able to operate such a system.

9.4 PACKAGING OF FROZEN FOODS

9.4.1 Manufacture of Packaging Materials

The majority of frozen fruit products are packed in bags or pouches made from PE; the manufacture of PE film, and the techniques used for the production of plastics film laminates are outlined, since they will be referred to throughout the remaining sections of this chapter. PE film is commonly manufactured by the blown film extrusion process; the PE granules are softened by pressure and heat, and the molten resin is then forced through a circular hole or die. Air is blown from the center of the die to expand the PE into a tube-shaped bubble, which is then pulled upward through a pair of rollers and emerges as a flattened tube. The two edges of the flattened tube are split, leaving two sheets of PE, which are separated and wound up individually.

With regard to the manufacture of plastics laminates, the three principal techniques used are as follows:

1. *Adhesive lamination:* Here, the base film is coated with an adhesive solution, dried, and combined with a second film by passing both films together through nip rollers.
2. *Extrusion coating:* Here, the base film is similarly coated with an adhesive primer on its upper surface, and is then pulled around a water-cooled steel drum called the chill roll. Molten resin, usually PE, is extruded in the same way as in film production above, except that the die here is in the shape of a thin slot. This layer of resin contacts the primed base film at the chill roll, and the two layers are joined together by pressure from a rubber-covered pressure roller.
3. *Extrusion lamination:* Here, a third layer of film is added to the laminate by introducing a second film, which uses the extruded PE as an adhesive.

9.4.2 Properties of Packaging Materials

Frozen food packaging has to protect the product principally from drying out by providing an effective barrier to moisture vapor loss from the food surface during refrigerated storage, thereby preventing freezer burn. PE is the best moisture barrier among the readily available packaging materials. Stability and abuse resistance at low temperatures is obviously important and, again, PE is better than other common resins.

9.4.3 Choice of Package

The choice of materials/containers for packaging frozen fruit products is wide, although PE film forms a component of the most successful ones, whether they be bags for individually quick-frozen soft fruits or lined/coated cartons for pies and similar products.

9.5 PACKAGING OF DEHYDRATED FRUITS AND FRUIT PRODUCTS

For dried fruits, packaging has to provide protection against loss of volatile components, against oxidation, against external attack, and physical protection against damage to the product itself. High resistance to moisture vapor with some fruit products, however, may be a disadvantage, since it may promote sweating of the product with consequent growth of molds and yeasts, microbial spoilage being basically controlled by reducing the free-water content below that necessary for growth. Where gas impermeability is required, such materials as heat-sealed aluminum foil laminates (PE/aluminum foil/PE) will provide the necessary protection, as will transparent nonfoil plastics laminates incorporating PVdC. Simpler structures include PVdC-coated cellulose film (coated on both sides) or one-side PVdC-coated cellulose film laminated to PE. Vacuum metallization of films also increases their impermeability to gases. In vacuum metallization, aluminum is vaporized by heating and then, under vacuum, becomes deposited onto the cold film. The oxygen transmission rate of some films is claimed to be reduced to one-fortieth by such a treatment. Apart from the necessity to protect the product from oxygen uptake during storage, it may be advantageous to remove as much oxygen as possible from within the pack at the time of filling. Vacuum/gassing equipment will handle both lined cartons and flexible packages; nitrogen gas packing is considered capable of doubling or trebling the shelf life of sensitive dried fruit products. The use of a pouch-in-carton system, or a lined carton, provides physical protection to the product.

9.6 PACKAGING OF FRESH/CHILLED FRUITS AND FRUIT PRODUCTS

Fresh fruits, by definition, continue to respire, that is, they take in oxygen and give off carbon dioxide and water vapor; some other gases, such as ethylene, are also produced. Plastics materials for the packaging of fresh fruit serve to reduce moisture loss and, for most products, microperforated LDPE is sufficient protection—too good a moisture vapor barrier can lead to microbial spoilage. Other fresh fruits may be packed in thermoformed PVC or expanded PS punnets or trays, overwrapped with LDPE or ethylene vinyl acetate (EVA) cling films.

Modified atmosphere packaging (MAP) is an increasingly popular technique for extending the shelf life of such perishable fruit products. The objective is to create the optimum equilibrium–modified atmosphere within the package (typically 3–10% oxygen and 3–10% carbon dioxide) by the use of a film of the appropriate permeability to these gases. The term *smart* or *intelligent* packaging has been coined for such applications. Having established the optimum combination of gases to add to the pack, knowing the respiration rate and maturity of the product being packed (5% carbon dioxide/5% oxygen/90% nitrogen is common), the packs themselves are sealed under the required atmosphere using a vacuum/gas packing machine. Vertical form-fill-seal units of the Multivac/Swissvac type, creating pillow packs, are very popular. Alternatively, a semirigid thermoformed tray may be used to hold the product, which is then hermetically sealed, in the correct gas atmosphere, with a heat-sealable semipermeable plastics film. Ideally, the gas:product ratio in all packs should be close to 2:1. Storage of the pack under refrigerated temperatures (0–4°C) is then required.

The choice of films for MAP is largely determined by their gas and water vapor transmission rates. Coating with an antifogging agent, which encourages the moisture released from the fruit to form a transparent film rather than unsightly water droplets, or a compound that selectively absorbs ethylene, thereby retarding the rate of fruit maturation, may be considered. In general, fruit with extremely high respiration rates needs to be packed in extremely high oxygen–permeable film.

Typical films for use as tray lidding materials or to produce packs on form-fill-seal machines are EVA, oriented polypropylene (OPP), or OPP/LDPE laminate. These materials have relatively high gas transmission rates. Trays for use with such products are normally manufactured from UPVC/LDPE laminate or HDPE. Typical film thicknesses for use with MAP are 20–50 μm, and typical shelf life extension may range from 50% to 250%. When using the MAP technique, it is recommended that the labeling regulations in each selected market are checked, since there may be a legal requirement to indicate that the packs contain added gas or a protected atmosphere. Gas sensor labels may be incorporated to confirm that the correct gas atmosphere has been maintained.

In summary, packaging makes available to the consumer a wider range of fruit products. It is used to extend the shelf life of fresh products, and it offers products that are both time consuming and uneconomic to prepare in the home. The traditional containers made of metal and glass still hold a strong position in the market, and developments with these materials are designed to maintain this position. However, the performance of plastics materials is improving with regard to their heat resistance and impermeability. Therefore, packaging for fruit products will continue to satisfy the increasing demands of both the food producer and the ultimate consumer.

SUGGESTED READING

Arthey, D., and Dennis, C., eds. (1991). *Vegetable Processing.* Blackie, Glasgow, Scotland.
Bennett, B., ed. (1995). *The Freshline® Guide to Modified Atmosphere Packaging.* Air Products PLC, Allentown, PA.
Guidelines for Modified Atmosphere Packaging (MAP). (1992). Campden & Chorleywood Food Research Association, Gloucestershire, UK.
HACCP: A Practical Guide. (1997). Campden & Chorleywood Food Research Association, Gloucestershire, UK.
Rees, J., and Bettison, J., eds. (1991). *Processing and Packaging of Heat Preserved Foods.* Blackie, Glasgow, Scotland.
Safe Packing of Food and Drink in Glass: Guidelines for Good Manufacturing Practice. (1998). Campden & Chorleywood Food Research Association, Gloucestershire, UK.
The Processing of Canned Fruit and Vegetables. (1980). Campden & Chorleywood Food Research Association, Gloucestershire, UK.

Chapter 10

The By-Products of Fruit Processing

Ruth Cohn and Leo Cohn

10.1 INTRODUCTION

The production of by-products from the waste of the citrus-processing industry has increased in recent decades for the following principal reasons:

- the increase in total quantity of fruit used for juice manufacture
- the economic importance of the by-products to the juice producer because they increase profitability
- the need to overcome the environmental problems caused by waste material accumulating in each plant

The statistics of the Brazilian and US citrus industry in 1991/1992 serve as an excellent example. The total fruit supply to the citrus industry in 1991/1992 in Brazil was 8,107,000 tonnes, of which about 4,000,000 tonnes was waste material; while in the United States, 7,155,000 tonnes of raw fruit was used, and gave rise to about 3,300,000 tonnes of waste material. There are large differences between the different fruit varieties in the amount of waste per tonne from processed fruit. In the apple industry, for example, waste is 10–15%, which compares with about 50% in the citrus industry and up to 70% with some tropical fruits. Nearly all the waste from the citrus industry can be successfully converted into animal food or fodder products.

10.2 BY-PRODUCTS OF THE CITRUS INDUSTRY

Citrus fruit production can be divided depending where it arises in the juice extraction process:

1. juice and juice cells, which form about 40–55% of the fruit
2. peel (flavedo) and rag (albedo), which constitute about 45–60% of the fruit (The flavedo contains the essential oils and the carotenoid pigments; the albedo contains cellulose, pectins, and flavonoids.)

Table 10–1 shows the different products that are produced from citrus fruit. The main by-products from the endocarp, or inner part of the fruit, are juice cells and pulpwash concentrate.

10.2.1 Citrus Premium Pulp (Juice Cells)

Juice cells are separated from the juice by sieving with variable pressure through a rotating drum finisher. Embryonic pips and small pieces of peel are separated by a cyclone installed in the juice production line. Typically, cells are pasteurized and cooled immediately after sieving in order to destroy pectic enzyme activity and to prevent fermentation. The cells contain natural juice and have the same Brix value as the juice. This product is used mainly to improve flavor and mouthfeel in single-strength juice made from reconstituted concentrate. Cells may be frozen or aseptically packed into drums or special cartons.

Washed Pulp and Pulpwash Concentrate

An alternative industrial possibility of utilizing the juice retained in citrus pulp is by extraction with water. This is usually done by a multistage countercurrent extraction system, using juice extractors for pressing the washed cells. Depending on the number of stages, it is possible to produce a low Brix 3–6% pulpwash liquid as well as washed pulp. The washed pulp is pasteurized, cooled, and packed in the same way as premium pulp. It has a low Brix, a high percentage of broken cells, and a weak flavor. This product is mainly used in drinks to increase "body" and improve appearance.

Because of its high pectin content, the liquid recovered from washed cells is very viscous and difficult to concentrate. Therefore, most producers treat the extract with pectolytic enzymes at 45°C batchwise or by a continuous process. After pasteurization

Table 10–1 Products of the Citrus Industry

Juice and Cells	Peel and Rag	Essential Oil
Concentrated juice	Pectin	Cold-pressed oil
Juice	Cloudy concentrate	Terpenes
Premium pulp	Hesperidin	Concentrated oil
Pulpwash concentrate	Naringin	Distilled oil
Dehydrated cells	Dried peel	
Water and oil-phase volatiles	Molasses	
	Alcohol	
	Natural color	

Table 10–2 Composition of Commercial Pulpwash Concentrate

	Orange United States	Orange Brazil	Grapefruit United States
Total soluble solids (Brix)	55.0	63.5	60.0
Citric acid (anhydrous) (% w/w)	3.2	4.5	4.3
Formol No. (100 mL)	23	21.8	31.2
Total sugar after inversion (% w/w)	39.8	45.3	41.4
Potassium (mg/kg)	1953	1700	1750
Pulp in juice (%) 1500 rpm/10 min	2.0	8.0	2.0

and centrifugation (for pulp reduction), the pulpwash extract is concentrated under vacuum to 58–64°B. Orange and grapefruit pulpwash concentrate is packed into drums and frozen. Composition of orange pulpwash concentrate from different origins is shown in Table 10–2.

The quality of the product is characteristic of the fruit as well as the technology used in its production. Low pulp content, good cloud stability, and a neutral (i.e., absence of bitter) taste are requisites for high quality. Valencia orange will give a product superior in color and taste to an early-season navel orange. Because the pulpwash concentrate is prepared by washing the cells with water, the legislation in some countries does not permit their subsequent use in natural juices, only in drinks.

Dehydrated Citrus Cells

The market for frozen citrus cells is limited, and there are large quantities of this raw material, about 10–15% of the extracted juice, depending on the equipment used. Therefore, the industry is looking for additional products that can be made from cells. Additionally, if the surplus cells are dried together with peel residues, part of them is lost as dust, leading to air pollution.

By drum drying the washed pulp, an edible dietary fiber product is produced that can be used in several ways. The composition of these fibers is shown in Table 10–3. The dried cells have a high pectin and crude fiber content, which makes them a valuable dietary product. Research carried out recently tends to support claims for their gastrointestinal function in preventing diseases of the colon, lowering cholesterol, and preventing heart diseases (Baker, 1993).

Table 10–3 Composition of Dehydrated Washed Cells

Moisture	Fat	Protein	Ash	Crude Fiber	Pectin	Carbohydrate	Total
11.7 %	2.1%	7.5%	2.5%	9.8%	29.0%	37.4%	100%

Source: Ferguson and Fox (1978).

Dried juice cells have high water and fat absorption ability. One part of dried cells can absorb 10–15 parts of water and 3–5 parts of fat. This opens opportunities for the baking industry for dried juice mixes and meat products (Kesterson and Braddock, 1973).

10.2.2 Products Prepared from Peel and Rag

The peel and rag of citrus fruit are used mainly as cattle feed, but as a result of considerable investment in industrial research, important edible products have been developed from them. Essential oils extracted from the peel are one of the sources for aroma development, and terpenes have valuable uses in the chemical industry. The composition of the peel (Table 10–4) shows the importance of sugar and pectin. The pectin industry uses increasing amounts of citrus peel as raw material instead of the traditional apple waste. New developments in this field, especially the process of debittering, have enabled the utilization of a large part of the sugars and other soluble ingredients, which can be used in the drink industry. The bioflavonoids, hesperidin and naringin, have found a place in the pharmaceutical industry. The following paragraphs describe the industrial production of these products.

Production of Special Concentrate from Peel Extract

Shredded peel and rag are used as raw material. Their extraction with water is carried out in a number of stages, either batchwise or with a continuous countercurrent diffusion process that includes heating. To reduce viscosity, most manufacturers add specific pectolytic enzymes during or after extraction and adjust the temperature, time, and pH to achieve optimal activity. The mixture of pectolytic enzymes may include polygalacturonase, lyase, transeliminase as well as cellulase (Pilnik and Voregan, 1991). An alternative process is to heat the peel–water mixture to 98°C prior to enzyme treatment; this eliminates the undesirable action of the pectin esterase present in the citrus peel, which results in loss of cloud stability. At the end of the extraction process,

Table 10–4 Composition of Israeli Orange and Grapefruit Peel

	Orange Mean	Orange Range	Grapefruit Mean	Grapefruit Range
Total soluble solids	12.5	9.4–17.7	10.9	8.4–13.4
Total sugar (% w/w)	8.0	5.5–10.5	6.7	4.7–8.4
Ash (% w/w)	0.54	0.44–0.9	0.57	0.49–0.73
Potassium (mg/kg)	940	600–1440	1077	815–1340
Total pectin (g/kg)	3.65	1.7–7.3	2.95	0.27–5.63
Hesperin (g/kg)	18.5	14.6–23.9		
Naringin (g/kg)			12.3	7.3–17.3
Formol No. (mL NaOH, 0.1 N/100 g)	27.5	18.3–36.7	22.3	12.0–33.6

Source: Cohen, et al. (1984).

the liquid is separated from the peel by a decanter. The extract varies from 3 to 6°B depending on the number of extraction stages. Extraction is followed by pasteurization to inactivate any enzyme activity and centrifugation to reduce pulp content as much as possible.

The peel liquid is finally concentrated *in vacuo* to 50 or 60°B. The color of the product depends on the carotenoid content of the peel used for extraction. Early fruit gives light-colored concentrate (30 ppm total carotenoids), Valencia about 50 ppm, and easy-peel varieties up to 80 ppm carotenoid. These special concentrates are all very bitter and can be added only in small quantities to a drink. They contribute a stable cloud and natural color and contain the bioflavonoids of the peel. The turbidity (cloud intensity) of the orange peel extract is a result of the interaction of the pectin, hesperidin, and protein (Ben-Shalom and Pinto, 1986; Kanner, Ben-Shalom, and Shomer, 1982). No interaction occurs between naringin and pectin and, therefore, grapefruit peel extract is not very turbid. The high naringin content of grapefruit peel extract often results in precipitation of naringin crystals in the concentrate. A typical flow diagram is shown in Figure 10–1. A similar process, but using different raw material, is described by Ragonese (1991).

Debittering of Citrus By-Products

All the citrus peel extracts are extremely bitter because of the very high limonin content. For example, orange peel contains 300–400 ppm limonin, and the capillary membranes about 300 ppm on a wet basis. The threshold of bitter taste resulting from limonin in orange juice is about 6 ppm. Some consumers even find 1 ppm quite bitter. Considerable research has been carried out to find ways of reducing the bitterness of orange and grapefruit products, and many commercial chemical companies now offer complete debittering plants. In the United States, the main effort has been concentrated on debittering navel orange juice; in Italy and Israel, on debittering cloudy peel concentrate extracted from peels. Grapefruit products can only be debittered by reducing both limonin and naringin. The structural formulas of limonin and naringin are shown in Figure 10–2.

The first material used for this debittering by the food industry was activated carbon. However, this method is no longer used because many vital juice components (such as ascorbic acid or the carotenoids) are also adsorbed by carbon.

The second method used specific enzymes. Naringinase free from pectinase can be used for reducing the bitter taste of grapefruit products. The optimum conditions are a temperature of 45–50°C, and the time varies depending on the amount of naringin present. The enzyme limonoate dehydrogenase has been evaluated for decomposing limonin (Hasegawa, Brewster and Maier, 1973).

The most effective debittering technique has been introduced only recently to the citrus industry and involves the adsorption of limonin and naringin onto special resins using ion-exchange equipment and techniques. The US Food and Drug Administration has permitted this method for debittering natural navel orange juice.

The first step is to reduce the pulp content of the peel extract by centrifugation to less than 1%; this prevents clogging of the resin. Resins adsorb limonin and naringin without affecting the other juice constituents.

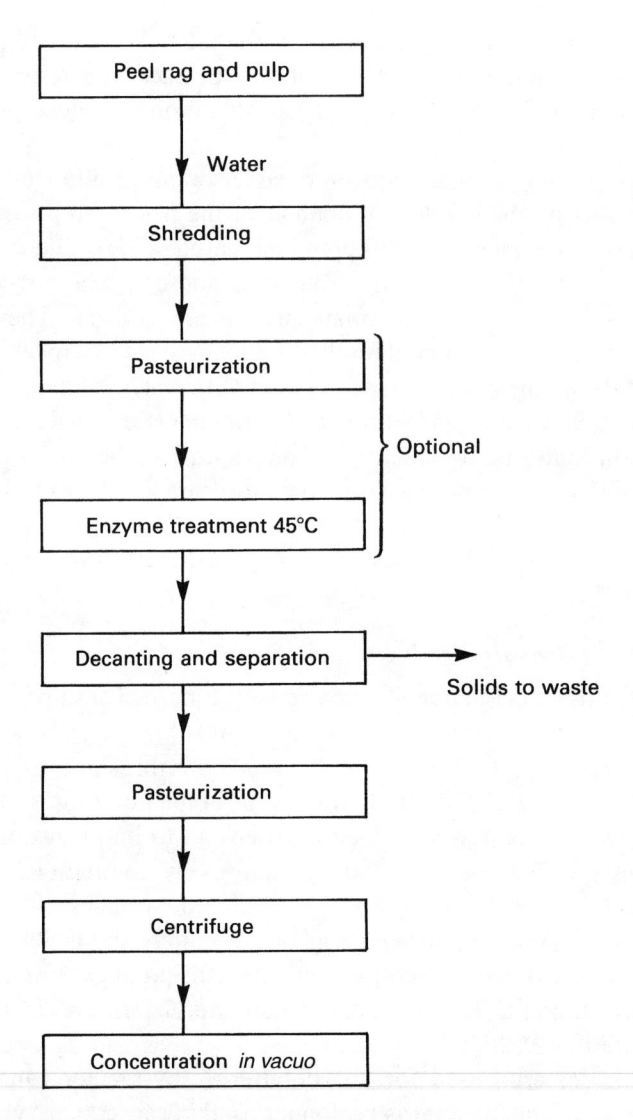

```
┌─────────────────────────┐
│   Peel rag and pulp     │
└───────────┬─────────────┘
       Water │
            ▼
┌─────────────────────────┐
│       Shredding         │
└───────────┬─────────────┘
            ▼
┌─────────────────────────┐
│     Pasteurization      │┐
└───────────┬─────────────┘│
            ▼              ├ Optional
┌─────────────────────────┐│
│  Enzyme treatment 45°C  │┘
└───────────┬─────────────┘
            ▼
┌─────────────────────────┐
│ Decanting and separation│──► Solids to waste
└───────────┬─────────────┘
            ▼
┌─────────────────────────┐
│     Pasteurization      │
└───────────┬─────────────┘
            ▼
┌─────────────────────────┐
│       Centrifuge        │
└───────────┬─────────────┘
            ▼
┌─────────────────────────┐
│  Concentration in vacuo │
└─────────────────────────┘
```

Figure 10–1 Production of cloudy concentrate from peel extract.

It is possible to use ultrafiltration for separating the fluid from the pulp, but this is an additional investment. However, without ultrafiltration the column must be backwashed from time to time to remove residual pulp from the resin.

Undebittered pulp can be added to debittered juice at the end of the process in order to recover the original cloud intensity of the peel extract. Most of the industrial equipment consists of alternating twin columns, allowing juice treatment and regeneration to proceed simultaneously in the different columns. The system can be fully automated, with regeneration using diluted caustic soda.

In big installations, up to 10,000 liters of extract per hour can be debittered. The final step of the operation is vacuum concentration. Two different concentrates are produced, one clear the other cloudy, and both have their uses in the drinks industry.

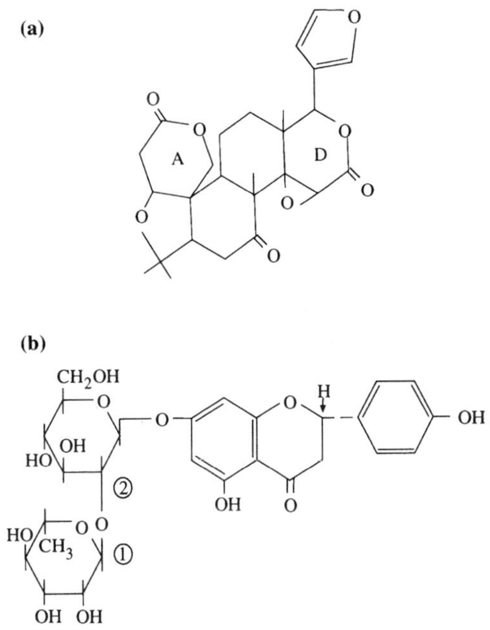

Figure 10–2 Structure of (a) limonin and (b) naringin.

In Italy, a special concentrate is produced from whole lemon or oranges of inferior juice quality. Torchiato press residue is diluted with water, heated, and enzyme treated. After pasteurization, the juice is centrifuged and debittered and forms a valued product for drink bases.

Hesperidin and Naringin Production

The production of citrus flavonoids began when it was realized that they possessed biological activity. Research in recent years has moved from a consideration of the effect of bioflavonoids on blood capillaries to a more central role relating to their general function in human health (Robbins, 1980).

Hesperidin is the bioflavonoid mainly present in oranges and mandarins, *Citrus sinensis* and *Citrus reticulata*, as well as in all lemon varieties (Figure 10–3). In Shamuti orange juice, the concentration of hesperidin is between 0.8 and 2.0 g/kg, and in the peel it is 13–24 g/kg. Hesperidin is practically insoluble in water at acid pH but is soluble in alkaline media. To extract hesperidin from orange peel, the shredded peel is mixed thoroughly with a dilute lime solution to a final pH of 8.0 by mixing in a tank or screw conveyor for up to 30 minutes. The peels are pressed in a screw press, as used in the production of dried citrus peel. The alkaline liquid is screened and then filtered with the assistance of filter aid in a filter press. The resulting clear liquid is acidified in a tank to a pH below 3.0, and hesperidin crystals are separated from the liquid by centrifugation.

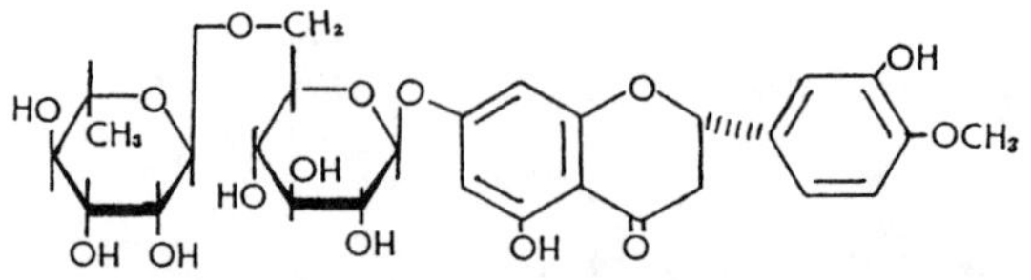

Figure 10–3 Structure of hesperidin.

The remaining peel extract of about 11°B and the pressed peels can be used in the production of molasses and dried cattle feed. The main market for hesperidin is in the pharmaceutical industry. The concentration of naringin in grapefruit peel is between 0.7 and 17.0 g/kg. Naringin is produced in a way similar to hesperidin, but the crystallization process is more difficult because of its greater solubility in water, even in acid solution. Therefore, concentration of the liquid and cold storage are recommended to enhance the process. Naringin is used in drinks to enhance the bitter taste, and it has been examined also as a raw material for further processing to an artificial sweetener.

10.2.3 Citrus Oils

Essential oils are extracted from all varieties of citrus, as prices are currently attractive and the production does not demand great investment. World production, according to Wright (1995), is about 16,000 tonnes of orange oil, 2,500 tonnes of lemon oil, and 180 tonnes of grapefruit oil. Lime and bergamot are among the highest-priced oils. The main market for essential oils is for the flavoring and drink industry.

Cold-Pressed Oils

Citrus oil develops in receptacles on the flavedo of the peel, which is covered by a protective wax. The quality and quantity depend on the variety and ripeness of the fruit, as well as the location of the grove. For example, oil from Valencia oranges is considered to be the best quality among orange oils, and Italian lemon and green mandarin oils are regarded as being superior to oils from other sources. The method of extraction is also of vital importance. The yield of essential oil from the fresh fruit is about 0.4% in lemon, between 0.2 and 0.4% in orange, and 0.2% in grapefruit. The main steps in the production of cold-pressed essential citrus oil are as follows:

1. extraction of the oil by mechanical pressure on the oil vesicles
2. separating the oil from the peel by water spray
3. centrifugation of the water–oil emulsion
4. clarification of the crude oil in a centrifuge
5. settling of the waxy crystals present in the oil by holding at low temperatures for several weeks (winterizing)
6. separating the wax from the oil by centrifugation before packing into drums

The extraction method used by each producer depends primarily on the type of juice extractor used in the plant. The FMC extractor, which simultaneously separates oil from the fruit during the juice extraction process, is used in most citrus-producing countries. The Brown juice extractor has two different oil extraction methods: (1) the Brown oil extractor is used before juice extraction, and (2) the Brown peel shaver is used after juice extraction. All three American systems produce good-quality oil. The different systems are described in detail by Redd and Hendrix (1993).

In Italy, different oil extraction systems have been developed and are adapted to the unique fruit varieties grown there. Two of them are still in use together with the American machines (Ragonese, 1991). The Pelatrice system is used by the Indelicato and Speciale oil extractor. The whole fruit is rasped by special rotating rollers that carry the fruit along a closed box, where water sprays wash the oil released by the rasping process. The Torchi system replaces the older Sfumari system. This oil extractor is used for peel as well as small whole fruit. The oil is extracted by two contrarotating helices with increasing pressure along the whole length of the oil press. Also, half lemon peels can be treated in this way after they have first passed through a Birillatrice juice extractor. Usually the process is in two stages, the first using low pressure, the second high pressure. The water–oil emulsion from the second stage is difficult to separate because of its high pectin content. The oil produced by this system is of good quality and the yield is high.

Clarification of the Oil and Waste Water Treatment

After extraction, the water–oil emulsion passes to a desludging centrifuge, producing crude oil and water still containing small amounts of oil, pectin, and sugars. In some extraction systems, such as the FMC, the water phase is recirculated into the water spray system to increase the oil yield and reduce the quantities of water that go to waste. Citrus oils act as bacteriocides and will interfere in the waste water treatment; they should be removed as far as possible by recirculation. The Brix of the water rises and can be added to the peel press water and used for the production of molasses. In this way the total solids of the waste water as well as the oil are removed in an environmentally sensitive way.

The crude oil is clarified in a desludger and transferred to settling tanks or drums for dewaxing. The lower the temperature, the faster the settling of the wax, which should take place below 0°C. After several weeks the oil is passed through a polishing centrifuge for dewaxing. Essential oils are inflammable and must be handled carefully. The oil is readily oxidized and, therefore, should be kept in full containers at low temperatures. Although cold-pressed oil contains natural antioxidants, some producers add butylated hydroxytoluene (BHT) or butylated hydroxyanisole (BHA) to prolong the shelf life of the oil.

Aroma and Oil from Juice

The excellent flavor of freshly pressed juice is caused by small quantities of oil (0.02–0.03%) and water-soluble aromatics (water phase) present in the juice. During the process of vacuum concentration, these are lost by evaporation. In modern juice plants, various aroma recovery systems have been installed, which strip the oil in the first step of concentration, usually after pasteurization. The oil and water phase are kept separately at low temperatures. They can be added back at a later date to the concentrate to enhance the flavor of the reconstituted juice. Recovered aromas are also used by flavor houses to produce soluble citrus aromas of excellent quality. A detailed explanation of aroma recovery equipment is given by Redd and Hendrix (1993).

Distilled Oils and Their Fractionation

Cold-pressed essential oils can be distilled to produce different aroma fractions for use in the drinks industry. The largest fraction is *d*-limonene, which comprises up to 95% orange oil and 90% lemon oil, but its contribution to flavor is limited. The important fractions are water-soluble esters and aldehydes, which contribute the fruity note to drinks. A great part of lime and lemon flavor is produced by distillation of cold-pressed oils. A detailed discussion of this subject is given by Redd and Hendrix (1993).

Concentrated Oil

Concentrated oils are produced by specialized flavor houses from different cold-pressed citrus oils. The terpenes are partly removed in the process, depending on the degree of concentration (1:5 to 1:20). Concentrated oils are less sensitive to oxidation and have a pleasant flavor; this is the result of the removal of a large proportion of the terpenes, which are insoluble in water. Therefore the formation of oil rings in the drinks is minimized. Concentrated mandarin oil (*Citrus reticulata*) can contribute to color as well as flavor, as it contains dark-colored carotenoids, such as cryptoxanthin, which are not present in *Citrus sinensis*. When concentrated up to 4000 ppm in the oil, these substances can be used as a natural colorant for drinks.

Terpenes

The press liquor produced from peel in the drying process contains a considerable amount of peel oil. In the first stage of molasses production, this oil is stripped and collected in tanks. The terpenes produced in this way have many uses in the chemical and cosmetic industry. Terpenes are excellent general-purpose organic solvents.

10.2.4 Comminuted Juices

Comminuted juices were introduced by the British Soft Drinks Industry soon after the Second World War. The main supplier of orange and grapefruit comminuted juices is Israel. Comminuted lemon juices are produced in Italy because of the superior quality of the lemons grown there. Originally the whole citrus fruit was crushed, passed through a finisher, milled, pasteurized, cooled, and stored, preserved. Only 3–4% of this mixture in a final drink is needed to contribute a fresh citrus flavor as well as a cloudy appearance to the drink. This cloud is caused by the presence of essential oil and peel substances in the comminuted juice.

To simplify standardization of the product, the industry now uses a different method of production. By blending different fruit components according to agreed formulae, many products can be produced to meet client requests. Oil and peel content can be changed and, by adding concentrated juice or cloudy peel, the concentration of the comminuted juice can be raised to 50°B. In this way, less packing material is used and shipping costs are reduced.

Most comminuted juices are packed in large plastic containers and usually preserved with sulfur dioxide, benzoic acid, or both. For aseptically packed comminuted juices, special plastic sacks with an aluminum foil barrier are used. Figure 10–4 gives the flow sheet of the production process.

Most whole citrus drinks on the British market contain up to 15% single-strength comminuted juice and are diluted to taste with water.

10.2.5 Dried Citrus Peel

Many of the citrus juice–processing plants have such a high throughput that disposal of the wet peel directly to farmers is not possible. In the United States and

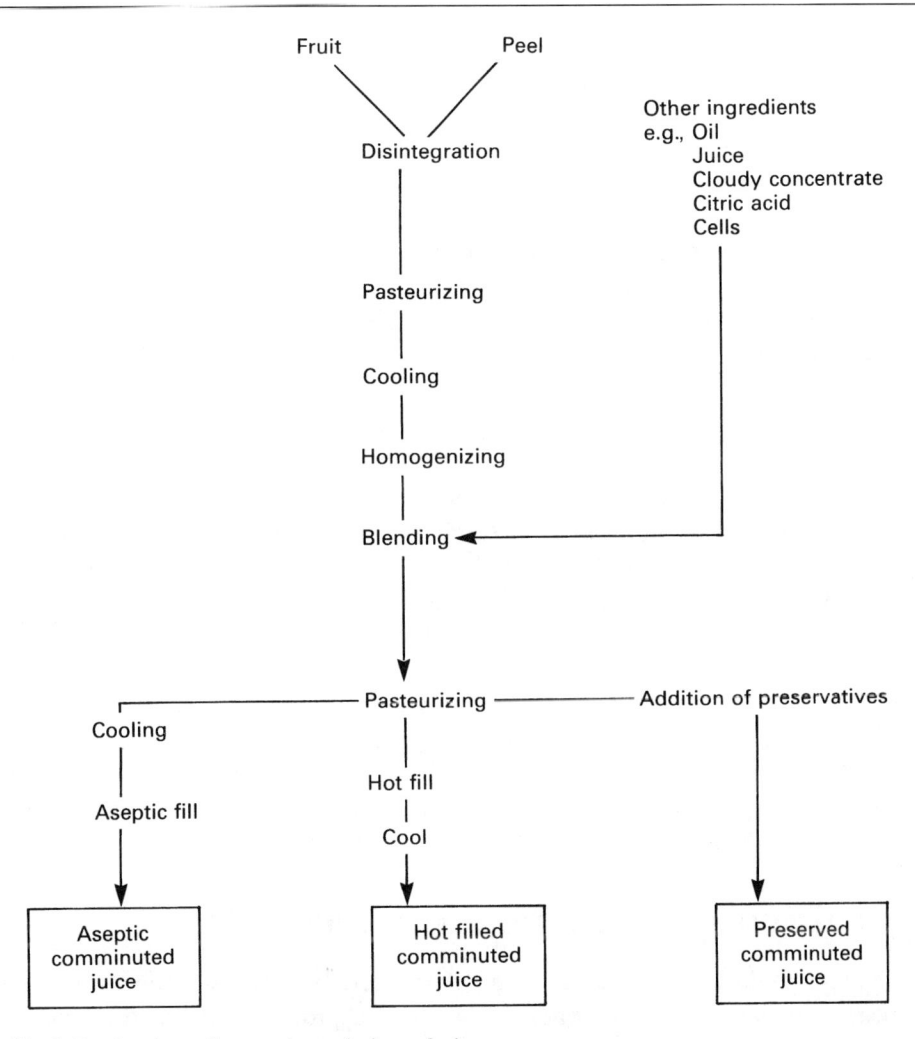

Figure 10–4 Production of comminuted citrus fruit.

Brazil, nearly all of the waste from the plant is dried and sold as animal feed pellets. Several manufacturers provide the industry with equipment and layouts for the production of dried peel for animal feed as well as for pectin. The general outlay of such a process is as follows:

1. The waste material is collected in silos and held there for not more than 4 hours. The moisture content at this stage is between 82 and 84%.
2. The peel is shredded.
3. Addition of 0.1% to 0.25% lime (calcium hydroxide) produces a pH of 5.5 to 7.0 and this is held for 10 to 20 minutes. The higher the pH, the shorter the retention time required, but more scaling in the evaporator of the press liquor should be expected. Lower pH is preferred, especially if the press liquor is to be used for fermentation to produce alcohol.
4. Pressing gives about 65% press cake with approximately 75% moisture content and 35% press liquor. The press liquor is concentrated in a multieffect evaporator to produce molasses, using the waste heat of the exhaust gases from the dryer. The molasses are either used as raw material for the production of fuel-grade alcohol or mixed with the press cake.
5. Either the press cake or its mixture with the molasses is fed into a rotating drum dryer, reducing the moisture content to 10–15%.
6. Finally, dried peel is passed through a cooling tunnel before being bagged or transported to a storage bin. Most peels are pelletized in order to reduce the volume during storage and transportation as well as to facilitate the feeding of animals. All dried peels preferably should have a moisture content not higher than 10% and be stored in a cool place that, if necessary, is ventilated with chilled air.

The nutrients of the dried citrus peel fluctuate according to the type and degree of ripeness of the fruit, as well as environmental conditions. Taking average figures of several available data, the values in percentage on a moisture-free basis are typically crude protein, 6.8%; fat, 4.7%; crude fiber, 13.4%; ash, 6.6%; and nitrogen-free extract, 68.5%. Comparative values for crude protein and crude fiber, respectively, in other meals are 49.6% and 7% for solvent-extracted soya bean seed, and 44.8% and 13% for cotton seed (Anon, 1987).

In comparison with other types of feeding material, dried citrus waste is of inferior quality as far as protein content is concerned and is consequently lower in price.

The production of fuel-grade alcohol, which during recent years has played an important role, especially in the Brazilian industry, assists in counterbalancing the economic disadvantages of disposal of the waste from the citrus industry.

10.3 NATURAL COLOR EXTRACTION FROM FRUIT WASTE

From ancient times, natural colors have been used to give food a more attractive appearance. The addition of artificial colors to food products has become a controversial subject in the last few years, and now many such colors have been restricted in use.

Even natural and nature-identical colors (such as β-carotene) must now be declared on product labels in most countries.

The extraction of color from fruit waste, therefore, has been intensively studied and ways of industrial application have been suggested. Two colors are of the utmost importance for the juice and drink industry; the orange-red color of the carotenoids and the dark red-blue color of the anthocyanins.

10.3.1 Extraction of Color from Citrus Peels

About 70% of the carotenoids of the whole fruit are concentrated in the flavedo. The main components are xanthophylls, but there are many other carotenoids in citrus varieties, and their dependence on ripeness, soil condition, and other factors has been discussed by Gross (1977). The extraction of the color from citrus peel is not easy, as the carotenoids are not water soluble. Therefore, only a few methods of extraction have reached industrial application.

Several research institutes have suggested different ways of color extraction from wet peel (Ting and Hendrickson, 1969; Wilson, Bisset, and Berry, 1971). A research team at the Volcani center in Israel (Kanner, Ben-Shalom, and River, 1984) has suggested the following process. Flavedo flakes are liquefied at pH 4.5 with the aid of pectolytic enzymes and cellulase at 40°C for 6 hours. The terpene *d*-limonene is added to the mixture in order to absorb the readily extracted carotenoids. After first centrifuging the solid particles, the liquid is again centrifuged to remove the highly colored *d*-limonene, which now contains 75% of the carotenoids present in the flavedo. The oil phase is dewaxed, concentrated, and purified. A 3.6% carotenoid solution can be achieved by this method.

Extraction of Color from Dried Orange Peel

An extraction process for color from dried citrus peel invented by the late M. Koffler has been carried out on an industrial scale in Israel during several production seasons. Peel, rag, and flakes are shredded and treated with lime to a pH of 8.0–9.0. They are then pressed and dried in a rotating dryer with a directly heated drum. Temperatures are carefully controlled to avoid overheating the peel. The dried peels are stored in a cool silo and extracted as soon as possible to prevent loss of carotenoids. Extraction is carried out in a commercial extractor, as used in the oil industry, with light petroleum ether mixed with alcohol. The extract is concentrated under vacuum, and *d*-limonene is added to the highly concentrated color solution. The petroleum ether is removed by evaporation at under 50°C, and a concentrated carotenoid solution in *d*-limonene of 4000 ppm results.

10.3.2 Extraction of Color from Grapes

Anthocyanins are present in many dark-colored fruits, such as red grapes, blue and red berries, dark cherries, and some tropical fruits. Commercially, it is generally the wastes of the grape juice and wine industries that are used for extraction. Anthocyanin powder and concentrated anthocyanin solution (typically 4 g/100 g) are prepared for the food industry. Recently a very detailed work on the distribution and analyses of anthocyanins in fruit and vegetables has been published (Mazza and Miniati, 1993).

The anthocyanins present in food products are very sensitive to heat, metals, pH, and air; therefore, it is not easy to produce attractive products without adding natural color. The concentration of anthocyanins in grapes is dependent on the temperature during ripening. The optimal temperature is about 20°C, and a temperature of 35°C will prevent color development. The most abundant anthocyanin in grapes is malvidin 3-acetylglucoside. Its concentration rises steadily during the ripening process, then declines slightly at the end of the process. Soil, water supply, light, and fertilizers are additional factors affecting the development of anthocyanins in grapes.

The anthocyanins in grapes are extracted mainly from the skin of dark-colored varieties after the juice is pressed. About 20% of the total fruit weight is skins, seed, and stems. In the extraction of the pomace, a diluted solution of sulfur dioxide is used, the same as in the maceration process of grapes in the wine industry. The sulfur dioxide acts as an antioxidant in the process of extraction. After color extraction is completed, the solution is filtered and concentrated under vacuum, causing a nearly complete evaporation of the sulfur dioxide.

10.4 APPLE WASTE TREATMENT

Apple juice production consists of several stages:

1. grinding to produce the mash
2. enzyme treatment
3. pressing
4. juice treatment
5. pomace disposal or its use

The pomace resulting from pressing mashed whole apples contains about 20–30% dry matter, 1.5–2.5% pectin, and 10–20% carbohydrates.

Because large quantities of apple juice are produced in the United States and Europe, the problem of waste disposal is a significant one. There are several ways of solving this:

- Feed the pomace directly to animals. Little investment is involved, but accurate planning of disposal during a 24-hour period has to be undertaken. Because the pomace is poor in protein, it must be mixed with additional nutrients.
- Ensilage the pomace to give a relatively stable product. Typical analyses are given by Steingass and Haussner (1988).
- Dry the pomace for pectin production (see section 10.5).
- Dry the pomace for feed; this is only possible in large plants.

An additional use of apple waste as dietary fiber and cake filling has been suggested. During recent years, the liquefying of the apple mash with special enzymes and the secondary extraction of the pomace with water have reduced the amount of waste material from apple juice production.

10.5 PRODUCTION OF PECTIN

Pectin represents an important component of the waste of two commercially important fruit types, citrus and apple. In the waste of citrus juice production, the pectin content is up to 4% of the fresh weight and in apples it is up to 2%.

Having acquired the technology to extract purified pectin solutions enabling their use in many food products, pectin has become a highly valuable by-product of the fruit juice industry. Pectin, being a natural plant component, has its use not only in food products but also in nutritional and medical applications with the great advantage of being legally acceptable. In recent decades, with increasing emphasis on the favorable functions of pectin in human metabolism, its economic value has increased considerably.

Many aspects of pectin have been studied, particularly regarding its role in plants throughout their life cycle and in products prepared with them. Also, many publications have appeared about pectin's interaction with the pectolytic enzymes present naturally in plants and in the microorganisms that degrade them.

10.5.1 Characterization of Pectin and Pectolytic Enzymes in Plants

The main pectin-containing tissue in the plant is the parenchyma. These cells are surrounded by primary walls and between them there is a middle lamella, both containing pectic substances. In the primary cell wall, the pectin, together with other materials, represents encrusting substances in between the cellulose fibrils, while the middle lamella (the main component) holds together the structure of the plant tissues (Schwimmer, 1981).

These pectic substances are polymers of d-galacturonic acid, partially esterified with methanol and coupled by $\alpha(1-4)$-glycosidic links (Kertesz, 1951; Doesburg, 1965). The degree of polymerization is so high that these pectic substances are designated as protopectin and are insoluble in water. The elaborate structure of pectin is summarized by Rolin and De Vries (1990) and Pilnik and Voragen (1991). The synthesis of pectin during cell formation is discussed by Northcote (1986).

Doesburg (1965), summarizing the data reported in the literature on the pectin content of apples during the harvest season, observed a nearly constant level with a slight decrease at the end of the season. Kertesz (1951) reported that in English apples the percentage of pectin based on fresh weight decreases during ripening and then becomes constant. Sinclair (1961) concludes that in orange peel during maturation and until senescence, the percentage of total anhydrogalacturonic acid on a dry weight basis first decreases and then levels off, but the relation of water-soluble pectin to total pectic substances rises until the peak of maturity and then gradually decreases.

Many data regarding the pectin content of citrus and apple can be found in the literature. Generally speaking, the pectin content in apple residue is about half that in citrus peels. Within the genus citrus, lemon, lime, and pomelo are the richest in pectin and mandarins the poorest; oranges and grapefruit have an intermediate position. As far as orange varieties are concerned, experience shows that Valencia is richer than Shamuti and Washington navel, but no figures have been published.

10.5.2 Pectin Enzymes in Plants Used for Production of Pectin

The pectolytic enzyme present in all plant materials used for pectin production is pectin methylesterase, which removes the methoxyl group from the carboxyl function. To a lesser extent, polygalacturonase, abundant in fungi, is present in higher plants but not in apple, and it is doubtful if it occurs in any citrus fruit. Polygalacturonase hydrolyzes glycosidic bonds next to free carboxyl groups. Two additional enzymes splitting glycosidic bonds by beta elimination are pectate lyase, mainly present in bacteria, and pectin lyase. The distribution of pectolytic enzymes is summarized by Pilnik and Voragen (1991).

The pectin manufacturer who needs to reach the required range of esterification in the final product is obliged to control the pectin methylesterase activity in the raw material. While natural polygalacturonase is not a problem, the presence of additional polygalacturonase and pectin lyase from fungi, yeasts, or bacteria, infecting decaying fruit (especially toward the end of the season), or from microbial contamination during production will cause concern. The activity of these enzymes of microbial origin may result in a loss of yield, deterioration of quality or, in extreme cases, even a complete failure of production.

10.5.3 Commercial Pectins and Their Production

To render the pectin suitable for use in different applications, the molecule has to be made soluble. This demands some shortening of the chemical chain. The degree of esterification (DE) also will determine the characteristics of the final product, and in the raw material this is very high and has to be adjusted in the process. The two processes, shortening of the polygalacturonic acid chain and partial deesterification, might interfere with each other. Success is based on the art of the manufacturer, who has to control the process to achieve the correct balance of these two requirements and obtain the maximum yield of the product. Obviously, manufacturers keep the details of their processes secret and, therefore, little has been published on this subject.

The commercial pectins consist of two basic groups according to their DE, with several variations in each group:

Pectins with DE above 50% (High-Methoxyl Pectin, HMP)

These pectins form gels only in the presence of a total soluble solids content (TSS) above about 57% or by replacing part of the water with an organic solvent at a pH below 3.5. These pectins are prepared by direct extraction, which in the traditional process is under acid conditions at elevated temperature, and are sequentially precipitated.

Pectins with DE below 50% (Low-Methoxyl Pectin, LMP)

These pectins form gels at lower TSS, depending upon the degree of methoxylation and require calcium ions for jellification. They are prepared either by direct extraction under more severe conditions or by modifying pectin previously extracted as HMP.

Each group of pectins includes various types, differing in chain length and degree of methoxylation and consequently have different characteristics. Each pectin manufacturer controls the production conditions in order to obtain the required type. Although they give the same name to each type, for instance, slow set, medium rapid set, these are similar but not exactly identical. In addition, pectin is made from a natural raw material that is influenced by climatic conditions, which vary for each season and area, by agricultural practices, and by conditions at fruit harvest and transport to the factory. Besides scrupulous control of production, the manufacturer must constantly counterbalance the influence of these factors. Even so, some slight differences may occur, which will have to be compensated for by the final user.

One of the major problems of pectin manufacture is the disposal of effluent. Pectin manufacture increases the value of the waste material from fruit juice production, but its traditional manufacturing process demands considerable quantities of water and its effluents often contain undesirable materials, such as soluble carbohydrates and neutralized acids.

Commercial Preparation and Sources of Raw Material for the Production of Pectin

Kertesz (1951) describes 50 different plants where details of the pectin content have been published in the literature. In order to obtain the desired quality of the final product, the protopectin should be esterified with methanol, and from the commercial point of view the percentage of protopectin should be high and sufficient quantities of raw material must be available at an economic price. In western Europe and the United States, raw material with these characteristics is limited to the residues of apples and citrus fruits after juice processing. During the 1980s, the use of apple pomace decreased considerably, with citrus peels being richer in pectin and more readily available in sufficient quantities. Also, changes in the processing of apples by the use of liquefying enzymes has rendered the pomace unsuitable for pectin production, and an increase in the value of apple pomace for cattle feed has diminished its attractiveness for pectin production. However, some food manufacturers still prefer apple pectin for some uses where specific properties are required, probably because of the presence of certain starchy fractions remaining in the final pectin. Citrus peels treated with enzymes are not suitable for pectin production. Other raw materials not originating from fruit juice manufacture have at certain times been used for pectin production. Examples are the residues from sugar beet processing and sunflower seed heads. As the pectin from these sources is partly esterified with acetyl groups, and sunflower contains undesirable terpenes, these raw materials are not popular for this purpose. Recently, however, a plant in western Europe began to produce low-methoxyl pectin from sunflower seed heads. It is also believed that a specific variety of watermelon is used in Russia as raw material for the production of pectin.

Because the period of waste availability after juice processing is limited and the plants of pectin manufacturers are often not located close to juice plants, a considerable proportion of the raw material has to be dried.

Certain points are essential in the preparation of dried peels for pectin production.

- The limitation, if not inactivation, of pectin esterase activity in the raw material and the avoidance of pectolytic enzymes from any type of microorganism is

essential, particularly if the raw material is transported long distances to the drying plant.
- The waste should be shredded and washed in order to remove soluble carbohydrates and to make it compressible. A countercurrent washing system may be used.
- Drying at the lowest possible temperature and storage under dry conditions using forced ventilation with cooled air prevents overheating by chemical reactions, which might cause spontaneous combustion (Rebeck, 1995).

A pectin plant situated next to a fruit juice plant has the advantage of being able to avoid the need for drying during the season. By using this procedure for citrus peels, it has been possible to increase the yield by 30–40% and improve the quality. The solution of pectin produced in this way is more cloudy.

Manufacture of Pectin

Until recently, the usual process in Western countries was to mix raw materials with acidified water for a certain time at elevated temperature. Each manufacturer selects his or her own parameters for preparing the different types of pectin. Although the principles are similar, the time of extraction may vary between 30 minutes and 45 hours, the temperature may vary between 40°C and 92°C, and the conditions of acidification differ. The change in the pectin molecule and its dissolution may be affected in one single or two separate steps, and in the latter procedure the ratio between the plant material and the acidified water may vary. The type and concentration of acid used will differ, not only according to the properties of the raw material and type of pectin required but also according to economic and environmental considerations. After separation of the crude pectin extract from the solid particles, it is clarified by passing through filter presses with a filter aid in one or two steps, by centrifugation, or by combining the two procedures.

In recent years Golubey and Gubanov of the Interbios company, in cooperation with the Russian Institute of Food Industry, have developed an essentially different process, using ultrasonics to split the protopectin. This process has been called "cold pectin technology." The process has successfully passed the pilot stage, and the first industrial trial, including filtration through a membrane system and spray drying, is now planned. If this method proves technologically and economically successful, it will contribute much to the solution of the environmental problems and may be a serious competitor to the traditional way of production.

Two alternative processes for the production of high-methoxyl pectin have been used until recently in the Western world. One is based on concentrating the purified pectin extract in a multiple-phase evaporator and precipitating the pectin with alcohol. The other precipitates the pectin from the purified extract with an aluminum salt. The different steps necessary to secure the final product for these two processes are represented in a flow sheet in Figure 10–5. In this figure the various stages where waste is created are indicated as (A), (B), and (C) for solid, liquid, and slurry consistency, respectively. These waste materials create serious environmental problems and, in spite of huge investments to try to solve them, have caused the closure of some pectin plants.

The solid waste may be used as fodder with or without such treatment, as may be required locally.

The flow diagram shown in Figure 10–5 has also been used for the production of low-methoxyl pectin, but with increasing time, acidity, and different conditions of precipitation. Usually it is more common to start with high-methoxyl pectin precipitated

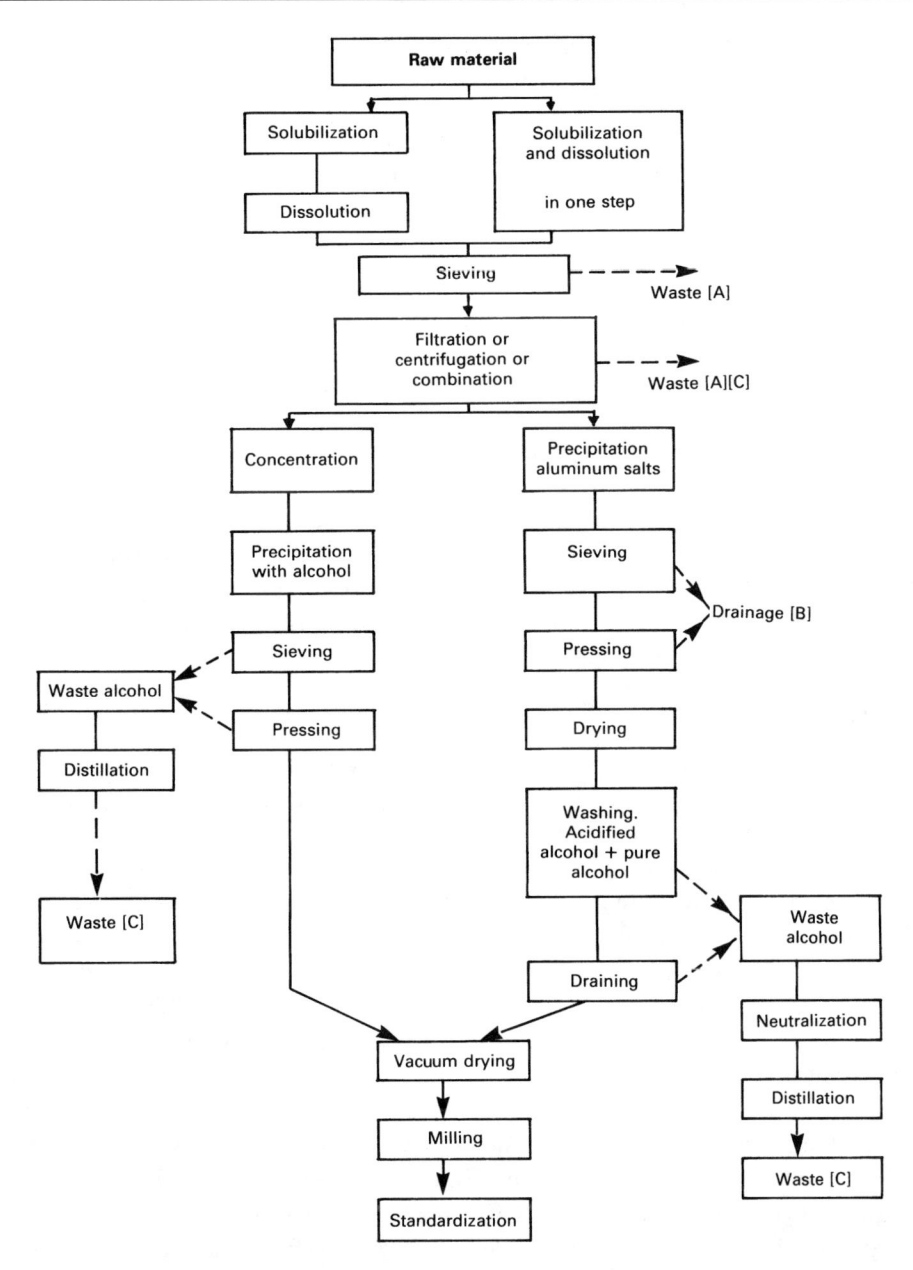

Figure 10–5 A flow diagram for two processes for pectin production. [A] solid waste; [B] liquid waste; [C] slurry.

either with alcohol or with aluminum salts and deesterification by acid or base in alcohol at reduced temperature. By deesterifying with ammonia, the methyl ester is partially substituted by amide groups. This type of low-methoxyl pectin is called amidated. A maximum of 25% amidation is permitted by law in some countries.

The pectins produced will have different gelatinization powers, according to their nature, the storage conditions of the raw material, and the production process. A number of instruments can be used for measuring the "jelly strength." As these devices do not measure the same parameter, it is impossible to correlate their results. The various methods have been discussed by Crandall and Wicker (1986) and a short description is presented by Rolin and de Vries (1990). Up to now the universally used method is the USA-SAG method (Committee for Pectin Standardisation, 1959). The standard for international trade is 150 grade USA-SAG, which means that under standard conditions one part of pectin causes jellification of 150 parts of refined sucrose at standard elasticity.

10.5.4 Different Types of Pectin and Their Application

The generally accepted nomenclature of pectic substances is that formulated by Baker et al. (1944). A slight change has been proposed by Doesburg (1965).

The main application of pectin in the food industry is as a gelling agent. The commercial types of pectin are divided into two main groups depending on the mechanism of gelatinization: those with esterification above 50%, which gelatinize only at TSS levels above 55% and in a restricted range of pH, and those with esterification below 50%, which gelatinize at a wider range of TSS and pH values in the presence of calcium ions. The theoretical explanation of the jellification process has been recently discussed by Crandall and Wicker (1986).

High-Methoxyl Pectins

Within this group, all other conditions being kept constant, the rapidity of setting is determined by the degree of esterification. The higher the percentage of esterification, within the range 57–82%, the more rapid the setting. Usually three types are offered in this range: slow set, medium rapid set, and rapid set, but sometimes ultra rapid set is also available. The different manufacturers do not indicate in their catalogues exactly the same range of esterification for these types. The rapidity of setting depends also upon the concentration of TSS, and if this exceeds about 74% even slow-set pectin might cause too rapid setting. Therefore, pectin manufacturers offer pectin mixed with inorganic and/or organic buffer salts and/or partially amidated pectin, which retard the speed of setting.

High-methoxyl pectins are also used for increasing the viscosity of liquids. The higher the degree of esterification, the higher the viscosity obtained. This characteristic can be used for stabilizing drinks and increasing mouthfeel, with the advantage that the additive is a natural fruit component. Another application is their ability to emulsify essential and, especially, vegetable oils. It is successfully used to prevent curdling of acidified milk products and, therefore, facilitates pasteurization of a mixture of milk and fruit juices.

Low-Methoxyl Pectins

At a lower degree of esterification, pectin can cause jellification at lower total solids content. Accordingly, different types of low-methoxyl pectin are offered for the production of dietary jams with different levels of sugar content, and milk products and aspics, especially those with a low sugar content. Theoretically, jellification can be obtained without any sugar present, but in practice a 20% sugar content is the lowest limit because below this syneresis occurs.

Because low-methoxyl pectins are produced by at least two different methods and several variations can be obtained by mixing with other organic polymers and other ingredients, the types are not so uniform. For any specific application, one of the manufacturers most specialized in this field should be consulted. After long discussions and much laboratory work, the safety of amidated low-methoxyl pectins has been approved by the international health authorities.

Intermediate Pectin

Pectin with a degree of methoxylation between 50% and 55% takes an intermediate position, combining to some extent the characteristics of both high- and low-methoxyl pectin. In the presence of calcium ions, it is less sensitive to the exact total solids content and pH than high-methoxyl pectin and, therefore, has found its application in household pectin, where a strict control of the cooking time to obtain jellification of jam is not vital.

Mixtures of Pectin with Other Polymers

Pectin may be mixed with other gelling or thickening agents, but usually the nature of the admixed material is not disclosed. Morris (1990) and Toft, Grasdalen, and Smidsrod (1986) have summarized the data concerning the characteristics of gel caused by mixing pectin and alginate compared with that of each component alone.

10.5.5 Application of Pectin in Medicine and Nutrition

In spite of some difficulties in formal definition, pectin is now considered to be within the group of fiber substances (Baker, 1980), and all theories regarding this group apply to pectin. The beneficial effects of pectin in cases of diarrhea have been known for many years. Several medicaments used against this ailment contain pectin. Also in the diets of patients, even babies, suffering from diarrhea, plants rich in pectin are used.

Diverse aspects of the interaction between pectin and several metal ions have been studied. One of these is the detoxification of poisonous metals, such as lead and arsenic, which are bound by the pectin molecule (Kertesz, 1951). A marked decrease in heavy metals in the blood has been observed after the administration of pectin solutions, and consequently pectin is regularly provided to people working under conditions liable to cause poisoning with harmful metals (Russian Institute of Food Industry, private communication). Behall and Reiser (1986), summarizing the literature on absorption of metals in the presence of pectin, report contradictory results. However, two patents were granted in the United States in 1967, where pectin was used as a carrier of metal ions for improved incorporation in the human body. One concerns bismuth (US patent

3,306,819), used in the treatment of gastrointestinal diseases, and the other is iron (US patent 3,324,109), used to cure its deficiency (Lawrence, 1973). Cerda (1988), however, states that pectin has no influence on absorbtion of iron, and, undoubtedly, Behall and Reiser (1986) are correct in suggesting that further research is required.

Since the early 1970s much attention has been paid to the level of cholesterol in human blood, and consequently the possible influence of pectin has become the subject of much research. Positive results may be of commercial value for the producer.

Cerda (1988) claims that the daily intake of 15 g of pectin causes the cholesterol level in human blood to decrease, and Baig and Cerda (1980) explain it by the formation of insoluble complexes of pectin with low-density lipoprotein. Unfortunately, this has been proved only in vitro. In most cases in which more than 6 g of pectin per day was administered, the cholesterol level decreased (Behall and Reiser, 1986).

Attention has been paid also to the hypoglycemic effect of pectin, and Behall and Reiser (1986) summarize the papers published between 1976 and 1984. In about 65 of the cases, administering pectin decreased the glucose level in blood.

REFERENCES

Anon. (1987). *Requirements of Dairy Cattle*, 5th ed. National Academy of Sciences, Washington, DC.
Baig, M.M., and Cerda, J.J. (1980). In *Citrus Nutrition and Quality*, ed. S. Nagy and J.A. Attaway. ACS Symposium Series 143. American Chemical Society, Washington, DC, p. 25.
Baker, G.L., Joseph, G.H., Kertesz, Z.I. et al. (1944). *Chemical Engineering News*, **22**, 105.
Baker, R.A. (1980). *Citrus Nutrition and Quality*, ed. S. Nagy and J.A. Attaway. ACS Symposium Series 143. American Chemical Society, Washington, DC, p. 109.
Baker, R.A. (1993). *IFT Annual Meeting—Book of Abstracts*. Chicago, pp. 43, 153.
Behall, K., and Reiser, S. (1986). *Chemistry and Function of Pectin*, ed. M.L. Fishman and J.J. Jen. ACS Symposium Series 310. American Chemical Society, Washington, DC, p. 248.
Ben-Shalom, N., and Pinto, R. (1986). *Lebensmittel Wissenschaft und Technologie*, **19**, 158.
Cerda, J.J. (1988). *Confructa*, **32**(1) 6.
Cohen, E., Sharon, R., Volman, L. et al. (1984). *Journal of Food Science*, **49**(4), 987.
Crandall, P.G., and Wicker, L. (1986). In *Chemistry and Function of Pectin*, ed. M.L. Fishman and J.J. Jen. ACS Symposium Series 310. American Chemical Society, Washington, DC, p. 88.
Doesburg, J.J. (1965). *I.B.V.T. Communication No 25*, Wageningen.
Ferguson, R.R., and Fox, K.I. (1978). *Transactions of the Citrus Engineering Conference*, Vol. 24, p. 23.
Gross, J. (1977). In *Citrus Science and Technology*, Vol. 2, ed. S. Nagy, P.E. Shaw, and M.K. Veldhuis. AVI Publishing, Westport, CT, p. 302.
Hasegawa, S., Brewster, L.C., and Maier, V.P. (1973). *Journal of Food Science*, **38**(7), 1153.
Kanner, J., Ben-Shalom, N., and Shomer, I. (1982). *Lebensmittel Wissenschaft und Technologie*, **15**, 348.
Kanner, J., Ben-Shalom, N., and River, O. (1984). *Proceedings of the International Federation of Fruit Juice Producers*, **18**, 219.
Kertesz, Z.I. (1951). *The Pectic Substances*. Interscience, New York.
Kesterson, J.W., and Braddock, R.J. (1973). *Food Technology*, **27**(2), 50.
Lawrence, A.A. (1973). *Edible Gums and Related Substances*. S. Noyes Data Corporation, Park Ridge, NJ.
Mazza, G., and Miniati, E. (1993). *Anthocyanins in Fruits, Vegetables and Grains*. CRC Press, Boca Raton, FL.
Morris, A.R. (1990). In *Food Gels*, ed. P. Harris. Elsevier Applied Science, London, p. 344.
Northcote, D.H. (1986). *Chemistry and Function of Pectins*, ed. M.L. Fishman and J.J. Jen. ACS Symposium Series 310. American Chemical Society, Washington, DC, p. 134.
Pilnik, W., and Voragen, G.J. (1991). In *Food Enzymology*, Vol. I, ed. P.F. Fox. Elsevier Applied Science, London, p. 303.

Ragonese, C. (1991). *Fluessiges Obst*, **58**(5), 222.
Rebeck, H.M. (1995). In *Production and Packaging of Non-carbonated Fruit Juices and Fruit Beverages*, ed. P.R. Ashurst. Blackie A&P, Glasgow, Scotland.
Redd, J.B., and Hendrix, C.M., Jr. (1993). In *Fruit Juice Processing Technology*, eds. S. Nagy, Shu Shen Chin, and P.E. Shaw. AgScience Inc, Auburndale, FL.
Robbins, R.C. (1980). *Citrus Nutrition and Quality*, ed. S. Nagy, J.A. Attaway. ACS Symposium Series 143. American Chemical Society, Washington, DC, p. 43.
Rolin, C., and de Vries, J. (1990). In *Food Gels*, ed. P. Harris. Elsevier Applied Science, London, p. 401.
Schwimmer, S. (1981). *Source Book of Food Enzymology*, Ch. 29. AVI Sourcebook and Handbook Series, AVI Publishing, Westport, CT, p. 511.
Sinclair, W. B. (1961). *The Orange, Its Biochemistry and Physiology*. University of California.
Steingass, H., and Haussner, A. (1988). *Confructa Studien*, **96**.
Ting, S.V., and Hendrickson, R. (1969). *Food Technology*, **23**(7), 947.
Toft, K., Grasdalen, H., and Smidsrod, O. (1986). In *Chemistry and Function of Pectin*, ed. M.L. Fishman, and J.J. Jen. ACS Symposium Series 310. American Chemical Society, Washington, DC, p. 117.
Wilson, C.W. Bisset, O.W., and Berry, R.E. (1971). *Journal of Food Science*, **36**(6), 1033.
Wright, J. (1995). In Food Flavourings, ed. P.R. Ashurst. Blackie A&P, Glasgow, Scotland, p. 24.

SUGGESTED READING

Bruernmer, R.B. (1976). *Proceedings Florida State Horticultural Society*, **89**, 191.
Cohn, R. (1985). *Confructa Studien*, **29**(3), 178.
Griffiths, F.P., and Lime, B.J. (1959). Food Technology, 13(7), 430.
Hofsommer, H.J., Fischer- Ayl off-Cook, K.P., and Radcke, H.J. (1991). *Fluessiges Obst*, **58**(2), 62.
Horowitz, R.M. (1961). In *The Orange, Its Biochemistry and Physiology*, ed. W.B. Sinclair. University of California.
IFT Committee on Pectin Standardisation (1959). Final report. *Food Technology*, **13**(8), 496.
Kimball, D. (1991). *Citrus Processing, Quality Control and Technology*. AVI Books, Van Nostrand Reinhold, New York.
Shaw, P.E., and Wilson, C.W. (1983). *Journal of Food Science*, **48**(2), 646.

Chapter 11

Quality Management System and Hazard Analysis Critical Control Points

Gerry Burrows

11.1 INTRODUCTION

All reputable processors need to control the quality of their finished products. Apart from the cost to the processor's business from any complaints or warranty claims, there is a danger of losing the confidence of the customer and future demand, which will, in turn, affect the success of the business.

In the past this was achieved by the application of quality control principles. The company would measure and record the quality of the raw materials, control the processing parameters, and determine the quality of the finished product. It produced a "snapshot" of what was happening at the time of testing. Over the past few years this has been superseded by quality assurance, which still incorporates the elements of quality control but also extends to incorporate all the activities of the business in an overall quality plan. Raw materials are sourced from reputable suppliers who have been assessed and approved by the company. This means that the raw materials are more likely to be within specification on arrival at the factory.

Quality assurance must be a companywide philosophy. It means working to a planned, regularly monitored system of controls and is applied to all activities that significantly affect or impact on the quality of the product and the efficiency of achievement of that quality. These activities require systematic control based on documented procedures.

The principles that enable a company to satisfy its customers' requirements in the most cost-effective way may be stated as: "Quality is everybody's business," where each management function and activity carry a specific quality-related responsibility; "Do it right the first time," when prevention is better than cure by planning ahead and anticipating problems; and "Communicate and cooperate," so that all employees know what to do in addition to understanding where they fit within the organization and with whom they interface. These basic principles are most effectively put into practice by the operation of a fully documented quality system.

11.2 QUALITY SYSTEM

A quality system may be defined as "The organization, structure, responsibilities, procedures, processes, and resources for implementing quality management." These are all detailed in the company's quality manual, which states the company's commitment to quality and is usually the first indication a potential customer has of this commitment. The manual also includes the operating procedures, which describe how activities are to be performed and by whom.

If the quality system is to be effective it must be organized in such a manner that clearly shows how responsibility and authority are delegated. This requirement is categorized by the structure, responsibilities, and role and position of the quality assurance function.

All employees need to be aware of their position within the business, and an organization chart should be compiled showing lines of reporting, responsibilities, and the position of the quality assurance function within the business. They must also be aware that, although the quality department has responsibility for monitoring quality, the attainment of that quality is everyone's responsibility.

The managing director or chief executive of the business carries the ultimate responsibility for the quality of all products manufactured by the company. Obviously he or she cannot directly oversee all of the activities necessary to manufacture the products and has to delegate responsibility to suitably experienced and qualified personnel. These authorities and responsibilities should be defined in written job descriptions and organization charts, and in clearly defined procedures that indicate the responsibilities for specific activities and describe their particular training needs.

11.3 ROLE OF THE QUALITY ASSURANCE FUNCTION

The quality assurance function is responsible for ensuring that the quality system is established and implemented and remains effective. Its principal activities are to ensure that the fundamental working methods are established and that approved procedures are in place to cover them. Also, it should see that all departments and personnel have access to current versions of these procedures and verify that those people responsible for checking and controlling an activity have done so in a systematic manner and that appropriate written records are maintained.

11.4 EXTERNAL ACCREDITATION

The documented quality system describes the established plan for achieving quality. However, there may be occasions when an organization needs to operate differently from the established systems in order to satisfy a specific customer requirement. In this circumstance an alternative quality system can be documented as a specific quality plan. This simply modifies the established quality system to reflect the specific requirements.

Perhaps the most widely used quality system standard is ISO EN BS 9000, which is divided into three parts, each of which has a series of requirements appropriate to a specific type of organization.

1. Part 1 applies to companies engaged in design/development, production, installation, and servicing.
2. Part 2 applies to companies engaged in production and installation.
3. Part 3 applies to companies engaged in final inspection and test activities.

The standard is written in such a manner that it can be applied to any type of business, including food production. The advantage of operating a formal quality system standard is that, in order to achieve and maintain accreditation, it has to be audited by an external body (e.g., British Standards Institute [BSI], Lloyds Register, SGS, etc.), who will assess compliance to the standard without any bias or commercial influence. The quality assurance function will ensure that all the procedures are reviewed and updated as necessary. It will also determine and report on the main causes of any quality losses and nonconformance. Finally, it should organize audits to verify that the quality assurance philosophy is being followed throughout the business and that effective procedures and work instructions are being implemented by all departments.

The quality assurance function need not necessarily carry out all of the audits but it should prepare the auditing schedule. The actual auditing can be decentralized and can be carried out by members of all departments. Ideally, they should audit departments other than their own.

There are some fundamental principles regarding the quality assurance function. It should not be regarded as a policing function or as the "super checker," nor is it a stopgap for all potential mistakes, errors, or carelessness. All employees should be encouraged to practice "ownership" of the department or workspace and to verify the quality of their own work by self-checking. The audit and review of the quality system should be undertaken by suitably qualified personnel who are independent of the function being audited.

The operation of quality assurance principles does not absolve management from needing to recognize apparently insignificant problems, sometimes dismissed as irrelevant, the cumulative effect of which can be much greater than apparently more important decisions taken at a higher level.

11.5 EXTERNAL AUDITING

Another route for external auditing is to follow the technical standard for "Companies Supplying Retailer Branded Food Products," compiled by the British Retail Consortium. It has been developed to help retailers in the fulfillment of their legal obligations to their consumers by providing a common standard for the inspection of companies supplying branded foods.

The standard is split into two levels: foundation and higher. All the requirements for the certificate of inspection at the foundation level must be met. Further specified criteria must be complied with to gain a certificate of inspection at the higher level. Ideally, companies should progress through the two levels of the standard and thus fully comply with all of its criteria. The frequency of inspection is determined by considering the type of operation and the level of certification sought.

In order to achieve certification, the company's quality management system is subject to scrutiny. Several elements are required, including the following:

- quality policy statement and quality manual
- organization, structure, responsibility, and authority
- document control
- procedures and record-keeping
- raw material and finished product specifications
- internal auditing and corrective action
- identification and traceability
- complaint handling and recall procedures
- supplier performance monitoring

Additionally, the environmental standards of the factory are considered, including the following:

- the location, including the condition of the perimeter and the grounds
- the fabrication of the factory
- the equipment and maintenance standards
- the staff facilities
- the layout and product flow to eliminate physical and chemical contamination of the product
- housekeeping and hygiene
- waste disposal
- pest control
- transport

The product must be controlled by considering the following factors:

- design and development, including the type of packaging
- product analysis
- stock rotation and shelf life
- metal and foreign body control
- segregation and release procedures and control of nonconforming material

Also, the process must be controlled by implementing the following controls:

- time and temperature control
- equipment and process validation
- calibration
- control of quantities
- any specific handling procedures

All staff must be suitable for work in a food factory and must have been subject to the following:

- personal hygiene standards
- medical screening
- access to suitable protective clothing
- training

Finally, the company must operate a hazard analysis critical control point (HACCP) system, which should be comprehensive and systematic. It is recognized that there are a number of ways of recording and documenting an HACCP system, but the company must be able to demonstrate that it has carried out a study and has implemented suitable controls with full records and corrective actions and has designated responsibilities.

11.6 HACCP

During recent times there has been an increasing awareness among the general public regarding microbiological, chemical, or physical food safety. This has created a greater demand for an effective mechanism for ensuring that all food supplied to the consumer is safe.

One of the requirements of ISO 9002 is to "establish, document and maintain an inspection system capable of producing objective evidence that material or services conform to the specified requirements, whether manufactured or processed by the supplier or procured from subcontractors." This can be substantially satisfied by the successful development and implementation of an HACCP system.

HACCP is an analytical tool that enables management to implement and maintain an ongoing system for ensuring food safety. It involves a systematic assessment of all the steps involved in a particular food manufacturing process and the identification of those steps that have a critical effect on food safety. An HACCP analysis results in a list of critical control points (CCPs), together with required parameters, monitoring procedures, and any corrective action for each CCP. Full records of each assessment should be kept, and the efficacy of the study should be reviewed at regular intervals, especially if any aspect of the operation should change.

The HACCP system can be used to identify microbiological, chemical, or physical hazards that could affect food safety. It should be applied to a specific product/process combination, either to an existing product or as part of the development brief for a new product or process. In order for the assessment to be carried out and for the subsequent successful implementation, the senior management of the company and the technical staff must give it their full commitment.

One of the many advantages of the HACCP concept is that it will enable a business to move away from quality control, which entails testing the final product, to a quality assurance approach whereby potential hazards are identified and controlled during manufacture. HACCP offers a greater security of control over product safety than would ever be possible by traditional final product testing and, when properly implemented, it can be used as part of the defense of "due diligence" in case of prosecution.

11.7 BENEFITS OF AN HACCP APPROACH

There are a number of benefits that can be derived from adopting an HACCP approach to food safety. The following are some of them:

- It is a cost-effective method for controlling food-borne hazards.
- It provides a systematic approach covering all aspects of food safety from raw materials to the final product.
- If correctly applied, it should identify all conceivable hazards, and these should include any that can realistically be predicted to occur.
- It will focus the technical resources into those parts of the process that are critical to food safety.
- It will reduce losses and costs by using a preventive approach and identifying hazards before the product is manufactured.
- It can be used with other quality management systems.

11.8 HACCP PRINCIPLES

As has been stated earlier, HACCP is a system that identifies hazards that can adversely affect food safety and specifies appropriate measures for their control. There are seven principles that should be applied when implementing such a system:

1. Conduct a hazard analysis. This is done by preparing a flow chart listing all the steps in the process and, from it, identifying and listing the hazards likely to occur at each step and determining the control measures.
2. Identify the CCPs in the process using the decision-tree approach (Figure 11–1).
3. Establish the parameters and the tolerances that must be met to ensure that each CCP is fully controlled.
4. Establish a system for monitoring the control of each CCP by scheduled testing or observation—frequency of testing can affect the degree of control.
5. Decide on the corrective action to be taken when the scheduled monitoring indicates that a particular CCP is moving out of control.
6. Ensure that all procedures, work instructions, and records pertinent to the HACCP system are fully documented.
7. Establish verification and validation procedures that confirm that the HACCP system is appropriate for the current process, has taken into account any changes, and is working effectively.

11.9 CONDUCTING AN HACCP STUDY

When conducting an HACCP study the seven principles listed above may be applied as 14 stages:

11.9.1 Determine the Terms of Reference

The HACCP study should be carried out on a specific product/process line, and it is essential that the terms of reference are clearly defined at the outset. Therefore it is necessary to define whether the study should consider microbiological, chemical, or physical hazards (or any combination of these) and whether product safety and microbiological spoilage are to be considered. The terms of reference must also clearly state whether the product is to be judged safe at the point of consumption, or at the end of manufacture with clear instructions for storage and use. When carrying out the initial study, priority should be given to product safety.

11.9.2 Select the HACCP Team

It is important that collection, collation, and evaluation of the technical data are carried out by a multidisciplinary team consisting of (1) a member of the quality control/assurance department who understands the types of hazard being considered and the associated risks for the particular product; (2) a member of the production staff who is closely involved with the process under study and understands the details of what actually happens on the production line; (3) an engineer who has knowledge of the engineering operation and hygienic design of the process equipment under study; and (4) other relevant specialists, for example, operators, packaging technologist, or hygiene manager who may be co-opted as necessary.

The team members should have a good knowledge and experience of HACCP techniques and ideally should be one of the above. A member of the team must be appointed chairperson, and another member of the team should be appointed as secretary to document the meetings so that it is clear who is to take action and what progress has been made.

When selecting the team it is recommended that the members have a working knowledge of the process and that they should not be so senior as to be divorced from it. The team members may need training in the principles of HACCP and how to approach the analysis systematically, its role in product safety, and the benefits of the system.

11.9.3 Describe the Product

A full description of the product under study should be prepared. It should be defined in terms of composition, structure, process, packaging, storage and distribution conditions, required shelf life, and any instructions for use.

11.9.4 Identify the Intended Use

The intended use of the product by the consumer and the consumer target groups should be identified.

11.9.5 Construct a Flow Diagram

The format of the flow diagram is a matter of personal choice, as long as each step in the process (including process delays) from the selection of raw materials through processing, storage, and distribution to retail and consumer handling is clearly outlined in sequence with sufficient technical data for the HACCP study to proceed.

Examples of the type of data to be included are as follows:

- all raw materials/ingredients and packaging used (include microbiological, chemical, and physical data)
- floor plans and equipment layout (to avoid cross-contamination)
- sequence of all process steps
- time/temperature history of all raw materials, intermediate and final products, including potential for delay
- flow conditions for liquids and solids
- product recycle/rework loops
- equipment design features (including presence of void spaces)
- efficacy of cleaning and hygiene procedures
- environmental hygiene
- personnel routes
- routes of potential cross-contamination
- high-care/low-risk area segregation
- personal hygiene practices
- storage and distribution conditions
- consumer use instructions

This is not an exhaustive list but will provide a good start to any study.

11.9.6 Establish On-Site Verification of the Flow Diagram

It is important that the flow diagram is verified by observation on the factory floor, including its operation on night shift or at weekends when, for one reason or another, procedures may vary. If it is a new process, the team must ensure that the flow diagram correctly represents the proposed process and then check the actual line during preproduction trials.

11.9.7 List All Hazards Associated With Each Process Step and List the Control Measures for Each Hazard

From the flow diagram, the HACCP team should list all the hazards that might reasonably be expected to occur at each process step, together with the control measures required for each hazard. The team should also consider the manner in which the process is managed and what could realistically occur that may not be covered by the flow diagram (e.g., process delays, temporary storage, etc.).

The hazards included in the study must be such that their elimination or reduction to acceptable levels is essential to the production of a safe foodstuff with the required attributes.

No attempt should be made at this stage to identify CCPs. This is in order to ensure that all conceivable hazards are identified.

Control measures are those actions or activities that are required to eliminate hazards or reduce their occurrence to an acceptable level. More than one measure may be required to control a specific hazard that occurs at different stages of a process. However, in other processes, one control measure at a single CCP may control more than one hazard. All control measures need to be detailed in specifications and procedures to ensure their effective implementation.

11.9.8 Apply an HACCP Decision Tree to Each Process Step in Order to Identify CCPs

Each process step listed in the flow diagram shall be considered for identification as a CCP, using the decision-tree approach (Figure 11–1). When the hazards and control measures for one particular step have been considered and agreed upon, the HACCP

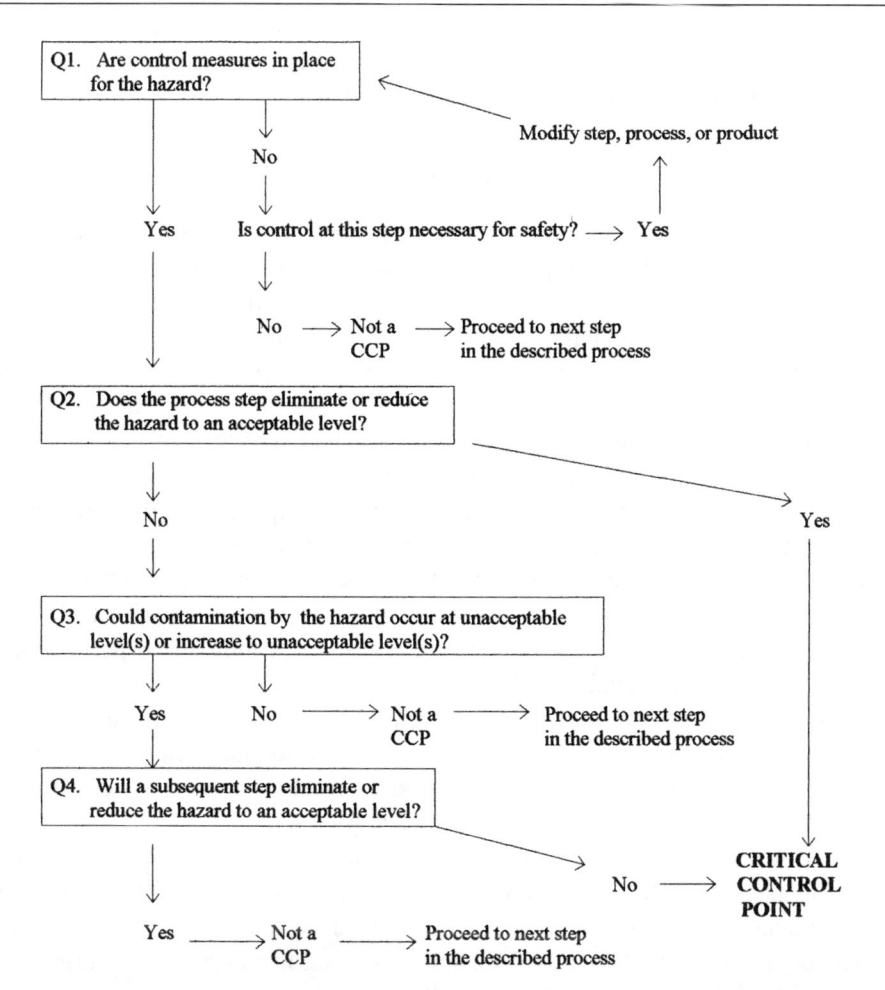

Figure 11–1 Critical control point decision tree. *Source:* Adapted with permission from Leaper (1992) *HACCP: A Practical Guide (Third Edition).* Published by Campden & Chorleywood Food Research Association. ISBN 0-905-942-05-1.

team should consider the hazards and control measures for the next step until the decision tree has been applied to all the process steps in the flow diagram. Application of the decision tree will determine whether the process step is a CCP for each specified hazard. There is no limit to the number of CCPs that can be identified.

The use of the decision-tree approach shall be flexible and requires common sense. This is particularly important when considering the impact of practices or procedures that could realistically occur but that may not be detailed in the flow diagram. Access to technical data will be necessary to answer the questions in the decision tree.

11.9.9 Establish Target Level(s) and Tolerance for Each CCP

Having identified all the CCPs in the product/process under study, the team should then identify targets and specified tolerances for the control measures at each CCP. The specific target level and tolerance must represent some measurable parameter related to the CCP. Target levels and tolerances that can be measured relatively quickly and easily are preferred (e.g., temperature, time, pH, water activity, chemical analyses, or visual assessments of product and operational practices).

11.9.10 Establish a Monitoring System for Each CCP

Monitoring is the scheduled measurement or observation of a CCP for compliance with the target level(s) and the specified tolerances determined for each control measure. The monitoring system describes the methods used to confirm that all CCPs are operating within specification and also produces an accurate record of performance for future use in verification. The selection of the correct monitoring system is an essential part of any HACCP study.

Monitoring procedures must be able to detect any loss of control at the CCP. Ideally, they should provide this information in time for corrective action to be taken to regain control of the process before there is a need to quarantine or reject any product. The monitoring systems may be on-line (time or temperature measurements), or off-line (measurement of salt levels or pH). On-line systems give an immediate indication of performance. As off-line systems by their nature require monitoring to be conducted away from the production line, a variable (sometimes long) time period may elapse before the results are available and action can be taken.

Monitoring systems may also be continuous, such as using a thermograph to record continuous process temperatures, or discontinuous, such as sample collection and analysis. The continuous systems provide a dynamic record of performance while discontinuous systems must ensure that any samples taken are representative of the bulk product. Whichever monitoring system is chosen, the HACCP team must ensure that the results obtained are relevant to the CCP and any limitations are fully understood. Physical, chemical, and sensory methods are preferred because of their speed of response. Microbiological testing takes time for the results to appear and then further time may be needed for interpretation of these results.

In addition to identifying the most appropriate monitoring system, the team should decide the following:

- Who is to act? The team must specify the job title of the operator who will carry out the monitoring. This person must have the knowledge and authority to carry out the corrective action if the specified tolerances are not met. If he or she does not have the authority to carry out this action there must be a clear instruction of who should be informed.
- When are they to act? If monitoring is not continuous, the frequency must be specified and shall be sufficient to ensure that the CCP is being controlled.
- How are they to act? A detailed description of the monitoring method is required, and the designated operators must be trained to understand the monitoring procedures and to carry them out correctly.

11.9.11 Establish a Corrective Action Plan

The HACCP team should specify the corrective action to be taken when monitoring results show that a CCP has deviated from its specified tolerance, or preferably what action should be taken when the monitoring results indicate a trend toward loss of control. In the latter case, action may be taken to bring the process back into control before the deviation leads to loss of control and becomes a hazard to product safety.

A disposition action will be needed to deal with any food that has been produced during the time that the CCP was out of control.

Both the corrective action and disposition action must be documented in the HACCP record with the responsibility clearly assigned.

11.9.12 Establish Verification and Validation Procedures

The HACCP team should implement procedures that can be used to demonstrate compliance with the HACCP plan. The members should examine the whole HACCP system and its records and should specify the frequency and methods used for verification. Examples of verification procedures include an internal and external (independent) review of the HACCP study and its records, together with a review of any deviations and product dispositions, and an audit of procedures and associated records at each CCP to observe whether the appropriate CCP is under control.

An important aspect of verification is the initial validation of the HACCP program to determine that the plan is scientifically and technically sound, that all hazards have been identified and that, if the HACCP program is properly implemented, these hazards will be effectively controlled.

Validation ensures that the HACCP program does what it was designed to do; that is, to be successful in ensuring the production of safe product. Food manufacturing plants are required to validate their own HACCP programs.

Information needed to validate the HACCP program often includes:

- expert advice and scientific studies
- in-plant observations, measurements, and evaluations

For example, validation of the refrigerated storage of fresh produce in pouches should include the scientific/technical justification for the required refrigerated temperature needed to prevent growth of pathogenic microorganisms, including *Listeria*, and studies to confirm that the conditions of refrigeration will prevent contamination.

Subsequent validations are performed and documented by the HACCP team or an independent expert as needed. For example, validations are conducted when there is an unexplained system failure; a significant product, process or packaging change occurs; or new hazards are recognized.

In addition, an unbiased, independent authority should conduct a periodic comprehensive verification of the HACCP program. Such authorities can be internal or external to the food operation. This should include a technical evaluation of the hazard analysis and each element of the HACCP program as well as on-site review of all flow diagrams and appropriate records from operation of the plan. A comprehensive verification is independent of other verification procedures and must be performed to ensure that the HACCP program is resulting in the control of the hazards. If the results of the comprehensive verification identify deficiencies, the HACCP team modifies the HACCP program as necessary.

11.9.13 Establish Documentation and Record Keeping

If the application of the HACCP system to a food process is to be successful, then efficient, accurate, and current records are essential. The food producer should be able to demonstrate that the principles of HACCP have been correctly applied and the documentation and records are appropriate for the size of the operation. The documentation should be assembled into a manual and integrated into a controlled quality management system. It should include the findings of the HACCP team (hazard analysis, CCP determination, control measures, and targets/tolerances), the records of the verification and validation procedures, and the procedures and work instructions. Each should be supported by current records.

Examples of the records required include the following:

- nature, scope, and quality of raw materials
- complete processing record, including storage and distribution
- cleaning and disinfection records
- all decisions reached relating to product safety
- deviations file
- corrective/disposition action file
- modification file
- verification and validation file
- review data

The length of time records should be retained is determined by the nature and shelf life of the product.

11.9.14 Review the HACCP Plan

The HACCP plan should be reviewed periodically, the frequency being determined by the "risk" of the product and its intended use. Also, it is necessary to have a system in place that will automatically trigger a review of the HACCP plan ahead of any changes in the process or equipment that may affect product safety.

Changes to any of the following must automatically be assessed to determine whether a review is needed:

- raw material or product formulation
- raw material supplier
- the processing system
- factory layout and environment
- process equipment (including modifications to existing equipment)
- cleaning and hygiene program
- packaging, storage, or distribution systems
- staffing levels and/or responsibilities
- customer or consumer use

A review may also be needed if information is received from the marketplace indicating a health or spoilage risk to the product or the emergence of food borne pathogens significant to public health.

All data arising from the HACCP review must be documented and form part of the HACCP records. Any changes resulting from the HACCP review must be included in the HACCP plan, as certain CCP control measures or targets/tolerances may have been changed or additional CCPs or control measures implemented.

11.10 IMPLEMENTATION OF THE HACCP PLAN

Once the HACCP plan has been developed by the team it must be implemented into the production operation. Essential to its successful implementation is the transfer of ownership to operators, supervisors, and managers, who should be suitably trained to operate and maintain the plan.

There are a number of approaches that can be taken, but they all rely on the continued commitment and support of senior management.

The responsibility for the everyday implementation of the HACCP plan lies with the line operators and supervisors, while the overall responsibility lies with the production management. These individuals must be encouraged to take ownership of the HACCP plan quickly, and this can be speeded up if the operators and supervisors have contributed to relevant aspects of the plan during its development.

In order to ensure successful implementation, the staff need to be informed why the HACCP plan was developed and by whom, the support given by management at all levels to maintain the system, and the implications on individual's roles and responsibilities in implementing the plan. They should be encouraged to suggest modifications for consideration by the HACCP team.

Successful implementation of the HACCP plan into a process needs to be accompanied by suitable training. The needs of the staff will vary according to their activity, and training should be ongoing rather than a one-off session. The training should ensure that all members of staff have an understanding of the following:

- sources of hazards and their effect on product safety
- critical control points and their effect on the assurance of product safety
- control measures at the critical control points for which they are responsible
- targets/tolerances that must be met
- monitoring procedures that they must operate
- corrective actions to be initiated and by whom, when the monitoring indicates that critical limits are exceeded or a trend toward loss of control is evident
- record-keeping requirements
- objective of verification procedures

The methods used for training are a matter of company preference, but it is essential that all staff understand their roles and responsibilities in the HACCP plan.

Once the training has been given, the HACCP plan should be implemented as quickly as available resources allow. Problems may occur during the early stages, and suitable adjustments will have to be made or further training given.

It is important that the HACCP plan is maintained if it is to remain effective. All staff need to understand that the team meetings, the verification audits, and changes arising from the findings of the audits all form part of the HACCP system and are aimed at achieving the objective of the study in the most effective (including cost-effective) manner.

11.11 EXAMPLES

If the company is operating a formal quality management system such as ISO 9002, then references should be made to work instructions, systems procedures, and the like. It is recommended in this case that written work instructions are not included in the HACCP documents, as these will result in a great deal of work in changing a number of instructions should a process change.

In order to demonstrate the application of HACCP principles, some examples of actual studies are given in Appendixes 11–A to 11–E. It should be noted that the details given for each example are not exhaustive, and verification activities are not included. The examples should not be taken as specific recommendations for similar processes or for direct use in the factory, but the information is intended only as a demonstration of how HACCP principles can be applied. The persons listed as having authority are only examples; obviously, staff in different factories will have differing job titles and responsibilities.

In this particular approach the control points have been categorized as follows:

S = Product safety. This includes microbiological, chemical, or physical hazards that can affect the safety of the food.

Q = Product quality. This includes items affecting the appearance, flavor, texture, or size of product but that will have no effect on the product safety.

T = Traceability. For any quality management system to be valid there must be full traceability from raw materials to the finished product. Without traceability there is no way that controls can be verified for a particular product on a specific day or shift.

It should be noted that full accountability is documented in the system so that there is no doubt where the responsibility lies.

The examples given in Appendixes 11-A to 11–E are intended to cover the various aspects of the canning process from the arrival of the raw materials at the factory to the dispatch of the finished product from the warehouse. The HACCP system is very versatile and can be used at other points of the food chain. It can be applied to the growing, harvesting, and packing of fresh fruit and it can also be applied to the distribution, storage, and retailing conditions of the finished product. It can also be used to detail those controls that may appear in every product/process, thus avoiding replication, or those that may not appear in any flow chart. These include staff selection, induction and training, protective clothing, hygiene and cleaning of the premises, control of foreign bodies (pests, glass, wood, plastic, etc.), door discipline, and control of waste.

An important point to remember is that, for the HACCP system to be successful, the material must not lose its identity so that batches are traceable throughout the process. This makes it possible to fully audit the system and verify the control of the identified batches.

It is worthy of note that although controls for different hazards may have been established in the past, the advantage of the HACCP system is its organized and formal approach.

SUGGESTED READING

British Retail Consortium. (1998). *Technical Standard for Companies Supplying Retailer Branded Products.*
British Standards Institute. (1994). *BS EN ISO 9002. Quality Systems—Model for Quality Assurance in Production, Installation and Servicing.*
British Standards Institute. (1994). *BS EN ISO 9000. Quality Management and Quality Assurance Standards—Part 1: Guidelines for Selection and Use.*
HMSO. (1994). *Guidelines for the Safe Production of Heat Preserved Foods.*
Institute of Food Science and Technology. (1991). *Food and Drink—Good Manufacturing Practice—A Guide to Its Responsible Management.*
Campden and Chorleywood Food Research Association. (1997, April). *HACCP: A Practical Guide*, 2nd ed.
Campden and Chorleywood Food Research Association. (1995). *Guidelines for the Prevention and Control of Foreign Bodies in Food. Guideline No. 5.*
Codex Alimentarius Commission. (1996). *Hazard Analysis and Critical Control Point (HACCP) System and Guidelines for Its Application.* Draft Report of the Twenty-ninth Session of the Codex Committee on Food Hygiene, Washington DC.
ILSI Europe. (1993). *A Simple Guide to Understanding and Applying the Hazard Analysis Critical Control Point Concept.*
Mortimer, S., and Wallace, C. (1994). *HACCP: A Practical Approach.* Chapman & Hall, London.
Royal Institute for Public Health and Hygiene. (1995). *HACCP Training Standard: HACCP Principles and Their Application in Food Safety.* RIPHH, London.

Appendix 11–A

Canned Rhubarb

This document covers rhubarb, both in syrup or water and solid pack, produced from fresh raw material.

Process Flow Chart

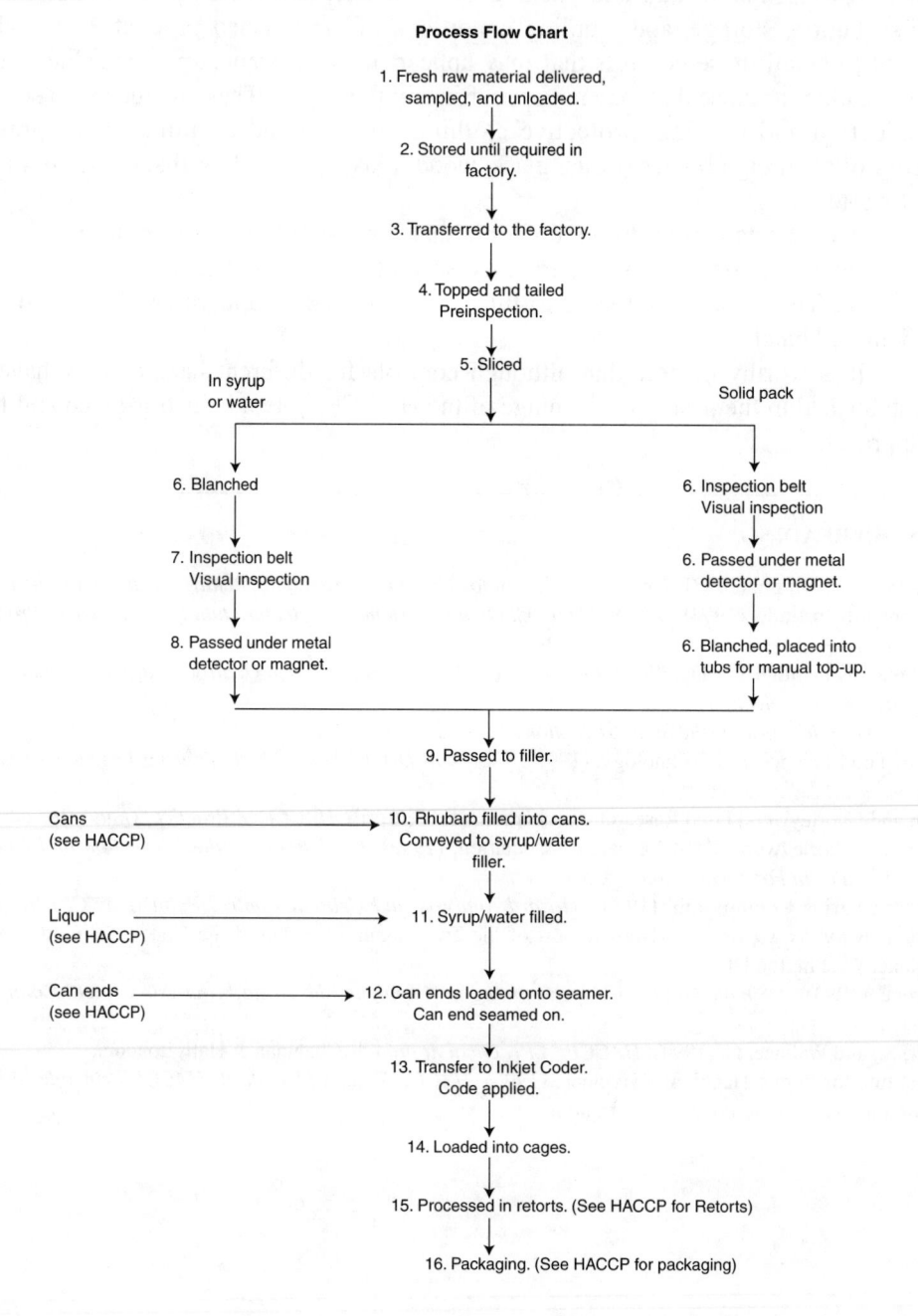

1. Fresh raw material delivered, sampled, and unloaded.
2. Stored until required in factory.
3. Transferred to the factory.
4. Topped and tailed Preinspection.
5. Sliced

In syrup or water:

6. Blanched
7. Inspection belt Visual inspection
8. Passed under metal detector or magnet.

Solid pack:

6. Inspection belt Visual inspection
6. Passed under metal detector or magnet.
6. Blanched, placed into tubs for manual top-up.

9. Passed to filler.
10. Rhubarb filled into cans. Conveyed to syrup/water filler. — Cans (see HACCP)
11. Syrup/water filled. — Liquor (see HACCP)
12. Can ends loaded onto seamer. Can end seamed on. — Can ends (see HACCP)
13. Transfer to Inkjet Coder. Code applied.
14. Loaded into cages.
15. Processed in retorts. (See HACCP for Retorts)
16. Packaging. (See HACCP for packaging)

Canned Rhubarb

CCP	Monitor/ Frequency	Hazard	Procedure/Control Measure	Parameter Range	Corrective Action	Authority
S1	Raw material intake. (Each load)	Foreign bodies (stones, metal, glass, etc.) in incoming rhubarb.	Each delivery sampled and checked against specification.	As determined in Raw Material Specification	Inform Buyer and Production Manager.	Quality Control
Q1	Ditto	Raw material unfit for canning (e.g., high blemish levels, damage, excessive EVM).	Ditto	Ditto	Ditto	Quality Control
T1	Ditto	Inability to trace grower and date/time of delivery	Details of each load to be entered onto Raw Material Sheet.	Traceability to be maintained.	Investigate and take appropriate action.	Quality Control
Q2	Ditto	Contamination or damage of raw material due to unsuitability of delivery vehicle.	All lorries and containers to be checked prior to unloading.	Absence of damage or foreign material or odors.	Quarantine delivery and inform Buyer and Production Manager.	Goods Inwards
S2	Storage (Continual)	Contamination with foreign material (e.g., plastic, wood, etc.)	Visual check of all plastic crates or wooden boxes.	All boxes and crates to be in good condition and raw material free from foreign bodies.	Do not use any defective crates or boxes.	Yard Foreman
S3	Transfer to factory. (Each load)	Contamination with foreign material.	Visual check of loads entering the factory.	Absence of foreign material.	Inform Supervisor.	Debox Operator
Q3	Topping/tailing. Preinspection. (Continual)	Tops and tails left in rhubarb.	Visual inspection. Manual top and tail.	Absence of tops and tails.	Inform Supervisor. Slow or stop inspection line.	Line Operators
S4	Ditto	String from tied bundles.	Manually remove string.	Absence of string.	Ditto	Ditto
Q4	Ditto	EVM in final product.	Visual inspection.	Absence of EVM.	Ditto	Ditto
Q5	Ditto	Blemished or damaged rhubarb in final product.	Visual inspection.	Absence of damaged or blemished rhubarb.	Ditto	Ditto
Q6	Slicing (Continual)	Incorrect slice size.	Engineers to set machine to give required slice size.	Size as required by specification.	Inform Q.C. Department and Management. Reset machine.	Engineers Line Supervisor
S5	Ditto	Broken slicer blades.	Check blades at start-up. Metal detection on line. (See procedure for operating metal detectors.)	No pieces of blade in final product.	Stop slicer. Engineers to replace blades.	Engineers Line Supervisor
Q7	Blanching (Continual) (Conditions recorded hourly)	Poor quality due to incorrect blanching conditions.	Blanch at specified time and temperature.	Blanching conditions laid down in manufacturing specification.	Adjust time and temperature. Quarantine any affected product.	Line Supervisor Quality Control
S6	Ditto	Cross-contamination of uncovered product in tubs.	Cover all tubs as they are filled.	No cross-contamination of product.	Quarantine any nonconforming product.	Line Supervisor
Q8	Inspection. (Continual)	Blemished or damaged rhubarb in final product.	Visual inspection at suitable product depth.	To meet final product specification.	Slow inspection belt or increase number of inspectors.	Line Supervisor

Approved (Production): _____ Approved (Q.C.): _____

continued

Canned Rhubarb *continued*

CCP	Monitor/ Frequency	Hazard	Procedure/Control Measure	Parameter Range	Corrective Action	Authority
S7	Ditto	Contamination from frayed or damaged belt.	Check condition of belt during cleaning procedures.	Inspection belt to be whole, sound, and without damage.	Report to Engineers for repair or replacement.	Line Supervisor Engineers
S8	Metal Detection. (Detectors checked prior to start of production and then hourly)	Metal in final product.	All rhubarb to be passed through a working metal detector. (See procedures for operation of metal detectors.)	Absence of metal in final product. Metal detectors to remove test pieces: Ferrous—2.5mm Nonferrous—3.0mm Stainless steel—4.0mm	If metal detector not working inform the Engineers and the Management. Quarantine product back to last good check.	Line Supervisor Quality Control Engineers
Q9	Passed to filler. (Continual)	Development of off-flavors if delays occur.	Stoppage times recorded.		Quality Control to check quality. Transfer to another line if necessary.	Line Supervisor Quality Control
Q10	Filling Rhubarb (Continual, visual) (Weight checks every 30 min)	Underfill. Failure to meet Code of Practice requirements.	Filler set up to meet specified requirements. Filled weight checks.	Weights as required by manufacturing specification.	Inform Engineers to adjust fill. Quarantine any nonconforming product.	Line Supervisor Quality Control Engineers
S9	Ditto	Overfill causing insufficient heat process.	Heat process determined on 10% overfilled cans.	Ditto	Ditto	Line Supervisor Quality Control Engineers Thermal Process Department
Q11	Filling liquid. (Weight and temperature checks every 30 min)	Low vacuum due to low filling temperature.	Liquid heated to sufficient temperature before being pumped to filler.	Minimum initial temperature as specified in manufacturing specification.	Inform management. Quarantine any nonconforming product.	Boiler Operator Line Supervisor Quality Control
Q12	Ditto	Underfilling leading to low net weight—failure to meet legislation. Excess headspace may cause cans to panel during process.	Filler set up to meet requirements. Net weight checks.	Weights as required by manufacturing specification.	Inform Engineers to adjust fill. Quarantine any nonconforming product.	Line Supervisor Quality Control Engineers
S10	Ditto (Headspace checks every 30 min)	Overfilling—lack of headspace causing peaking during process.	Can tipper or spacers to ensure adequate headspace. Headspace checks.	Target headspace according to manufacturing specification.	Inform management. Quarantine any nonconforming product.	Line Supervisor Quality Control
S11	Can ends seamed on. (Continual, visual.) (Seam checks every 2 hours)	Faulty seams allow ingress of pathogenic and spoilage microorganisms during cooling.	Seamer set up to specified parameters.	Absence of visual defects, e.g., spurs, skidders, cut-overs, etc. Seam dimensions to be within critical limits.	Stop line. Inform Engineering. Reset. Quarantine any affected product.	Seamer Operator Line Supervisor Quality Control Engineers

Approved (Production): _____ Approved (Q.C.): _____

continued

CCP	Monitor/ Frequency	Hazard	Procedure/Control Measure	Parameter Range	Corrective Action	Authority
S12	Ditto	Insufficient or excess vacuum causing peaking/paneling during heat process.	Use of steam-flow closure where necessary to achieve required vacuum.	As laid down in manufacturing specification.	Adjust steam flow and/or liquid temperature. Quarantine any suspect product.	Seamer Operator Line Supervisor Quality Control
T2	Ink Jet Coding. (Code checked prior to start of production and then hourly)	Loss of identification and traceability and failure to meet customer and legislative requirements.	Code applied to can end as determined by Production Program. Inkjet code includes Best before end date. Checked by Seamer Operator and Supervisor at start of production.	Code to be legible and as specified in the Coding Manual.	Stop line if Ink Jet Coder is not working. Inform Electrician to reset. Quarantine any nonconforming product.	Line Supervisor Seamer Operator Quality Control Electrician
S13	Cans loaded into cages. (Continual)	Damage to cans, causing leaks.	Correct procedures implemented.	All cans to be free from damage.	Remove all damaged cans.	Cage Loader Cage Unloader
S14	Ditto	Microbiological spoilage due to incorrect heat process caused by using wrong number of layer pads in cages.	Ditto. Remove all layer pads from empty cage prior to filling it.	No layer pad in the bottom of the cage and only one between each layer of cans.	Quarantine any suspect stock.	Cage Loader Cage Unloader

Approved (Production): _____ Approved (Q.C.): _____

Appendix 11–B

Cans and Ends Flow Chart

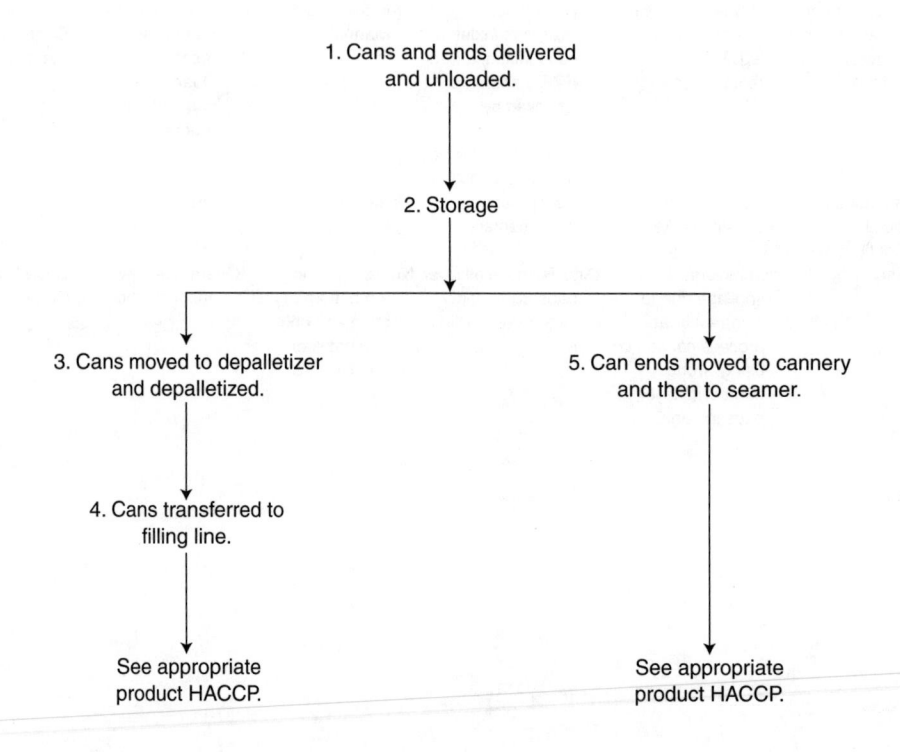

1. Cans and ends delivered and unloaded.//↓//2. Storage

3. Cans moved to depalletizer and depalletized.

4. Cans transferred to filling line.

See appropriate product HACCP.

5. Can ends moved to cannery and then to seamer.

See appropriate product HACCP.

Cans and Ends

CCP	Monitor/ Frequency	Hazard	Procedure/Control Measure	Parameter Range	Corrective Action	Authority
S1	Cans and ends delivery and unloading. (Every load)	Can or end faults due to problems in can or end manufacture.	Approved Supplier and compliance with specification. Assess supplier.	To meet agreed supplier specification.	Appropriate action taken with supplier.	Purchasing Manager Technical Manager
Q1	Ditto	Incorrect specification, i.e., lacquer, tin coating.	Ditto. Identified on pallet ticket.	Ditto	Ditto	Purchasing Manager Technical Manager
T1	Ditto	Inability to trace cans/ends back to point of manufacture.	Details on supplier's pallet label. Loads checked against delivery note.	Pallet label to be present on all pallets of cans and ends when delivered.	Reject pallet.	Goods Inwards
S2	Ditto	Foreign bodies in cans getting into final product.	Visual check on incoming pallets.	Absence of evidence of rodent, insect, or bird activity.	Reject any affected pallets.	Goods Inwards
S3	Cans and ends storage. (Continual)	Contamination with metal, glass, wood, etc.	Good housekeeping. Housekeeping audits. Absence of glass from building. Glass breakage procedure. Cans covered on delivery. All part pallets covered by layer pad.	Adherence to Good Manufacturing Practice.	Progress any action points raised by housekeeping audit. As detailed in glass breakage procedure.	Stores Foreman Technical Manager
S4	Ditto	Infestation by rodents, insects, or birds.	Suitable building. Good proofing. Pest control contract. Housekeeping audit.	Absence of infestation.	Progress action raised by pest control inspection and by housekeeping audit.	Stores Foreman Technical Manager
T2	Ditto	Loss of identification and traceability, particularly of part pallets.	Check for presence of suitable identification.	All pallets and part pallets to be identified.	Investigate and take appropriate action.	Production Manager
T3	Depalletizing. (Each pallet)	Loss of traceability.	Record details from pallet label and date/time used.	Traceability to be maintained.	Investigate and take appropriate action.	Depalletizer Operator Production Manager
Q2	Ditto	Wrong cans used.	Cans taken to unloader as detailed in production program.	As in manufacturing specification.	If correct cans not available, acceptable substitutes may be used. Quarantine suspect product.	Stores Foreman Depalletizer Operator Production Manager
S5	Ditto	Damaged or faulty cans reaching seamer.	Damaged or faulty cans to be removed by Depalletizer Operator.	No damaged or faulty cans to reach the seamer.	Pallet removed and quarantined.	Depalletizer Operator Production Manager

Approved (Production): _____ Approved (Q.C.): _____

continued

Cans and Ends *continued*

CCP	Monitor/ Frequency	Hazard	Procedure/Control Measure	Parameter Range	Corrective Action	Authority
S6	Cans transferred to canning lines. (Continual)	Foreign bodies from cans in final product.	Cans inverted and air blown to remove foreign bodies. Lines covered before cans are seamed. Check that air blower is working prior to start of production. Visual check of cans.	Absence of foreign bodies in final product.	If air blower not working report to Engineers for repair. If foreign bodies present, quarantine and inform management.	Seamer Operator Line Supervisor Engineers Production Manager
S7	Can ends transferred to seamer. (Continual)	Wood from pallets.	Ends transferred onto metal pallets or plastic containers for transfer to seamer. Wood audit.	No wooden pallets to enter production areas.	Remove from line.	Line Supervisor Seamer Operator Technical Manager
T4	Ditto	Loss of traceability.	Record pallet ticket information with date/time used.	Traceability to be maintained.	Investigate and take appropriate action.	Seamer Operator Line Supervisor
Q3	Ditto	Incorrect ends supplied.	Ends delivered to line as required by production program. End specification checked by Seamer Operator.	As in manufacturing specification.	Do not use. Report to Line Supervisor.	Seamer Operator Line Supervisor

Appendix 11-C

Liquor Preparation Flow Chart

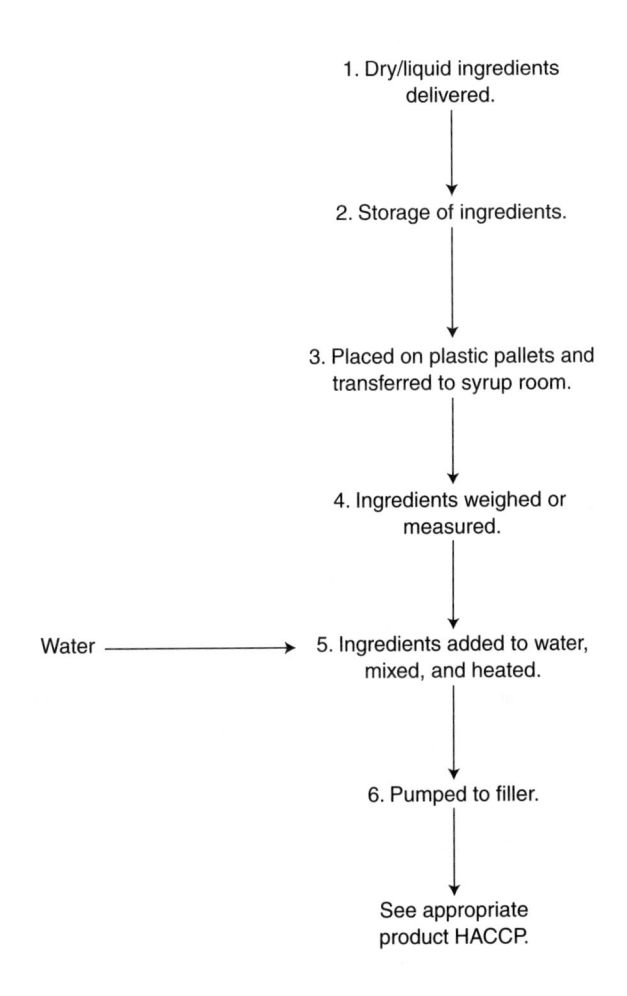

1. Dry/liquid ingredients delivered.

   ↓

2. Storage of ingredients.

   ↓

3. Placed on plastic pallets and transferred to syrup room.

   ↓

4. Ingredients weighed or measured.

   ↓

Water ⟶ 5. Ingredients added to water, mixed, and heated.

   ↓

6. Pumped to filler.

   ↓

See appropriate product HACCP.

Liquor Preparation

CCP	Monitor/ Frequency	Hazard	Procedures/Control Measures	Parameter	Corrective Action	Authority
S1	Intake of ingredients. (Each delivery)	Foreign bodies in final product.	Approved supplier. Supplier assessment. Compliance with specification. Certificate of Conformance from supplier.	As in appropriate raw material specification.	Appropriate action taken with supplier.	Purchasing Manager Technical Manager
Q1	Ditto	Incorrect specification making unsuitable for use.	Ditto	Ditto	Ditto	Purchasing Manager Technical Manager
T1	Ditto	Loss of identification and traceability.	Lot/batch numbers given each raw material.	Traceability to be maintained.	Material without lot/batch number to be identified before use.	Goods Inwards Technical Manager
Q2	Ingredients stored. (Continual)	Deterioration due to poor storage conditions.	Dry, ambient warehouse. Housekeeping audits.	Adherence to Good Manufacturing Practice.	Progress action points raised in housekeeping audit.	Stores Foreman Technical Manager
S2	Ditto	Contamination with odor, metal, glass, etc.	Good housekeeping and cleaning. Housekeeping audits. Absence of glass. Glass breakage procedures.	Ditto	Ditto	Stores Foreman Technical Manager
S3	Ditto	Infestation by rodents, birds, or insects.	Suitable building with adequate proofing. Pest control contract. Housekeeping audits.	Absence of infestation.	Progress action points raised by housekeeping audits and pest control inspections.	Stores Manager Technical Manager
S4	Transferred to plastic pallets and moved to syrup room. (Continual)	Wood contamination of final product. Plastic contamination from broken pallets.	Use of plastic pallets for ingredients entering syrup room. Wood audits. Housekeeping audits.	No wooden pallets to enter syrup room. All plastic pallets to be sound.	Progress action points raised by housekeeping and wood audits.	Stores Foreman Syrup Room Operator Technical Manager
Q3	Ingredients weighed and dispensed. (Continual) (Each batch)	Incorrect ingredient weights affecting final product quality.	Weighed/measured with calibrated equipment. Weights and measures according to recipe sheet. Recipe sheet checked against production program. Check each batch.	All equipment to be calibrated to required accuracy. As in manufacturing specification.	Adjust and recalibrate equipment. Check. Isolate suspect product.	Syrup Room Operator Quality Control Engineers
T1	Ditto	Loss of traceability of specified ingredients.	Lot/batch number of specific ingredients recorded. Audits.	Traceability to be maintained.	Investigate and take appropriate action.	Syrup Room Operator Technical Manager

Approved (Production): _____ Approved (Q.C.): _____

continued

CCP	Monitor/ Frequency	Hazard	Procedures/Control Measures	Parameter	Corrective Action	Authority
S5	Water. (Daily filter check. Weekly analysis)	Foreign body contamination of product. High bacterial counts.	Filter all incoming water. Check that filter is in place and is sound. Send weekly sample to Public Analyst.	No foreign bodies present. T.V.C. levels on incoming water <100/mL.	Take appropriate action, i.e., remove foreign material and dump any prepared liquor. Quarantine any suspect product.	Production Manager Syrup Room Operator Technical Manager
Q4	Ingredients added to mixing tank and made up to required volume. (Each batch)	Incorrect additions affecting final product appearance, flavor, and texture.	Additions recorded on sheet. Brix, salt, taste, and color checks on each batch.	As in manufacturing specification.	Adjust water/ingredients to meet requirements or dump.	Syrup Room Operator Quality Control
S6	Ditto	Insufficient heat process due to low liquor temperature.	Heat until specified temperature reached. Monitored regularly at filler.	Minimum 60°C.	Adjust temperature to meet requirements.	Syrup Room Operator Filler Operator Quality Control
S7	Pumped to filler. (Hourly)	Foreign bodies in final product.	All liquor passed through a filter. This is checked for presence and soundness.	No foreign bodies present.	Quarantine suspect product.	Line Supervisor Filler Operator Quality Control

Approved (Production): _____Approved (Q.C.): _____

Appendix 11–D

Retorts Flow Chart

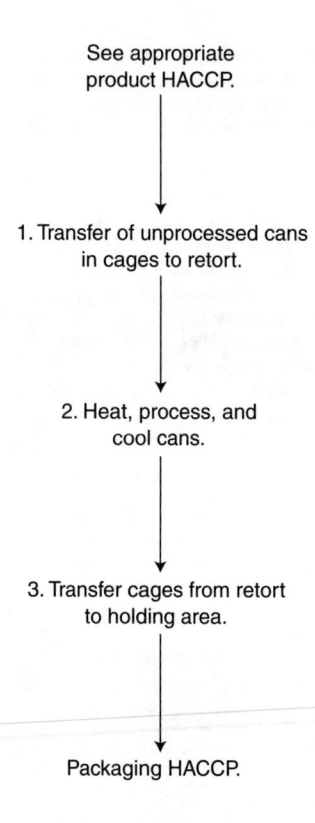

See appropriate
product HACCP.

↓

1. Transfer of unprocessed cans
in cages to retort.

↓

2. Heat, process, and
cool cans.

↓

3. Transfer cages from retort
to holding area.

↓

Packaging HACCP.

Retorts Processing

CCP	Monitor/Frequency	Hazard	Procedures/Control Measures	Parameters	Corrective Action	Authority
T1	Transfer of unprocessed cans in cages from canning lines to retorts. (Each cage)	Loss of identification.	All cages to carry red sequentially numbered disc with line number and can code. Checked at retorts. All cans are coded. Checked on line.	All cages to be identified. All cans to be coded.	Unidentified cages reported to Quality Control. Quarantine if necessary.	Cage Loader Retort Operator Quality Control
S1	Cans heat processed in static retorts then cooled. (Each cage)	Insufficient heat process, leading to survival of spoilage or pathogenic organisms.	Heat process established by heat penetration studies and scheduled process sheets issued. Process time and temperature set up as on production program and retort operated correctly. Audit procedure and check records.	Process to be established before first production run. Process details as in manufacturing specification.	If necessary carry out heat penetration study on simulation. Quarantine any suspect product.	Retort Operator Technical Manager Production Manager
S2	Ditto (Each process)	Survival of *Clostridium botulinum* spores.	Calibration of thermometers and other process instruments.	Instruments to be calibrated to correct level of accuracy.	Repair/replace/recalibrate. Assess effect on previous production. Quarantine suspect production.	Retort Operator Technical Manager Production Manager Engineers
S3	Ditto (Each process)	Poor heat distribution, causing inadequate process.	Heat distribution, checks. Retort properly vented. Condensate bleeds to be open.	As detailed in Retort Work Instructions.	Inform Technical Manager and quarantine any suspect product.	Retort Operator Production Manager Technical Manager
Q1	Ditto (Each process)	Overprocessed cans, leading to poor quality product.	Record cook start and finish times.	As detailed in manufacturing specification and Retort Work Instructions.	Technical Manager to check quality and quarantine nonconforming product.	Retort Operator Production Manager Technical Manager
S4	Ditto (Each cage)	Wrongly processed cans.	All cages identified by line and product. Each retort bank only processes one product.	All cages to be fully identified and processed correctly.	Unidentified cages reported to Quality Control. Cages at wrong bank removed to correct bank.	Retort Operator Quality Control Production Manager
S5	Ditto (Hourly chlorine check)	Microbial contamination from cooling water.	Chlorination of cooling water. Cooling water tanks and pits to be emptied and cleaned regularly. Chlorine levels checked.	Free chlorine target 4–12 ppm. TVC target <100 mL. Coliforms not detectable.	Adjust chlorine level. Quarantine any suspect product.	Quality Control Production Manager Technical Manager

Approved (Production): _____ Approved (Q.C.): _____

continued

Retorts Processing *continued*

CCP	Monitor/ Frequency	Hazard	Procedures/Control Measures	Parameters	Corrective Action	Authority
S6	Ditto (Continual)	Mixing of processed and unprocessed cans.	Destruction of dropped cans. Layout and procedures designed to prevent mixing of processed and unprocessed cans. Check heat-sensitive tape for color change.	Physical separation of processed and unprocessed cans.	Quarantine any suspect product. Dump if required.	Retort Operator Quality Control All Staff
S7	Ditto (Hourly)	Insufficient cooling causing stack burn. Excess cooling preventing drying—rust.	Cooling times set to give adequate cooling. Postprocess temperatures checked.	Can temperatures as in manufacturing specification.	Adjust cooling times as necessary.	Quality Control Retort Operator
T2	Ditto (Each cage)	Loss of identification.	Cages identified with tickets showing line and product. All cans are coded. Tickets and codes checked.	All cages to be identified. All cans to be coded.	Quality Control to be informed of any cages without identification.	Quality Control Retort Operator Cage Unloader
S8	Ditto (Continual)	Postprocess microbial spoilage of product while cans warm/wet.	Avoid handling wet cans. Supervisors to ensure correct procedures followed.	No wet cans are to be handled.	Quarantine any suspect product.	Line Supervisor All Staff
T3	Transfer of processed cage from retort to holding area. (Each cage)	Loss of identification.	All cages to carry green sequentially numbered disc carrying line/code. All cans are coded. Presence of disc and code checked.	All cages to be identified. All cans to be coded.	Unidentified cages reported to Quality Control. Quarantine if necessary.	Quality Control Retort Operator
S9	Ditto (Continual)	Wet cans may cause postprocess contamination.	Every cage tipped and drained of water before transfer. Do not handle wet cans.	No postprocess contamination.	Cages retipped if necessary.	Retort Operator Cage Unloader

Approved (Production): _____ Approved (Q.C.): _____

Appendix 11–E

Packaging Flow Chart

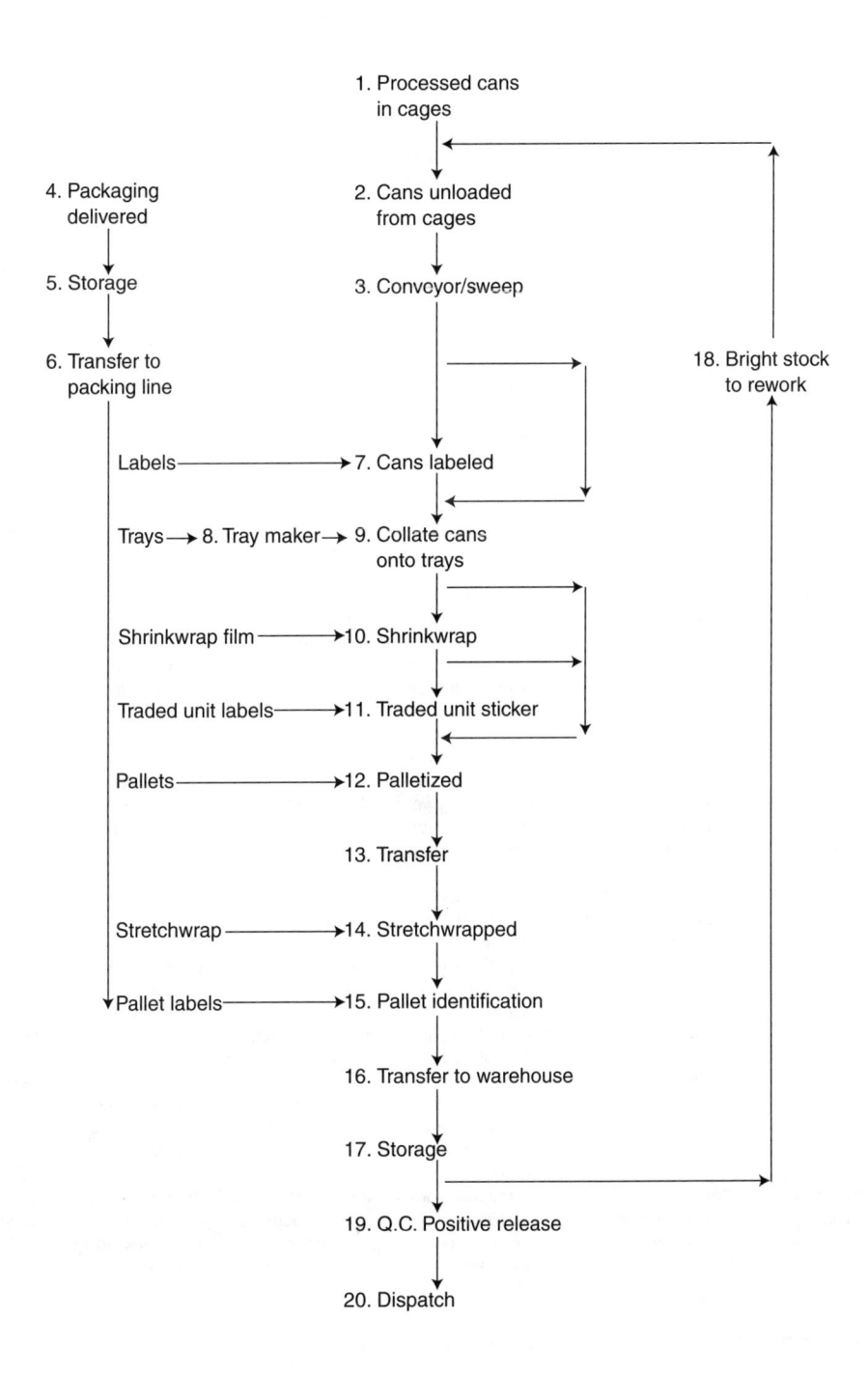

1. Processed cans in cages
2. Cans unloaded from cages
3. Conveyor/sweep
4. Packaging delivered
5. Storage
6. Transfer to packing line
7. Cans labeled (Labels)
8. Tray maker (Trays)
9. Collate cans onto trays
10. Shrinkwrap (Shrinkwrap film)
11. Traded unit sticker (Traded unit labels)
12. Palletized (Pallets)
13. Transfer
14. Stretchwrapped (Stretchwrap)
15. Pallet identification (Pallet labels)
16. Transfer to warehouse
17. Storage
18. Bright stock to rework
19. Q.C. Positive release
20. Dispatch

Packaging

CCP	Monitor/ Frequency	Hazard	Procedure/Control Measures	Parameters	Corrective Action	Authority
S1	Processed cans in cages transferred from retorts to packing lines. (Continual)	Postprocess microbial spoilage while cans wet and warm.	Avoid handling wet cans. On-line can driers.	No postprocess contamination.	Quarantine any suspect product.	Cage Unloader Quality Control
T1	Ditto (Continual)	Wrong cans to packing line. Loss of identification and traceability.	Cages identified with green discs with line/code. All tickets checked.	No unidentified cages to be used. Green indicates processed OK.	Unidentified cages reported to Technical Manager.	Cage Unloader Technical Manager
S2	Ditto (Continual)	Unprocessed cans to packing lines.	Heat-sensitive tape must have changed color. Q.C. inspection of each cage positive release.	No unprocessed cages through system.	Any suspect cage quarantined. Technical Manager informed.	Cage Unloader Technical Manager
S3	Unloading of cages. (Continual)	Postprocess spoilage while cans wet and warm.	Avoid handling wet cans. On-line can driers.	No postprocess contamination.	Quarantine any suspect product.	Cage Unloader Quality Control
T2	Ditto (Continual)	Wrong cans to packing line. Loss of identification and traceability.	Cages identified with green discs with line/code.	No unidentified cages to be used. Green indicates processed OK.	Unidentified cages reported to Technical Manager.	Cage Unloader Technical Manager
S4	Ditto (Continual)	Unprocessed cans to packing line.	Heat-sensitive tape must have changed color. Q.C. inspection of each cage positive release.	No unprocessed cans through system.	Any suspect cages quarantined. Technical Manager informed.	Cage Unloader Technical Manager
Q1	Ditto (Continual)	Can damage at point of unloading.	Care when unloading. Checked on label line.	No damaged cans.	Damaged cans removed.	Cage Unloader Labeler Operator
S5	Conveying cans to point of unloading. (Continual)	Postprocess microbial spoilage while cans wet/warm.	Avoid handling wet cans. Conveyors and sweeps are sanitized.	No postprocess contamination.	Quarantine any suspect product.	Cage Unloader Labeler Operator Quality Control
Q2	Packaging material delivery. (Each delivery)	Poor condition of packaging—dirty or damaged.	Approved supplier. Compliance with specification. Certificate of conformance. Visual check.	To meet appropriate specification.	Report to Packaging Buyer. Take appropriate action with supplier.	Goods Inwards Packaging Buyer
Q3	Ditto (Each delivery)	Incorrect information on printed packaging.	Approved supplier. Compliance with specification. Visual check. Proofs checked prior to placing order.	To meet appropriate specification and approved artwork.	Report to Packaging Buyer. Take appropriate action with supplier.	Goods Inwards Packaging Buyer Technical Manager
Q4	Storage of packaging. (Continual)	Damage or deterioration of material.	Dry, ambient store. Good housekeeping. Housekeeping checks.	Adherence to Good Manufacturing Practice.	Progress action raised by housekeeping checks.	Stores Foreman Technical Manager

Approved (Production): _____ Approved (Q.C.): _____

continued

Quality Management System and Hazard Analysis Critical Control Points

CCP	Monitor/ Frequency	Hazard	Procedure/Control Measures	Parameters	Corrective Action	Authority
Q5	Ditto (Continual)	Infestation by rodents, insects, or birds.	Suitable building with good proofing. Pest control contract. Housekeeping checks.	Absence of infestation.	Progress action raised by housekeeping checks and pest control inspections.	Stores Foreman Technical Manager
Q6	Transfer of packing material to line. (Each delivery)	Incorrect packaging supplied to line.	Supplied according to production program. On-line check.	As in Packaging Instructions and to meet customer requirement.	Nonconforming product quarantined and reworked. Wrong packaging removed.	Labeler Operator Packing Supervisor Stores Foreman
Q7	Cans labeled (Continual)	Misplaced or damaged labels. Unlabeled cans.	Correct operation of labeler. On-line checks.	Good presentation.	Mislabeled cans removed and reworked.	Labeler Operator Quality Control
Q8	Ditto (Continual)	Wrong labels used.	Labels from previous run removed.	Correct label used.	Fault reported and appropriate action taken.	Labeler Operator Quality Control
Q9	Ditto (Continual)	Wrong cans used.	Check can code. Remove cans from previous run.	To meet customer requirements.	Fault reported and product quarantined for rework.	Labeler Operator Line Supervisor Quality Control
Q10	Trays formed. (Continual)	Wrong trays used.	Trays used as in Production Program.	As in packaging specification.	Remove wrong trays.	Tray Erector Line Supervisor
Q11	Ditto (Continual)	Badly made trays.	Tray erector set up to produce good trays.	As in packaging specification.	Remove all bad trays.	Tray Erector Line Supervisor
Q12	Cans collated into trays. (Continual	Incorrect packaging format.	Trays packed according to Production Program.	As in packaging specification.	Nonconforming product quarantined for rework.	Tray Packer Line Supervisor
Q13	Ditto (Continual)	Wrong cans included in pack.	Cans cleared from previous production run.	To meet customer requirements.	Nonconforming product quarantined for rework.	Tray Packer Line Supervisor
Q14	Trays shrink-wrapped. (Continual)	Wrong width film used.	Film supplied in accordance with Production Program.	To meet customer requirements.	Nonconforming product quarantined for rework.	Shrinkwrap Operator Line Supervisor
Q15	Ditto (Continual)	Badly labeled cans included in pack.	Unlabeled and badly labeled cans removed.	To meet customer requirements.	Nonconforming product quarantined for rework.	Shrinkwrap Operator Line Supervisor
Q16	Ditto (Continual)	Shrinkwrap loose or poorly sealed.	Correctly set machine. Regular checks on shrinkwrap quality.	As in packaging specification.	Nonconforming product quarantined for rework.	Shrinkwrap Operator Line Supervisor
T3	Trade unit labels printed and applied. (Continual)	Incorrect details printed onto trade unit labels.	Format programmed into computer. BBE date checked against can code. Bar code verified.	As in packaging specification.	Nonconforming product quarantined for rework.	Label Printer Line Supervisor
T4	Ditto (Continual)	Wrong trade unit label applied to pack.	Set up and check that labels are being correctly applied. Regular checks.	As in packaging specification.	Nonconforming product quarantined for rework.	Label Printer Pallet Stacker Line Supervisor

Approved (Production): _____ Approved (Q.C.): _____

continued

Packaging *continued*

CCP	Monitor/ Frequency	Hazard	Procedure/Control Measures	Parameters	Corrective Action	Authority
Q17	Stacked on pallet. (Each pallet)	Incorrect stacking pattern.	Stacking pattern set out in Packaging Work Instruction.	As in packaging specification.	Restack pallet.	Pallet Stacker Line Supervisor
Q18	Ditto (Each pallet)	Badly presented packs included on pallet.	Poorly presented packs to be removed prior to stacking.	As in packaging specification.	Remove all poorly presented packs and restack pallet.	Pallet Stacker Line Supervisor
Q19	Ditto (Each pallet)	Badly stacked pallet.	Care taken when stacking.	As in packaging specification.	Restack pallet.	Pallet Stacker Line Supervisor
T5	Transfer of pallet to stretch-wrap. (Each pallet)	Loss of identification and traceability.	Bar-coded pallet ticket applied. Check for presence of ticket.	Traceability to be maintained.	Investigate and apply appropriate ticket.	Pallet Stacker Line Supervisor Forklift Driver
Q20	Pallet stretch-wrapped. (Each pallet)	Contamination of pallet with dirt. Also unstable pallet.	Operator checks pallet is fully covered with film.	See packaging specification.	Poorly wrapped pallets must be rewrapped.	Stretchwrap Operator Forklift Driver
T6	Bar-coded pallet tickets printed and applied. (Each ticket)	Incorrect details printed, causing loss of traceability.	Details held in computer. Bar code verified.	See packaging specification.	Apply correct ticket and cancel incorrect ticket.	Label Printer Pallet Stacker Line Supervisor
T7	Ditto (Each ticket)	Label applied to wrong pallet.	Can code checked against pallet ticket.	See packaging specification.	Apply correct ticket and cancel incorrect ticket.	Label Printer Pallet Stacker Line Supervisor
T8	Pallet moved to warehouse. (Each pallet)	Loss of identification.	Check pallet ticket before moving into warehouse.	Traceability to be maintained.	Investigate and apply appropriate ticket.	Stores Foreman Forklift Driver
Q21	Storage (Continual)	Damage or deterioration.	Dry, ambient storage. Good housekeeping. Housekeeping checks.	Adherence to Good Manufacturing Practice.	Progress action points raised by housekeeping checks.	Stores Foreman Technical Manager
Q22	Ditto (Continual)	Infestation by rodents, insects, or birds.	Suitable building with good proofing. Pest control contract. Housekeeping checks.	Absence of infestation.	Progress action points raised by housekeeping checks or pest control inspection.	Stores Foreman Technical Manager
T9	Bright stock moved for rework. (Each pallet)	Loss of traceability.	Pallet ticket checked and removed as pallet emptied.	Traceability to be maintained.	Quarantine any unidentified product.	Stores Foreman Line Supervisor
Q23	Ditto (Each pallet)	Wrong cans depalletized onto line.	Check can code, pallet tickets, and production program. Clear all cans from previous run.	All cans to be correct.	Quarantine any suspect stock for checking.	Line Supervisor Stores Foreman Forklift Driver
S6	Q.C. Positive release. (Each batch)	Release of spoilt stock.	Incubation checks and microbial testing (pH and vacuum).	All stock released to be sound.	Quarantine any suspect stock for investigation.	Stores Foreman Technical Manager
T9	Dispatch to customer. (Each pallet)	Loss of traceability.	Pallet label details recorded together with customer/destination.	Traceability to be maintained.	Investigate and take appropriate action.	Dispatch Manager

Approved (Production): _____ Approved (Q.C.): _____

Chapter 12

Water Supplies, Effluent Disposal, and Other Environmental Considerations

Michael J.V. Wayman

12.1 INTRODUCTION

As water is the principal component of fruit and fruit products, it is surprising that only relatively recently its sourcing and disposal have been given proper consideration. The more settled climate of past decades has given way to increasingly erratic weather patterns worldwide. This change has far-reaching implications for crop husbandry and, not least, for environmental concerns. In periods of drought or in areas of low rainfall, the recycling or at least the serial use of water, even if only for irrigation, has become pressing.

Attention to the usage and disposal of water is therefore claiming more management time and investment in technology than ever before. This is not without its benefits. Improved control can be exercised over both the quality and the economics of production. Indifferent attention has, in the past, resulted in off-flavors, spoilage, and frequent extravagance in the volumes of water used. This has been particularly true of primary processes. Improved instrumental and computer technologies have become available that facilitate close monitoring of key parameters. Appropriate software now makes it easy to manage the water sector closely and efficiently.

Environmental auditors, purchasers of produced goods, shareholders, and, not least, potential investors are taking an increasing interest in the ways in which companies use raw materials, natural resources, and energy. The advent of environmental assessment accreditations, such as British Standard (BS) 7750, the ISO and EN series, and EMAS (Eco-Management and Audit Scheme), means that comparisons between operations and operators can now readily be made.

There is an international dimension, too. Long-distance distribution has resulted in the development of more generally agreed standards, arguable though some of these still are. The European Union is a clear example of the move toward universal objectives. Compliance with these, rather than with local, national, or traditional standards, will become the norm in the third millennium.

12.2 WATER SOURCING

Fruit processing has historically made use of considerable amounts of water. Rising costs of acquisition and disposal have prompted an interest in reducing this use. Hydraulic discharge flumes have been replaced by conveyor belts, and solid wastes collected dry instead of being flushed to drain. The trend from canning to freezing has not had as much impact in this area as might have been expected.

Town water requires the minimum of capital investment on the part of the fruit processor. The supply is usually dependable and its characteristics not subject to great variation. As the costs are not under the control of the consumer, however, they can rise to levels that threaten profitability.

In order to minimize spoilage from handling and delays occurring in transportation, many fruit-processing plants are located in the more rural and remote areas where the contributing crops are grown. Here, recourse has to be made to a nearby river, borehole, or spring. Provided that appropriate necessary treatment is applied, all these can be made entirely satisfactory for the intended use. River water varies the most widely in composition as a rule, and so calls for the longest sequence of unit treatments to render it suitable for use.

Unless the water source is a major one, storage will be required at some point. This may be prior to any treatment, combined with a settlement/clarifying stage, or later on as partially or fully treated water. The capacity of any such reservoir can readily be calculated. Due allowances may be made for extreme demands (firefighting, large-scale wash-down), and transfer pumps and pipework rated accordingly.

12.3 PRIMARY TREATMENT

Only rarely can water be used as it is acquired. Some boreholes and conduited springs may seem to qualify, but even if innocent of unwanted components, the water from them may still need adjustments in regard to pH value, alkalinity, and biological stabilization.

The range of characteristics requiring attention before raw waters comply with fruit industry standards includes the removal of suspended solids; compounds conferring color, taste, and/or smell; dissolved minerals; metal ions; salts; and, not least, microorganisms: animal, vegetable, fungal, bacterial, and viral. The minimum standard to be aimed for is that for potable water. In this connection, a most useful and readable introduction will be found in Volume 1 of *Guidelines for Drinking-Water Quality*, published by the World Health Organization (WHO).

12.3.1 Screening

Water intakes from rivers should be carefully sited so as to minimize the ingress of silt and floating matter. Suitable screens will keep out larger debris and aquatic life.

Fine strainers fitted to pump suction pipes will keep out smaller particles. Vigilance has to be exercised to maintain these in good order; self-cleaning devices are available for these duties. Downstream grit channels can easily be arranged to allow sand and small stones to drop out under gravity.

12.3.2 Color Removal

Surface waters are often brownish as the result of vegetative breakdown products, such as humic acid derivatives and/or iron species. The standard response to this is to apply a coagulant treatment followed by a polyelectrolyte flocculant. Coagulants now available include ferric and aluminum complex sulfates and silicates. Synthetic polyelectrolytes are cationic, anionic, or nonionic. Laboratory treatability studies will soon establish the most appropriate regimen for a particular water.

The extensive surface area produced by the above combination adsorbs colored elements, clay, chalk, and so on. After treatment, a clear and bright water results. The selected reagents should be monitored from time to time so that adjustments can be made to correspond to seasonal or other variations in raw water quality.

Borehole water may require particular treatment to deal with dissolved metals (iron, manganese) and/or anionic species (bicarbonate, sulfide). Depending upon the amounts present, chemical dosing, aeration towers, and catalytic filters are used to remove these components (Figure 12–1). All reagents and materials in contact with the system should be approved for use with potable water.

12.3.3 Adjustment of pH Value

Fruit products will tolerate only a limited range of pH values before changes of hue or color become evident, or before hydrolysis results in irreversible impairment of flavor, vitamin content, or physical structure. Natural waters may be acidic or alkaline: town water is usually well corrected to around neutrality. Local primary treatment with metal salts and/or lime may result in a water with a pH value that needs further adjustment before process use. Granted a retention time of a few minutes and fully mixed conditions, a pH meter-controller can add reagent acid or alkali automatically to achieve and maintain any desired target range.

pH adjustment is the usual way in which dissolved toxic metals may be removed by precipitation as their hydrated oxides. Some metals (e.g., nickel) require high pH values to achieve minimum solubility; others again may be present in anionic forms (e.g., plumbate, chromate), which are not directly precipitatable.

12.3.4 Filtration

Unless extended time is available, settlement will not usually result in exhaustive removal of suspended solids. Natural turbulence and thermal eddy currents will always conspire to give residual suspensions of a few parts per million. Filtration can reduce these to whatever low limits may be specified for particular applications. Municipal waterworks use gravity sandbeds, but a more appropriate industrial option is that of pressure filtration. There are many configurations. There are fixed or moving beds of

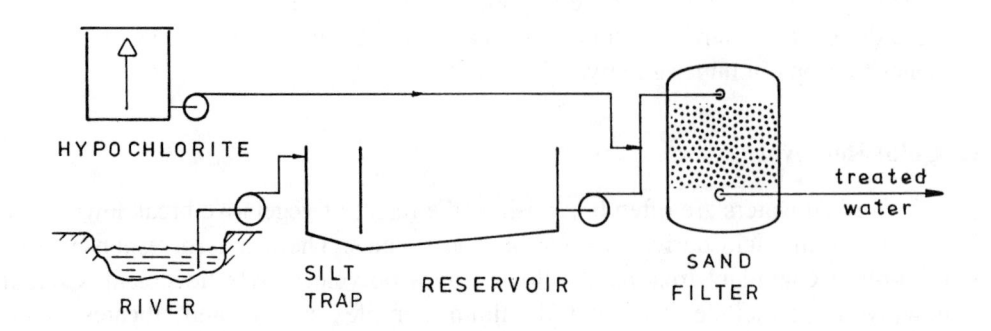

Figure 12–1 A river water treatment plant in northern Nigeria using a reservoir, sand filtration, and hypochlorite disinfection.

sand, gravel, sometimes anthracite, depth-wound cartridges and supported membranes. All filters eventually become occluded with accumulated solids. These are disposed of either by backwashing or by changing the filtration element. It is desirable, therefore, to install filters in duplex working mode so that on-line production can be maintained. On occasion, filters can become breeding places for microorganisms, so it is important to allow for cleaning-in-place (CIP) using a sterilizing regimen (Figure 12–1).

12.3.5 Carbon Adsorption

Certain organic compounds, present in only trace amounts, impart unacceptable odors and flavors to water. These stem largely from human sources: materials of construction (e.g., bitumen joint sealants in water mains), or the direct pollution of an aquifer or watercourse. The latter may be contaminated by the upstream outfall of a mine of some sort or effluent treatment works.

Techniques are now available to detect and monitor submicrogram concentrations of solvents, oils, phenols, hormones, and pesticides. If not dealt with at this point in the treatment sequence, later chlorination may result in more intractable problems (e.g., the conversion of phenols to more highly flavored chlorinated derivatives.

Most of the above can be removed satisfactorily by passing the water through a fixed bed of activated carbon. Granular forms of this medium are derived from coal or a range of vegetable materials—coconut shells and particular softwoods (vine wood). All these give charcoals having open microstructures and are sometimes pretreated chemically to impart mechanical or enhanced absorptive properties.

As with sand filtration, means have to be engineered for occasional backwashing, bed expansion, and/or regeneration. It may be possible to effect the last in situ, or the carbon may have to be removed for stoving. Again, duplex units are commonly used so that forward flow is maintained.

12.3.6 Primary Disinfection

Water may appear to be of good quality but still carry bacteria, viruses, yeasts, and mold spores. These carry through to the fruit products and often lodge in pipework to become a source of chronic infection.

Conventional biocidal agents include halogen species (chlorine, bromine, iodine), silver, oxidants (peroxide, ozone, ultraviolet radiation), and reducing agents (sulfur dioxide and derivatives). On grounds of applicability and costs, chlorine is the preferred initial consideration. Its most effective germicidal form is hypochlorous acid (HClO). The proportion of this present in any given reactive environment is related to the net pH value therein. Above pH 9 or so, the hypochlorite becomes dissociated and is ineffective; below pH 7.5, acceptable activity is secured (Figure 12–2).

Chlorine may be applied as the gaseous element dissolved in recirculated water, as sodium or calcium hypochlorite (bleaching fluid/bleaching powder), or as part of an organic carrier such as trichloroisocyanuric acid (or its sodium salt). The ultimate choice will depend upon the scale and engineering of a particular operation. A standard plant is available for a wide range of demands.

Sufficient reagent is added to maintain a small residual at the farthest point of distribution. For normal drinking, cooking, and washing purposes in temperate climates, this may be only 0.2 mg of chlorine per liter. In the tropics, a higher concentration (0.5 mg chlorine per liter) is desirable. For the washing of fruit suspected of having been contaminated, for example, with irrigation water of poor quality, even higher levels may have to be employed. Otherwise, potable water will be acceptable for all noncritical purposes: normal fruit preparation and incorporation into blended fruit juices and beverages. Further treatment (especially disinfection) will be a statutory requirement for certain applications.

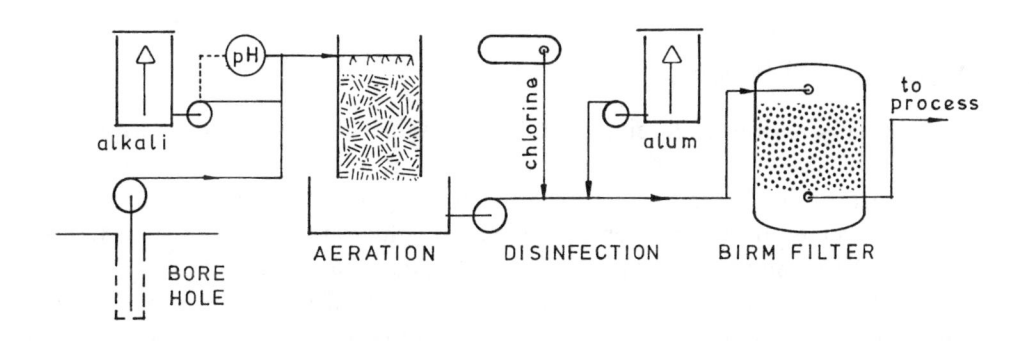

Figure 12–2 A borehole water treatment plant in West Malaysia using aeration, iron removal, and chlorine disinfection.

12.4 SECONDARY TREATMENT

12.4.1 Boiler Feedwater

Considerable quantities of low-pressure steam and high-pressure hot water (HPHW) are used in fruit processing. If boilers are to continue to function at optimum efficiency, then precautions must be taken to protect them from corrosion and scaling. Often, it is found sufficient to add metered amounts of appropriate chemicals to the feedwater: phosphates, phosphonates, tannins, alkalis, and so on. Sometimes, partial softening or demineralization will have to be applied. There should be regular monitoring of the boiler contents, and prompt and sufficient blowing-down practiced when indicated to control the concentrations of dissolved salts that build up.

12.4.2 Cooling Water

The requirements here parallel those for boiler feedwater. Precautions have to be taken to prevent scaling and corrosion. Additionally, biocides must be used to control fouling and consequent blockage of heat exchangers and pipework.

The health hazards posed by legionnaires' disease and *Cryptosporidium* spp. have prompted moves toward air-cooled systems and sometimes to the installation of refrigerated chilled-water facilities. As a rule, the suppliers of proprietary treatment chemicals will advise and monitor the use of appropriate reagents for a specific plant.

12.4.3 Water for Bottle Washing

Hard water leaves unacceptable films of dried salts on glassware, so at least the final rinsing should be performed with soft or softened water. The base-exchange process is simple and relatively cheap to operate. The calcium and magnesium ions responsible for hardness in the water are replaced by sodium. Subsequent heating does not then give rise to any precipitates. Regeneration of the ion-exchange resin that effects this substitution is achieved by means of a saturated solution of common salt (brine), with the waste stream comprising the chlorides of the calcium and magnesium. This treatment also allows the reduced use of detergents/surfactants.

12.4.4 Water for Fruit Dressing

Virtually all fruit as received will be contaminated to some extent with soil, plant debris, juice, pesticide, and/or other husbandry residues. After primary sorting, the greater part of this foreign matter may be removed by appropriate washing equipment. Medium- to high-pressure water is sprayed through arrays of nozzles so as to reach all surfaces. As already cited, potable water is acceptable for this operation.

If the cost of water is significant, or if water itself is in short supply, then some degree of recirculation may be practiced. A common sequence is gravity or mechanical screening, followed by pressure filtration. In any case it is desirable to fit a meter or meters to the water distribution system so that proper audits can be carried out periodically.

For cut fruit, additional precautions must be taken to guard against spoilage. Water for this purpose is chlorinated to a higher level: a residual of between 25 and 30 mg/L is often adopted, but there must be careful monitoring of the fruit to ensure that excessive chlorine is not causing its own problems of tainting, decolorization, or loss of flavor essentials.

12.4.5 Water for Process Use

The characteristics required of process water may be determined by the particular use to which it is to be put. Canning, fruit-juice production, jams, wine making, or pharmaceutical preparations exhibit very different intrinsic physicochemical environments.

A water satisfactory for one product may well not suit another. pH value, alkalinity, and salts content can all bear noticeably on flavor and color. Tolerances of these may vary not only from product to product, but also between harvests, fruit varieties, and between fruit grown on different types of soil or under differing seasonal conditions. The subtleties of wines demonstrate this point well.

Trials should be conducted to determine whether a particular water supply is acceptable for the process use intended. If some unwanted effect becomes apparent, then it will be necessary to investigate and pursue a remedy. Depending upon the problem, the action required may be a simple chemical adjustment—of pH value, for example—or some more radical further treatment may be found to be necessary.

In some areas, the mains water supply may be overchlorinated to an extent that affects the fruit/products adversely. The color may be bleached and/or the flavor diminished. In this case, the water can be presented to a dechlorinating filter, based on activated carbon. Water treated in this way loses its residual protection against infection, so dechlorination is carried out no earlier than at the point of use. Otherwise, the risk may be underwritten by the later inclusion of a local disinfecting stage. Almost always, this will be by ultraviolet irradiation. This comprises a high-actinic discharge tube mounted within a silica sheathing tube, placed in turn in an in-flow casing. Means can be included for monitoring the continuing performance of the unit (Figure 12–3).

12.4.6 Water for Special Applications

For certain uses, the process water may be required to comply with various national or international pharmaceutical or other standards. Generally, these relate to the dissolved salts content and the microbial count. The former may be controlled by a specific property (conductivity), a net parameter (hardness, dissolved solids), or a specific limitation (nitrate, chloride). Contributing minerals may be removed down to the specified levels by several means: by a sufficient degree of deionization/demineralization, by a membrane process (reverse osmosis), or by distillation. These unit processes may be used alone or in some combination. The choice of system will depend upon the amount of water being demanded and upon the requirements of the specification being applied (Table 12–1).

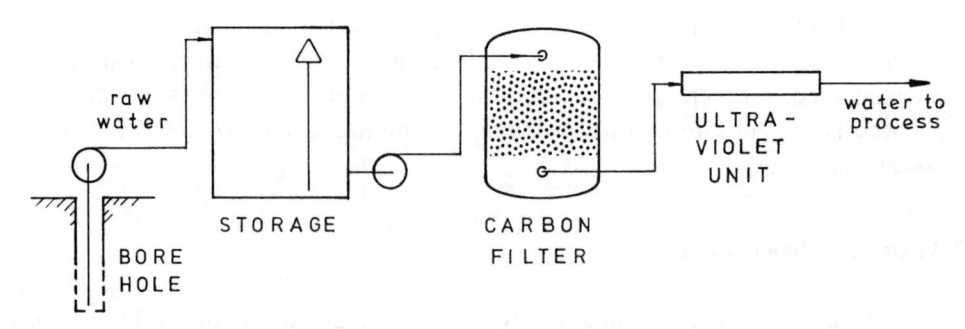

Figure 12–3 Process water for fruit beverage production in South Wales comprising storage, activated carbon filtration, and ultraviolet irradiation for disinfection.

12.5 EFFLUENT PLANNING

While the collection and disposal of surface (rain) water is usually given proper consideration at an early stage in laying out a factory site, too often comparable arrangements for handling process effluents are neglected. At worst, it may be assumed that a single drainage scheme will suffice for both. This is inviting trouble. Extremes of rainfall can overload any finite effluent plant and result in the escape of untreated wastes to the receiving water course. High oxygen–demanding material will then asphyxiate aquatic life. It may also affect downstream uses for irrigation or potable supplies to others.

Factory layouts should be arranged so that receipt of fruit and primary handling are as remote as possible from finished goods production and storage. This may be achieved in a notional way by the provision of bunds or other means whereby separation can be rigorously observed.

The natural slope of the terrain, if any, can be a significant benefit. Floor gullies, drains, catch-pits, and sumps all need adequate relative inverts for proper operation. Falls need to be steeper if solids are likely to be present. Due allowance should always be made for maintenance access. This may include rodding and/or flushing with jetted water.

Table 12–1 Typical Water Quality Standards

Category	Standard	pH value	TDS	S/S	T.H.	M/O	Conductivity
Potable	WHO	6.5 to 8.5	500	1.0	500	100	5
B. Pharm	BS 3978	6.0 to 8.5	1.0	0.1	0.1	10	0.1

Note: TDS, total dissolved solids; S/S, suspended solids, T.H., total hardness; MO, alkalinity to Methyl Orange (indicator); WHO, World Health Organization; B. Pharm, British Pharmacopoeia.

Source: Data from British Pharmacopoeia, 1980 and *Guidelines for Drinking Water Quality*, 1984–1985, World Health Organization.

Underground pipework should be buried at such depths as to comply with local bylaws and be adequately marked, both on site and on drainage layout drawings. Information regarding bore, invert, and materials of construction should be included on the latter.

Underneath roads, yards, and other trafficked areas, buried pipes will need additional protection, usually by being sheathed in concrete.

Ideally, all the above factors should be considered in the course of initial site planning. The later addition and construction of effluent collection systems can be very much more difficult and costly.

12.5.1 Segregation

Apart from the desirability of installing independent rain-water collection, consideration should be given to technical and economic arguments for arranging some degree of segregation of the effluent contributors themselves. Combined systems may appear easier and cheaper in the first place, but in practice may suffer from problems of conflict and inflexibility. Segregation, on the other hand, can facilitate the recovery of energy, of materials, and of water itself. For instance, sensible heat from condensate and used bottle-washing rinses can easily be imparted to incoming cold water by appropriate heat exchangers. Water discarded by reverse osmosis plants will be acceptable for other, noncritical operations.

12.5.2 Effluent Transfer

It often happens that the ideal of a fully gravitational flow system cannot readily be attained. There will then have to be sundry reception sumps with associated pumps to transfer the waste-waters to the distant treatment facilities. Every sump should be provided with at least a simple screen and the pumps (duty and standby) fitted with strainers. Debris of all sorts finds its way into factory drains; some can cause considerable trouble if it comes to lodge in vulnerable places.

Preferably, local collection points should be fitted with their own devices to separate fruit solids from the aqueous component. These are often gravity parabolic or cylindrical screens of proprietary design. Mechanical clearance is frequently installed. Macerators are generally not to be approved, as they serve to increase the load of dissolved oxygen–demanding material, which burdens later treatment processes.

The transfer pumps can be controlled automatically by one of several level-monitoring systems. A flood-level alarm may be interlinked with the main site control room. Pipe runs should be chosen intelligently to avoid too many changes of direction. Swept bends are to be preferred to elbows. Care should be taken to size the pipework appropriately. If too large, the lower velocity may allow silt and heavier solids to settle and build up obstruction. In temperate countries, there may be a requirement for heat tracing and insulation against freezing. Adequate support, physical protection, and labeling are essential in all cases.

Pipework and fittings may most conveniently be of polymeric material. Unplasticized polyvinylchloride (uPVC) is the cheapest overall; acrylonitrile-butadiene-styrene copolymer (ABS) is stronger and has better thermal resistance; polypropylene resists solvent effects best, but requires welding installation. For hot water and pressurized duties, stainless steel is the industry standard.

Throughout the site, the greatest care should be exercised to ensure that all fruit juices are comprehensively collected and routed only by the designated drainage and transfer system. Leakages and spillages to the foundational subsoil will contaminate ground water and may have grave and certainly unwanted consequences.

- Acids may corrode concrete and reinforcing steelwork.
- The fruit juice may ferment and loosen the soil structure.
- Solutes may migrate into local aquifers, boreholes, and watercourses.
- The organic material may form a substrate for sulfate-reducing bacteria, which draw their oxygen supply from sulfate (present in concrete). The consequence of such activity can be the evolution of hydrogen sulfide gas. This is toxic at a parts-per-million level and has the property of deadening the human olfactory sense. If the smell is overlooked or tolerated, collapse and possibly death may ensue.
- Lengthy and costly remediation may involve replacement of civil foundation structures, as not only will withdrawal of sulfate weaken concrete, but hydrogen sulfide itself rapidly corrodes steel and cast iron.

12.5.3 Effluent Reception

Some of the effluents will be fairly consistent in flow and composition. Others, for example, machinery wash-downs, will be intermittent and carry strong chemical loads (sugar, caustic soda). Measures to handle these in the reception and/or transfer stages are essential to the economic operation of any subsequent treatment system. The reception tanks for all streams should be made large enough to provide not only sufficient flow and load balancing, but also spare capacity to allow short-term problems with treatment facilities to be addressed.

On an existing site, a survey will soon indicate what balancing capacity is appropriate. Samples should be drawn from representative points and over a sufficient period of time to embrace the range of variations likely to be experienced. For a projected fruit factory, best estimates should be made and an agreed-upon notional effluent schedule compiled. This should allow also for any future foreseeable developments.

12.5.4 Treatment Objectives

Treatment objectives may be less or more onerous, depending upon the site and subsequent destination of the used and treated water. At the very least, some limitations will be imposed on the pH value, on the solids content, and on the demand for oxygen exhibited by the final effluent. Increasingly, maximum levels are being set to govern other parameters such as color and temperature.

As with the supply water, standard analytical reference methods are published. These should be used by the controlling laboratory. Occasional short-cuts may be adopted for internal use (e.g., Brix measurement for sugar content). Discharge consent results may be categorized or banded according to the nature of the receiving water

Water Supplies, Effluent Disposal, and Other Environmental Considerations 291

course or other disposal. A small works adjacent to a major river or the sea may, by virtue of the dilution available, have relatively easy limits to attain (Table 12–2).

12.6 EFFLUENT CHARACTERIZATION

For the purposes of effluent treatment, the nature of the wastes from fruit processing is not so important as its effects. The pH value can have a critical influence on chemical and biochemical reactions. In addition to the effluent pH values, the contributing acidities and alkalinities must be determined because of buffering effects. Fruit acids in particular can be responsible for giving very misleading pH results if taken on their own. The measurement of pH values is best made using a metering instrument, as color indicators do not work well with fruit wastes.

12.6.1 Suspended Solids

The assessment of suspended solids is not as straightforward as may at first appear. At the simplest level, the figure is the weight of insoluble material per volume of effluent. A sample is filtered through glass-fiber sheet and dried to constant weight or for a specified time at an agreed-upon temperature (usually 105°C) before being weighed. In practice, it may be important to distinguish between "settleable" solids, related to a certain time scale, and nonsettling or colloidal matter. Also, besides the weight/volume ratio, there is occasionally a requirement for the volume/volume information for the same solids.

12.6.2 Oxygen Demand

Apart from a small minority of specialized microorganisms, all life on earth requires access to oxygen in some form or other in order to function. In the aquatic environment, this is available in solution at a concentration in equilibrium with the atmosphere. In cold waters, this is about 10 mg/L. In warmer conditions, this falls to 7 mg/L or so.

In a watercourse, any solute that competes for this oxygen threatens whatever life is already established there. Examples include inorganic species having reducing properties, such as bisulfite, and all soluble organic compounds. The measure of the threat

Table 12–2 Typical Discharge Limits

Parameter	Watercourse	Local Drainage
pH value	6.0 to 9.0	6.0 to 10.0
BOD5*	max. 20 mg/L	—
COD†	—	max. 600 mg/L
Suspended solids	max. 30 mg/L	max. 400mg/L

*Biological oxygen demand
†Chemical oxygen demand

that all these represent is termed the *oxygen demand* (OD) and is cited in milligrams of oxygen per liter (in parts per million).

The original analytical method for determining this property uses a culture of microorganisms and takes 5 days of incubation. It is thus known as measuring the biological oxygen demand (BOD) or BOD5. Results can vary with the source of the culture used for the inoculation.

Two, more rapid, techniques developed subsequently use chemicals as reagents. Potassium permanganate for the permanganate value (PV) or oxygen absorbed (OA) test, and chromic acid for the chemical oxygen demand (COD) test. There are no universal interconversion factors for the results obtained by the three above methods, but for a particular effluent, it may be possible, by experience, to establish some rough and ready ratios.

Because some compounds react only partially (or not at all) with these oxidizing agents, a further analysis is sometimes performed. This determines the total organic carbon (TOC). The sample is pyrolyzed in a stream of oxygen and the carbon dioxide evolved is trapped and measured. This test has the merit of avoiding errors caused by the inclusion of inorganic reducing substances, as well as functioning in the presence of inhibitors to the other methods.

12.6.3 Other Parameters

As suggested previously, there may be particular local interest in other characteristics such as temperature or color. The quantitative measurement of the latter is not easy, even using instrumentation, and several contingent standards are in use. As always, it is important to agree at the outset with the regulating authority what referee methods are to be adopted.

12.6.4 Effluent Monitoring

A database of effluent analytical records is invaluable. An appropriate program of sampling, testing, and flow recording, allied to production figures, will enable optimization of the use of water and the minimization of waste. pH values, temperatures, and flow rates are easily logged automatically and down-loaded when required to computer spreadsheets. Abnormalities and trends can then easily be represented graphically, recognized, and responded to.

Nowadays, there are several excellent on-line programs that allow remote computer terminals to display treatment plant functions and operating parameters. The latter can be altered at will in response to variations in effluent characteristics or local operating developments.

12.7 EFFLUENT TREATMENT

Fruit-processing effluents can present a wide range of characteristics. A treatment process or sequence of processes appropriate to one may not suit others. The several discharge-consent conditions should be considered item by item against each scheduled contributor and an initial table of contraventions drawn up.

It will then be necessary to produce integrated design variants to minimize the engineering ultimately to be installed. Opportunities for mutual treatment may well present themselves (for example, alkali peeling liquor to neutralize citrus juices, or chlorinated water to satisfy part of an oxygen demand).

It may be wise to hold back relatively clean streams until the end of another treatment sequence, so as to take advantage of the dilution thus available. The possibility of partial or side-stream treatment should also be considered. Some processes result in a product water so well within some limiting parameter that only a corresponding portion needs to be treated. An example here might be the use of pressure filtration to trim suspended solids down to the level required.

Eventually, an appropriate overall scheme will be arrived at. As a rule, this will comprise a sequence of unit processes or stages, each dedicated to the limited adjustment or removal of particular parameters. The selection and specification of these processes are a matter of judgment based on the effluent schedule, the discharge limitations, and any incidental recoveries that may prove feasible.

12.7.1 Solids Removal

In a gravity-fed system, all manner of plant debris may be expected. The simplest way of removing this is by passing the flow through a run-down parabolic screen. These are constructed of specially profiled elements to obviate blockaging. Larger installations will require mechanically raked or brushed devices of varying sophistication. Collected material should be removed regularly to approved disposal.

Finer solids are generally more reluctant to settle. It may be helpful, particularly if colloids are present, to introduce a chemical addition stage at this point. The judicious use of ferric and/or aluminum salts, perhaps in conjunction with a synthetic polyelectrolyte flocculant, can coagulate and absorb a significant proportion of polluting loads. Laboratory treatability studies will show whether or not this step would be worthwhile. Both ferric chloride and ferric sulfate are available; the former tends to be slightly more expensive but can give superior results. Poly-aluminum silico-sulfate (PASS) also has its advocates. It should be noted that all these are acidic in nature and that some corresponding pH correction may be needed.

Further, induced or assisted flotation could be installed. Microbubbles are generated within a tank containing the effluent, by electrical or mechanical means. In the latter, the technique is known as dissolved air flotation (DAF). The bubbles nucleate on dispersed particles and increase their buoyancy so that migration to the surface is accelerated. A mousse-like layer builds up and is removed progressively by a mechanical skimmer. The air is derived from a small compressor and contacted with a percentage of the recirculated water at a pressure of some 6 bars. The amount of air capable of being dissolved at this pressure is about 10 times ambient.

12.7.2 pH Adjustment

The net balance of acids and alkalis will depend on the type and variety of fruit being processed, the end-product in view, and the use of any ancillary chemical (e.g., for cleaning, sterilizing, demineralization, and so on).

The pH of each effluent will have an initial value (field pH), which may change with time as the result of continuing chemical reaction (hydrolysis or oxidation) or biological activity (fermentation, enzymatic transformations). It may be necessary to adjust the pH value not only initially, but also subsequently, to allow a process to proceed optimally. Biooxidation performs best in slightly alkaline conditions, between pH 7.0 and 8.0. Acidic fruit juices must be neutralized, therefore, before being presented to this type of treatment. Further, biodegradation itself results in the production of (carboxylic) acid residues. There will thus be a corresponding need for ongoing pH control.

The measuring electrode of the pH meter system is usually made of a special glass. If fluorides are present, from the surface etching of aluminum fruit juice/beverage cans, then an antimony electrode has to be substituted. It is important to locate pH probes so that they monitor a representative portion of the whole. Adequate agitation will ensure this. They require regular inspection and cleaning, together with occasional recalibration, if they are to continue to function reliably. Since pH reference cells have a finite life, it is advisable to replace these elements annually as a matter of course.

The neutralizing reagents most readily used are sulfuric acid and sodium hydroxide. The former may be acquired as 50% battery acid, while the latter is available at various concentrations as caustic soda liquor. Both can be delivered in IBC tanks containing 1000 L, from which they can be dispensed as needed. To avoid wasteful overcorrection of pH value, both these should be diluted to between 5% and 10% strength before application. This can be carried out automatically in a day-tank of suitable size. Other reagents in use include slaked lime slurry (milk of lime: calcium hydroxide) and carbon dioxide.

The neutralization reaction is virtually instantaneous, but because of statistical "leakage," it is usual to allow a retention for this stage of at least 15 minutes. The transfer of reagent can be by actuated valve, if inverts permit; otherwise, pumps have to be installed. For larger installations, a ring main with branches to solenoid-dosing valves may be better. Duty, boost, and alarm functions should always be included. The tank or pit should be provided with an electric stirrer matched to the duty. Direct-coupled, marine propeller types are suitable for vessels up to 5000 L; above this capacity a geared turbine of some sort will be required.

12.7.3 Biological Oxygen Demand

This is by far the most significant of effluent properties needing attention. It will require the greatest capital investment and entail the highest running costs. Fruit-processing wastes are made up of considerable amounts of biodegradable solids and dissolved matter: fruit acids and proteins, carbohydrates, sugars, colloids, and esters. All these contribute to differing extents to the net BOD loading.

Naturally occurring microorganisms are used to absorb then digest substances dissolved in the effluent. Organic compounds are then converted into solid and mineral forms as a biomass, which can be separated by conventional means. Most of the carbon is lost to the atmosphere as carbon dioxide, while organically bound sulfur and nitrogen are converted to sulfate and either nitrate or nitrogen itself.

Traditionally, the seed for a new biological system is drawn from local sewage works or allowed to develop naturally over a period of weeks. There are now available, however, cultures particularly developed and acclimatized to more specific duties, and it might be worthwhile investing in these.

Optimum conditions for biooxidation can be achieved in several ways, depending upon the scale of the exercise, the type of effluent being handled, the availability of space and, not least, the target quality required for the fully treated discharge. The metabolic requirements are oxygen, appropriate pH and temperature ranges, and the necessary retention times for contact with the effluent for subsequent stabilization and for the removal of solids by settlement.

It is essential, too, that due balance is maintained between the nutrients necessary for growth (ammoniacal nitrogen, phosphate, and trace minerals) and the carbon content of the organic material present. In round figures this should be in the ratios of 100:5:1 for BOD:N:P, respectively. Fruit-processing effluents are usually deficient in both nitrogen and phosphate to the extent that supplementation is required. Stock solutions of the necessary salts are added by dosing pump, the rates of application being monitored by the site control laboratory and altered from time to time if need be.

All biocidal compounds *must* be excluded. This should be understood by everyone on site. These include janitorial disinfectants, herbicides applied to pathways, insecticides used to control *Drosophila*, timber preservatives, and the like. There can also be machinery maintenance proprietaries (oils, solvents, sterilizers, lubricant/hydraulic silicones), rodent control baits, paint, and fuel. Alternative means of disposal should be mandatory for all these. Washing them down the drain may so interfere with biological treatments that the plant could take weeks to recover fully.

12.8 FORMS OF BIOLOGICAL TREATMENT

12.8.1 The Activated Sludge Process

The activated sludge process is the simplest form of biooxidation. The effluent stream is brought into contact with an appropriate culture of microorganisms and the whole is agitated with air. Organic matter is digested and converted. After several hours, this process reaches completion and the effluent is said to have become stabilized. The solids component augmented thereby (surplus sludge) is then allowed to separate under the influence of gravity and the treated effluent discharged as permitted. Some sludge is returned to inoculate the next batch of effluent. There are three principal ways in which air can be introduced:

1. *Mechanical agitation of the surface:* Various designs of rotor are marketed, which are mounted in some way just below the surface. Electric motors drive these at speed so that water is picked up and dispersed over the adjacent area. At the same time, air is driven into the bulk of the effluent. Some concomitant noise is generated and odors may be emitted. The plant is thus usually sited remotely from habitation. It is not possible to modulate mechanical aerators to any great extent; that is, they are either on or off.

2. *Tank aeration:* A remote air blower supplies a header-and-lateral manifold mounted across the bottom of the tank or pit. At calculated intervals are fitted diffusers, cap-pieces, bubble-guns, helix tubes, or other devices for breaking up the air and inducing a degree of turbulence. By including valves in the distribution system, it is possible to economize when less than the maximum amount of air is needed.
3. *Venturis:* This simple system is useful for smaller installations. A proportion of the effluent is recycled through a suitably specified venturi element. Atmospheric air is drawn in and intimately mixed with this side-stream, which then rejoins the aeration tank. The effluent enjoys lateral mixing, while energy requirements are competitive with the previous methods.

Although batch treatment is sometimes practiced, most activated sludge plants are designed on a continuous-flow basis. Dimensions are related to between 6 and 8 hours' passage. A proportion of the separated biomass, the activated sludge, is recirculated to seed the influent stream. Since aeration costs money, some form of control is engineered. Measurements of dissolved oxygen (DO) are made at strategic points, and the system is adjusted accordingly.

To avoid the capital costs of a large plant at a seasonal site that could lie idle for much of the year, a technique has been developed with smaller plants whose performance can be boosted by adding oxygen gas to the air supply as and when required. This has the effect of increasing their treatment capacity by a significant factor. Very careful attention needs to be given to the cost effectiveness of every candidate plant.

There are three established developments of the activated sludge process, each offering certain refinements and advantages. In the extended-aeration version, the retention time is lengthened to a day or more. This requires a plant that occupies more space, of course, but that copes better with variations of load and gives improved performance via exhaustive digestion and solids consolidation (Figure 12–4).

If the long channel above is made into a closed loop, an "oxidation ditch" is formed. This is often oval, rather than circular, and is provided with some way of inducing the effluent to circulate. Horizontally mounted brushes or paddles are used.

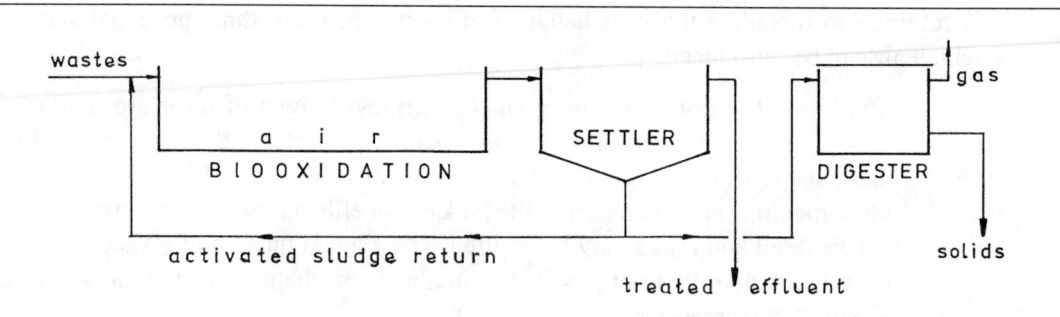

Figure 12–4 Conventional biooxidation treatment system: extended-aeration variant.

Air is introduced by these and/or a surface agitator or blower. A section of the ditch is fitted with a baffle, behind which surplus solids/biomass settles out and is removed at intervals. It works well with consistent and balanced effluents, but once installed, its range of adjustment is very limited. If the retention time is long enough, reliable results are consistently attained.

The third development, contact stabilization, is more highly engineered still, but more adaptable and controllable. The process is separated into compartments dedicated to the different stages of absorption, digestion, and separation. Incoming effluent is mixed immediately with a considerable proportion of well-aerated returned sludge. This rapidly takes up dissolved organic material. After a relatively short time, the effluent passes on to a settlement stage, from which the clarified portion runs directly to outfall. The settled sludge has not by then completed its digestion of the absorbed material, so it is pumped to a separate tank for aeration. The latter can then be much smaller than for aeration of the entire flow and thus saves energy (Figure 12–5).

This version is efficient in its use of space and can cope well with peak loads. Operational variables include the rate of return of sludge and its degree of aeration. Long-term sludge reduction can be included. A large installation will require full-time oversight for sampling, testing, and making corresponding adjustments to particular plant functions.

Attempts to improve the efficiency of the activated sludge process have been made from time to time by adding some dispersed substrate to the aeration stage sponges, and magnetized particles, for example, both being readily removable. These have not been as effective as envisaged or hoped for.

12.8.2 Percolating Filters

The activated sludge techniques depend upon the managed admixture of effluent, air, and suspended biomass. The critical and limiting factor is the surface area of the latter, across which exchanges of organic matter and oxygen take place. An alternative way of engineering this area is to support the biomass on some fixed, solid medium. Nineteenth-century studies resulted in the development of beds of mineral material (clinker, road-stone) onto the top surface, of which effluent is trickled or sprayed. The

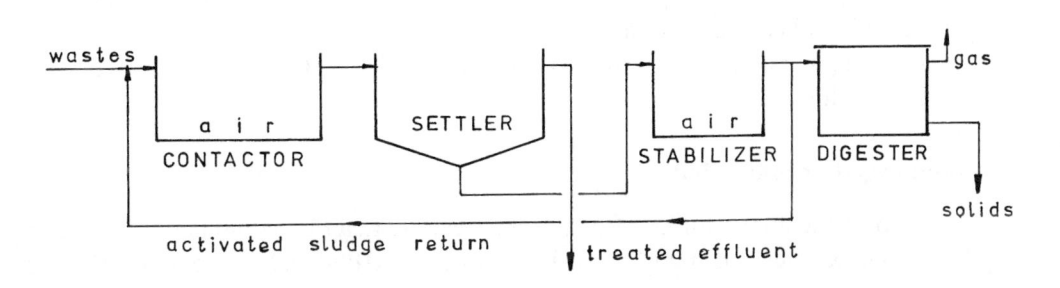

Figure 12–5 Contact stabilization treatment system: flow diagram.

beds may be circular in plan or rectangular. Biomass forms on the surfaces of the mineral, absorbs organic solutes, grows, and sloughs off from time to time for gravity separation. The rate of presentation must be related to the BOD loading and to the capacity of the installation.

Trickling filters enjoy widespread use. They are frequently installed in tandem, with periodic alternation of the sequence in which they are used. Known as alternating double filtration (ADF), this practice is capable of very low residual levels of BOD. Although cheap to run, such filters do not adjust well to variations in load or balance and are somewhat at the mercy of weather conditions.

12.8.3 High-Rate Filtration

The advent of specially designed plastics elements has allowed a more intensive use of space for biofiltration. A simple enclosed tower structure is filled with either random or ordered plastics modules, and the effluent is again distributed over the top layer. Biomass establishes itself on the exposed surfaces and grows as described previously. Because of the open nature of the filling media, the effluent travels rapidly down to a collection sump underneath. From here, some of it is recirculated, while the remainder runs forward to a settlement stage. Running and maintenance costs are very modest. Units have been installed in a wide variety of food and beverage situations all over the world and found to operate very successfully.

This technology has been found in practice to cope well with high initial loadings and rapid fluctuations of characteristics. Its performance is relatively modest, however, being between 65 and 75% BOD removal per stage. It is usual, therefore, to place two towers in series to achieve some 90% removal. This may suffice for some discharge requirements, but more often some secondary treatment is called for. This can be either activated sludge (the contact stabilization mode) or ADF.

12.8.4 Mechanical Contacting

A more intensively engineered form of high-rate filter is the biocontactor. This is composed of an array of circular elements, perhaps 3 or 4 m in diameter, mounted along a slowly rotated shaft. The elements may be either rigid corrugated plastics discs or caging enclosing random-fill plastic packing pieces. Their lower portions are immersed in a trough containing the effluent. Biomass develops on the elements and acts upon the dissolved organic matter as previously described. Biocontactors offer the best utilization of space, if that is at a premium.

Another version is the moving-bed bioreactor (MBBR), which is particularly effective for high-strength wastes (Sunner et al., 1999).

12.8.5 Submerged Aerated Media

Two of the above forms of biological treatment can be combined to give a third, the submerged aerated media (SAM) modification. Plastics elements (both random and ordered) as designed for high-rate filters are installed in an activated sludge tank (as in Figure 12–4). Effluent fills the tank and is recirculated locally via a venturi aerator (Figure 12–6). As before, biomass grows on the support medium, surplus being entrained in the overflow and removed by subsequent settlement, flotation, and/or filtration (Bigot et al., 1999).

This modification has been applied successfully to the up-rating of existing aeration tanks at modest capital cost. It may also be considered where space is particularly limited. It is less suitable, however, for higher loadings of BOD, when a combination of high-rate and percolating filters offers greater resilience and capacity (Hodkinson et al, 1999).

12.9 TERTIARY TREATMENT

12.9.1 Filtration

If very low discharge limits have to be attained, further "polishing" may be necessary. Gravity sand filters can be used to remove residual solids, as in waterworks practice. For organic traces, if sufficient land is available, constructed wetlands or reed beds may commend themselves. There are several variations of the latter, the most popular being the horizontal flow, root-zone format. Shallow trenches are lined with an impermeable membrane and filled with large aggregate. Inlet and outlet arrangements are made using gabions. Seedlings of the common reed, *Phragmites australis*, are planted at 1-m intervals. This plant has the unique property of releasing oxygen from its extensive root/rhizome system. Biomass builds up on the aggregate and functions in combination with the reeds. Higher forms of life graze on the biomass. Removal of up to 98% of the applied BOD has been reported. Two growing seasons are needed for full efficiency to be reached, but thereafter many years' usage may be expected, with very little required in the way of maintenance.

12.9.2 Solids Removal and Disposal

The settlement stages associated with the foregoing depend upon the effluent's being allowed quiescent conditions so that the upward flow vector of the effluent is

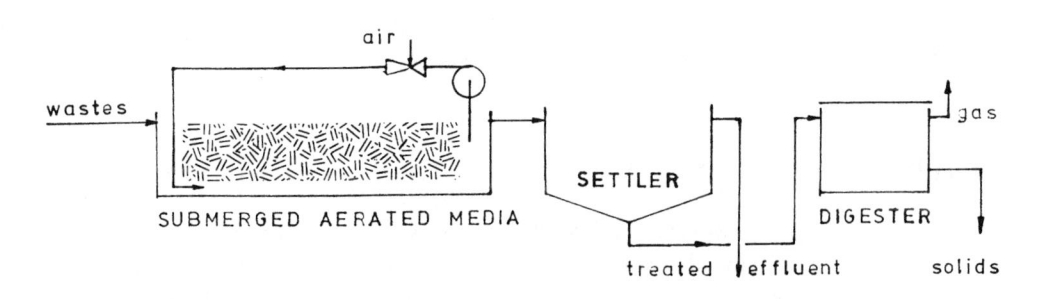

Figure 12–6 Submerged aerated media modification of conventional system.

less than the downward fall rate of solids particles present. This is achieved through the dimensional ratios and other aspects of their design. Vertical, horizontal, and radial patterns are all well known, the last two being applicable more to larger duties. Fittings to settlers include inlet distribution launders and still-wells, intermediate baffles, and outlet weirs. Accumulated solids are drawn off by scrapers of half- or full-bridge design.

A more rapid and complete clarification is often achieved if a small dosage of synthetic polyelectrolyte flocculant is added just prior to settlement. This practice can help any later dewatering exercise, too, and is inexpensive to institute and run.

Accumulated solids can be digested and greatly reduced in volume by anaerobic reaction, but it is doubtful whether the investment in the necessary plant could be justified unless it were to be in use the year round. It is unlikely that the resultant products (principally methane) would have much market value. Methane is not compressible, so it is usually utilized on site for low-energy heating and power.

Most crudely, surplus activated sludge/slurry can simply be sprayed onto whatever land is available. Since the accompanying odors may be detectable a mile or more away, this practice is rapidly becoming socially unacceptable. There are doubts, too, about the hazards to health posed by any pathogens or bioactive compounds that may be present and that might find their way back into potable supplies. There is, therefore, a trend toward the former practice of drying beds in which moisture is lost by evaporation.

It may well be preferred, however, particularly on larger sites, to dewater the slurry mechanically. For this, the filter-press is a robust and reliable device. A horizontal stack of plates and frames is interleaved with filter cloths (usually woven polypropylene twill). The slurry is pumped in at high pressure to result eventually in a cake having a solids content of between 15 and 30%. This is of spadable consistency and may be disposed of ultimately to landfill or, increasingly, to incineration.

12.9.3 Recovery and Reuse Options

Attempts have been made from time to time to utilize waste fruit tissue and surplus biomass as a culture medium for growing single-cell protein for incorporation into animal feedstuff. Apart from the fact that low world protein prices have tended to render such attempts commercially unattractive, the experience of the bovine spongiform encephalitis (BSE) epidemic has led to great circumspection latterly about what is now fed to livestock.

There is considerable potential, though, in drier regions for the further use of spent water for irrigation purposes. Proper consideration must be given, of course, to crops appropriate to the situation and season(s).

12.10 ENVIRONMENTAL AUDITING

The term *environmental auditing* refers not only to the straightforward quantification of the uses of natural resources and utilities, but also to judgments as to whether these uses have been optimized to achieve minimum environmental impact.

12.10.1 Baseline Assessment

For a new fruit-processing plant, comparisons of its later performance in regard to the environment will be made with reference to the parameters obtaining prior to its construction and commissioning. It is not always easy to anticipate which of these may come to be of public interest.

Inquiries should be made, therefore, of parties having an actual or potential interest in the fruit-processing enterprise or its surroundings. These will include the supplier or suppliers of process and mains water (which may come from different sources) and the authorities regulating effluent and atmospheric discharges and waste disposal. Advice as to prevailing or pending legislation should be sought also from the health and safety executive, food industry inspectorate, local planning authority, and any relevant trade body. Another sector to be informed, if not consulted, is that embracing community, amenity, and political interests.

Limited surveys covering disposals, fumes, noise, traffic, and so on should be drawn up with the knowledge of the relevant interests. Ideally, to maintain and demonstrate objectivity, such work should be undertaken by an outside specialist company.

12.10.2 Period Review

At suitable intervals, determined by experience and discussion, the baseline surveys should be revisited and key elements repeated. Significant changes will show up straightaway. More gradual trends will become apparent only as data build up over longer periods of time. Graphical representations are always helpful, and statistical analysis sometimes reveals information not otherwise readily discerned.

12.10.3 Quality and Environmental Standards

After a long period of development, during which individual companies experimented with their own quality-control schemes, the British Standards Institute (BS) and American Standards Association (ASA) have published consolidated guidelines: BS 5750 and ISO 9000/EN 29000, for example. Unfortunately, accreditation to these does not imply any objective compliances. Quality targets are entirely self-set. Of much greater importance are the later BS 7750 standard and the European Union's Eco-Management and Audit Scheme (EMAS). Assessments for awarding certification for these are obliged to take into account the effects of the business operation on the global environment, and to state whether these effects are likely to have the least feasible impact. Considerations include the use of energy and water, disposal of service and manufacturing wastes (solid, liquid, and gaseous), and the recycling potential of products and their packaging.

Environmental standards have been covered largely in the previous sections. It remains to reiterate that these are certain to become ever more limiting and more closely monitored. For new establishments, the regulating authorities will need to be assured of best environmental practice before overall planning permission is sought and granted. In the long run, anyway, such are sure to return dividends on the initial investments made in them, as well as confirming the reputation of the operating company as being environmentally conscious and responsible.

REFERENCES

Bigot, B., Le Tallec, X., and Badard, M. (1999). A new generation of biological aerated filters. *Journal of the Chartered Institution of Water and Environmental Management*, October, **13**, 363.

Hodkinson, B., Williams, J.B., and Butler, J.E. (1999). Development of biological aerated filters: A review. *Journal of the Chartered Institution of Water and Environmental Management*, August, **13**, 250.

Sunner, N., Evans, C., Siviter, C., and Bower, T. (1999). The two-stage moving-bed/activated-sludge process: An effective solution for high-strength wastes. *Journal of the Chartered Institution of Water and Environmental Management*, October, **13**, 353.

FURTHER READING

Water Supplies and Treatment

Camp, T.R. (1963). *Water and Its Impurities*. Reinhold, New York.

HMSO. (1976–1999). *Index of Methods for the Examination of Waters*. HMSO, London.

Houghton, H.W., and McDonald, D. (1978). In *Developments in Soft Drinks Technology*, ed. L.F. Green. Applied Science Publishers, London.

Sykes, G. (1965). *Disinfection: Theory and Practice*, 2nd ed. E. & F.N. Spon Ltd., London.

Waste Treatment and Disposal

American Chemical Society Division of Agriculture and Food Chemistry. (1988). *Quality Factors of Fruits and Vegetables*. American Chemical Society, Washington, DC.

Besselievre, E.B. (1952). *Industrial Waste Treatment*. McGraw-Hill, New York.

Gurnham, C.F. (1965). *Industrial Wastewater Control*. Academic Press, New York.

Hayes, G.D. (1991). *Quality in the Food Industry*. Department of Food Manufacturing, Manchester Polytechnic, Manchester, UK.

Herzka, A., and Booth, R.G, eds. (1981). *Food Industry Wastes: Disposal and Recovery*. Applied Science Publishers, London.

Hulme, A.C., ed. (1970). *The Biochemistry of Fruits and Their Products*, Vol. I. (Vol. II, 1971). Academic Press, New York.

Koziorowski, B., and Kucharski, J. (1972). *Industrial Waste Disposal*. Pergamon Press, New York.

Manual of British Practice in Water Pollution Control: Unit Processes in Biological Filtration. (1988). The Chartered Institute of Water and Environmental Management, London.

Index

A

Activated sludge process, 295–297
Air separators, 69
Albedo, 5
Alfa-Laval Centritherm, 106
Alternating double filtration (ADF), 298
Aluminum
　cans, 213
　foil trays, 218
Anthocyanins, 28–32
Antioxidants, content in fruits, 49
Apple juice
　apple waste, use of, 238
　fermented. See Cider
　microorganisms found in, 127, 131–132
　processing of, 91–93
Apples
　canning, 159–161
　fermentation, 4
　pectin production, 239–246
　regions grown, 3
　varieties, 3
　waste treatment, 238
Apricots, canning, 161
Ascorbic acid, 10
　benefits of, 46–47
　content of fruits, 44–45, 49
　and fruit processing, 49, 50
Aseptic packaging, 218–221
　bulk aseptic systems, 220–221
　meaning of, 174
　packaging containers, 175
　retail-size containers, 218, 220
　sterilization/pasteurization process, 174–175
Aseptic processing, fruit juice, 101–102
Aspergillus organisms, types of, 131
Astringency, and flavor of fruit, 15–16
Atmosphere
　changes, effects on fruit, 66–67
　controlled. See Controlled atmosphere technology
　and fruit storage, 66–70
Auditing, external auditing, 251–253

B

Bananas
　ripening sensitivity, 9
　volatiles of, 12–13
Benzoic acid, 10
　as preservative, 90, 102

Berry fruits, 6–7, 53
Berry juices, processing of, 94
Bilberries, canning, 161
Biocontactors, 298
Biofungicides, 76
Biooxidation. *See* Effluent treatment
Black cider, 113, 124
Black currants, canning, 162
Blackberries, canning, 161–162
Blanching fruit, for canning, 152
Blast freezing, 171
Botrytis cinerea, 74
Bottling
   fruit juice, 102
   fruit processing for, 170
Brettanomyces organisms, 132
British Standard (BS) 7750, 281
Browning reactions, 25–27
Buffer salts, in preserves, 180
Bulk aseptic systems, 220–221
Byssochlamys organism, 131

C

Calcium, 42–44
Cancer, protective effect of fruit, 48–49
Candida oleophila, 76
Canning
   acceptance of fruit at factory, 151
   apples, 159–161
   apricots, 161
   bilberries, 161
   black currants, 162
   blackberries, 161–162
   blanching fruit, 152
   can vacuum, 157
   cherries, 162–163
   closing cans, 154–155
   cut out, 153
   exhausting, 153–154
   filling cans, 152
   finished pack pH values, 158
   fruit cocktail, 164–165
   fruit juice, 100–101
   fruit salad, 164
   gooseberries, 163
   grapefruit, 163–164
   heat treatment/processing, 155–157
   hygiene standards, 150–151
   loganberries, 165
   nutritive changes and processing, 50–51
   oranges, 165–166
   peaches, 166
   pears, 166–167
   peeling, 151–152
   pie fillings, 161, 165
   pineapple, 167
   plums, 167–168
   prunes, 168
   raspberries, 168–169
   rhubarb, 169
   solid pack, 159–160
   strawberries, 169–170
   syrup, 152–153
Cans, 152, 213–216
   aluminum, 213
   corrosion of, 215
   lacquered cans, 213
   manufacture of, 213–215
   tin-free steel, 213
   tinplate, 213
Carbohydrates, content in fruit, 38, 41
Carbon, and water supply, 284
Carbon dioxide, and fruit storage, 66–69, 71
Carbonated beverage bottles, 217
Carbonated beverages
   fruit juice, 107
   fruit wines, 142
   sparkling ciders, 133
Carbonyls, and flavor formation in fruit, 10–11
Carotenoids, 49
   and fruit processing, 49–50
Catalytic O_2 burner, 69
Centrifuge extraction, 92
Chemical hazards, sources of, 80
Chemical injury, causes of, 79
Cherries
   canning, 162–163
   glacé preparation, 193
   storage, 6
Chlorophylls, and color of fruit, 46
Cider, 4, 112–135
   black cider, 124
   chemical composition of, 133–134
   cider apples, types of, 113–114
   cider flavor, 134–135
   cider orchards, 114–115

cidre nouveau, 125
code of practice, 123
fermentation process, 121–122
fermentation substrate, 121
fermentation vessels, 117–121
fermentation yeasts, 127–128
final processing, 122–123
harvesting and pressing, 115–117
history of, 112–113
ice cider, 124
low-alcohol cider, 124
maturation process, 129, 131
microorganisms found in, 127, 131–132
organic cider, 126
secondary fermentation/maturation, 122, 132–133
sparkling ciders, 133
sulfur dioxide, 126–127
traditional ciders, 113–114, 125–126
varieties of products, 113
vintage cider, 124
white cider, 124
Cidre nouveau, 125
Citric acid, 10, 44, 46, 180
Citrus byproducts, 5, 225–236
   citrus oils, 232–234
   citrus peel, dried, 235–236
   citrus premium pulp, 226–232
   comminuted juices, 234–235
Citrus fruit, 5, 53
   regions grown, 5
Citrus oils, 232–234
   aroma recovery, 233–234
   clarification of, 233
   cold-pressed oils, 232–233
   concentrated oil, 234
   distilled oils, 234
   extraction systems, 233
   terpenes, 234
Citrus peel
   color extraction from, 237
   concentrate from peel extract, 228–229
   debittering, 229–231
   dried, 235–236
   glacé preparation, 195
   in marmalade, 186–187
   in preserves, 180
Citrus premium pulp, 226–228, 226–232
   dehydrated citrus cells, 227–228
   peel extract, concentrate from, 228–229
   pulpwash concentrate, 227
   washed pulp, 226–227
Clarification
   fruit juice, 99–100
   fruit wines, 140
Climacteric
   climacteric fruit, 54
   fruit ripening, 7–8, 54–56
   nonclimacteric fruit, 54, 56
Climateric rise, meaning of, 8
Cold-pressed oils, citrus oils, 232–233
Collapsible tubes, 218
Color extraction, from fruit waste, 236–238
Comminuted juices, 234–235
   production of, 234–235
Concentration, fruit juice, 90, 106, 108
Controlled atmosphere technology, 68–71
   air separators, 69
   analysis of gases, 69–70
   catalytic O_2 burner, 69
   fruit respiration for, 68
   liquid N_2, 69
   modified atmospheres, pros/cons, 71
   open-flame hydrocarbon burner, 68–69
   safety hazards, 69
   scrubbers, 69
Cool storage, 63–65
   coolroom, requirements, 65–66
   heat removal methods, 60–63
   high humidity room, 64–65
Countercurrent extraction, fruit juice extractors, 92–93, 96
Crystallization. *See* Glacé fruits
Crytosporidium, 131, 286
Currants, 197
Cut out, canning, 153
Cyclic esters, 12
Cyclospora, 80

D

Dates, drying process, 200
Debittering, citrus fruit for preserves, 229–231
Diabetic jam, 187
Dietary fat, content in fruit, 41–42
Dietary fiber, content in fruit, 41, 48
Dimethyl pyrocarbonate (DMPC), added to fruit wine, 140

Discoloration, and processing, 32
Disease, 73–74
   and environment, 73
   fungal diseases, 73–74
   host susceptibility, 73–74
   and petolyic enzymes, 74
Disease control, 74–76
   biofungicides, 76
   fungicides, 74–76
   heat treatment, 76
Disinfection
   water supply, 285
   See also Disease control
Disorders, 76–78
   atmosphere stress, 77
   control methods, 77–78
   high-temperature/solar injury disorders, 76
   low-temperature injury, 77
   nutrient imbalance, 77
Dried citrus cells, 227–228
Dried fruit, 50, 51, 196–200
   citrus peel, 235–236
   fruit tree fruits, 199–200
   packaging for, 222
   sulfur dioxide in, 51, 200
   vine fruit, 196–199

E

Eco-Management and Audit Scheme (EMAS), 281, 301
Effluent characterization, 291–292
   oxygen demand, 291–292
   suspended solids, 291
Effluent planning, 288–291
   effluent reception, 290
   effluent transfer, 289–290
   rainwater collection, 288–289
   segregation, 289
   treatment objectives, 290–291
Effluent treatment, 292–295
   activated sludge process, 295–297
   alternating double filtration (ADF), 298
   biocontactors, 298
   biological oxygen demand, 294–295
   filtration, 299
   high-rate filtration, 298
   percolating filters, 297–298

   pH adjustment, 293–294
   recovery/reuse, 300
   solids removal, 293
   solids removal/disposal, 299–300
   submerged aerated media (SAM), 298–299
Environmental auditing, 300–302
   baseline assessment, 301
   meaning of, 300
   period review, 301
   quality/environmental standards, 301
Enzyme-catalyzed reactions and processing, 23–25
Enzymes, content in fruit, 46
Escherichia coli, 80, 131
Essence recovery, fruit juice, 103–104
Esters, and flavor formation in fruit, 10–11
Ethylene, and fruit ripening, 8–9, 56–57
European Union
   agricultural policy of, 16
   food safety directives, 81
Exhausting, canning, 153–154
Extra jams, 182–183, 185

F

Factory farming, meaning of, 2
Fermentation process
   cider, 121–122
   fruit pulp fermentation, 141–142
   fruit wines, 139
   products of, 134
   secondary, 122, 132–133
   yeasts for, 127–128
Fermentation vessels, types of, 117–121
Fermented fruit beverages
   alcohol content, 112
   cider, 4, 112–135
   fruit spirits/liqueurs, 142–145
   fruit wines, 136–142
   perry (*poiré*), 135–136
Figs, drying process, 200
Filtration, effluent treatment, 299
Filtration sterilization, fruit juice, 102–103
Fining, fruit juice clarification, 99–100
Flavor of fruits
   analytical data, 9–10
   flavor characteristics, 15–16
   formation of flavor, 10–14

and fruit juice processing, 103–104
physiological/biochemical aspects, 14
taste and aroma, 10
Flavorings from fruits, 200–201
Flower petals, crystallized, 195–196
Fluidized belt freezing, 171
Food Machinery Corporation (FMC), fruit juice extractors, 87, 89
Foodborne illness
pathogens, types of, 80
risks and fruit, 80
Forced-air cooling, 61–63
Freezing, fruit juice, 102
Freezing fruit
blast freezing, 171
block-freezing, 2
fluidized belt freezing, 171
liquid nitrogen freezing, 172
packaging for, 221–222
packaging materials, 173
room freezing, 171
spiral belt freezing, 171
storage of frozen fruit, 172–173
washing fruit, 171
Freezing point, meaning of, 58
Fresh fruit, packaging of, 222–223
Fruit cocktail, canning, 164–165
Fruit juice
adulteration of, 108
apple juice, 91–93
aseptic processing, 101–102
berry juices, 94
bottling, 102
canning, 100–101
carbonated beverages from, 107
clarification of, 99–100
concentration, 90, 106, 108
essence recovery, 103–104
filtration sterilization, 102–103
freezing, 102
fruit juice drinks, 106
fruit nectars, 107
guava pulp, 98–99
high-pressure processing, 103
mango pulp, 95–96
nutritive value and processing, 51
orange juice, 87, 89–90
papaya purée, 95
passion fruit juice, 96–98

pear juice, 93
pineapple juice, 95
preservatives, 89–90, 102
and quality of fruit, 85–86
stone fruit juice, 93–94
therapeutic value of, 48
thermal treatment, 100
Fruit juice drinks, 106
Fruit juice extractors
centrifuge extraction, 92
citrus skin oil removal, 87, 88
countercurrent extraction, 92–93, 96
Food Machinery Corporation (FMC), 87, 89
Polyfruit extractor, 99
presses, 92, 118–119
rack and cloth method, 91
reaming type, 87, 88
Fruit processing
and anthocyanins, 28–32
browning reactions, 25–27
discoloration and processing, 32
enzyme-catalyzed reactions during, 23–25
flavor problems and processing, 27–28
minimal processing, 19–23
nutritive changes during, 49–51
Fruit processing byproducts
apple waste treatment, 238
citrus byproducts, 225–236
color extraction from fruit waste, 236–238
increase in, 225
pectin, 239–246
Fruit quality
external affecting factors, 15
postharvest factors, 19
preharvest factors, 19
and variety of fruit, 14–15
Fruit salad, canning, 164
Fruit spirits/liqueurs, 142–144, 142–145
production of, 142–143
types of, 144, 145
variations on recipes, 143–144
Fruit types
citrus fruits, 5
classification of, 53
pome fruits, 3–4
soft fruits, 6–7
stone fruits, 5–6
Fruit wines, 136–142
alcohol-fortified wines, 142

clarification, 140
composition of, 137
defects of, 140
fermentation process, 139
fermentation of pulp, 141–142
final processing stage, 140
fruit processing for, 138–139
maturation/mellowing, 140
sparkling wines, 142
styles of, 137
types of fruits in, 137–138
Fumaric acid, 10
Fungal diseases, 73–74
    of apple cider, 131–132
    control with fungicides, 74–76

G

Ginger, glacé preparation, 194–195
Glacé fruits, 192–196
    cherries, 193
    citrus peel, 195
    flower petals, 195–196
    ginger, 194–195
Glass containers, 207–213
    closure of, 208, 210–212
    manufacture of, 208
Global market
    international standards, 3
    and price, 16
    quality management practices, 81–82
    and SNIF NMR methods, 16–17
Gluconic acid, 10
Gooseberries, canning, 163
Grape juice, processing of, 94
Grapefruit, canning, 163–164
Grapes
    color extraction from, 237–238
    raisins from, 196–199
Green life, meaning of, 58
Growth pattern, 5–6
Guava pulp, processing of, 98–99

H

Hazard analysis critical control point (HACCP), 253–263

benefits of, 254
decision tree, 257
examples of, 262–280
flow diagram, 255–256
implementation of plan, 261–262
principles of, 254
purpose of, 253
record keeping, 260
survey, stages of, 254–261
team for, 255
verification/validation procedures, 259–260
Heat treatment
    canned fruit, 155–157
    disease control, 76
    fruit juice, 100
Hesperidin, in preserves, 231–232
High-performance chromatography (HPLC), 17
High-pressure processing, fruit juice, 103
High-rate filtration, effluent treatment, 298
Histamine, in fruit wines, 141
Humidity, obtaining in cool storage, 64–65
Hydrocooling, 63

I

Ice cider, 124
Impact bruise injury, 78
Injury, 78–79
    chemical injury, 79
    insect injury, 78–79
    mechanical injury, 78
    packaging/mechanical handling, 79
Insect injury, 78–79
    control of, 78–79
Irradiation, and nutritive value of fruit, 51
ISO EN BS 9000 standard, 3, 251, 253

J

Juice. *See* Fruit juice

L

Lacquered cans, 213
Lactic acid, and fermentation process, 129
Lactones, and flavor formation in fruit, 10–11

Laxative effects, of fruits, 48
Limonoids, 48
   for debittering, 229–230
Liqueurs. *See* Fruit spirits/liqueurs
Liquid N$_2$, 69
Liquid nitrogen freezing, 172
Listeria monocytogenes, 80
Loganberries, canning, 165

M

Malic acid, 10, 46, 180
Malonic acid, 10
Mango pulp, processing of, 95–96
Marmalade, production of, 186–187
Maturity of fruit
   for fruit juice, 86
   maturity standards, 57–58
   meaning of, 53–54
Mechanical contacting, effluent treatment, 298
Mechanical handling, and injury, 79
Mechanical injury, forms of, 78
Medicine, use of pectin, 245–246
Microorganisms
   in apple cider, 131–132
   in apple juice, 127
   in foodborne illness, 80
   in fruit wine, 141
Minerals. *See* Vitamins and minerals
Minimal processing, 19–23
Mixed storage, 66
Modified atmosphere packaging (MAP), 222–223
Mold rot, 74, 80
Monolayer plastic containers, 217–218
Muscatels, 199
Mycotoxins, types of, 80

N

Naringin
   for debittering, 229–230
   in preserves, 231–232
Near-infrared (NIR) spectroscopy system, fruit juice verification, 108
Nectars, from fruit juice, 107
Nutrition and fruit
   nutritional components of fruit, 37–46
   nutritional importance of fruit, 46–49
   processing and nutritional changes, 49–51

O

Open-flame hydrocarbon burner, 68–69
Orange juice, processing of, 87, 89–90
Oranges, canning, 165–166
Organic acids
   content in fruit, 44, 46, 47
   and flavor of fruit, 10–11, 15–16
Organic cider, 126
Oxygen, and fruit storage, 67, 68–69, 71
Oxygen demand, effluent, 291–292, 294–295

P

Packaging
   aluminum foil trays, 218
   aseptic packaging, 218–221
   cans, 213–216
   collapsible tubes, 218
   of dried fruit, 222
   of fresh fruit, 222–223
   frozen foods, 221–222
   functions of, 205–206
   glass containers, 207–213
   plastic packaging, 216–218
   recycling of, 205
Packing, and injury, 79
Papaya purée, processing of, 95
Passion fruit juice, processing of, 96–98
Pasteurization process, aseptic packaging, 174–175
Patulin, 80, 131–132
Peaches, canning, 166
Pear juice, processing of, 93
Pears
   canning, 166–167
   perry (*poiré*) production, 135–136
   varieties, 3
Pectin, 239–246
   cold pectin technology, 242
   commercial pectins, production of, 240–242
   extraction from fruit, 38, 41
   high-methoxyl pectins, 240, 242, 244
   intermediate pectin, 245

low-methoxyl pectin, 240–241, 245
for making preserves, 181–182
manufacture of, 242–244
pectic substances in plants, 239
and pectolytic enzyme, 240
raw materials for, 241–242
use in nutrition/medicine, 245–246
Pectolytic enzyme, 74, 240
Peeling fruit, for canning, 151–152
Penicillium pathogens, types of, 73, 80, 131
Percolating filters, effluent treatment, 297–298
Perry (*poiré*)
pear varieties, 136
production of, 135–136
Pharmacological effects, of fruits, 48
Phenols, and flavor formation in fruit, 14
pH value
adjustment of water supply, 283
canned fruit, 158
effluent treatment, 293–294
Physical contamination, sources of, 80
Picking fruit, and disease control, 74
Pie fillings, canning, 161, 165
Pineapple, canning, 167
Pineapple juice, processing of, 95
Plastic packaging, 216–218
carbonated beverage bottles, 217
closure of, 217
fruit juice, 102
hot-filled barrier containers, 217
manufacture of, 216–217
materials used, 216
monolayer plastic containers, 217–218
Plums
canning, 167–168
ripening, 6
Polyfruit extractor, 99
Polymer beads, to remove bitterness from juice, 90
Pome fruits, 3–4, 53
storage, 15
Potassium carbonate, dried fruits, 198–199
Precooling, for field-heat removal, 60–61
Preservatives
fruit juice, 89–90, 102
See also Sulfur dioxide
Preserves
acids/buffers, 180
boiling, 188–191

citrus peel addition, 180
containers, filling of, 191–192
diabetic jam, 187
extra jams, 182–183, 185
fats, 180
formulation of recipe, example of, 183–186
fruit, preservation for, 178–179
fruit preparation, 187–188
gelling agents, 181–182
marmalade, 186–187
preprocessing of fruit, 187–188
reduced-sugar products, 187, 188
standard books on, 177
sweetening agents, 180, 183, 184, 185
types of fruits for, 179–180
variations of product, 182–183
Presses, fruit juice extractors, 92, 118–119
Produce safety initiative, 81
Proteins, content in fruit, 38
Prunes, canning, 168
Pseudomonas syringae, 76
Pulp wash, 90

Q

Quality assurance, functions of, 250
Quality management practices, 81–82
Quality system
components of, 250
external auditing, 251–253
hazard analysis critical control point (HACCP), 253–263
ISO EN BS 9000 standard, 3, 251, 253
Quarantine, pest control, 78–79
Quinic acid, 11

R

Rack and cloth method, fruit juice extraction, 91
Rainwater collection, 288–289
Raisins, drying process, 196–199
Raspberries, canning, 168–169
Reaming type, fruit juice extractor, 87, 88
Reduced-sugar products, preserves, 187
Rhubarb, canning, 169
Ripening of fruit, 7–9, 54–58
changes to fruit during, 54

climacteric fruits, 54
climacteric rise, 8
and ethylene, 8–9, 56–57
nonclimacteric fruits, 54, 56
respiration climacteric, 7–8
room for controlled ripening, 56–57
and susceptibility to disease, 73
Room freezing, 171

S

Saccharomyces
resistance to sulfur dioxide, 132
strains for fermentation process, 127
types of organisms, 90, 132
Salicylic acid, 10
Salmonella, 80, 131
Scrubbers, 69
Shikimic acid, 10
Site-specific nuclear isotopic fraction by nuclear magnetic resonance (SNIF NMR), capability of, 16–17
Sludge. See Effluent treatment
Snacks, fruit as, 48
Soft fruits, 6–7
storage, 15
structure of, 6–7
Solid pack, apples, 159–160
Sorbic acid, as preservative, 90, 102
Sparkling ciders, 133
Sparkling fruit wines, 142
Spinning cone column, 104–105
Spiral belt freezing, 171
Standards
cider making, 123
for environmental assessment, 281–282
environmental standards, 301
international standards, 3
ISO EN BS 9000, 3, 251
water quality standards, 288
Staphylococcus aureus, 131
Sterilization process, aseptic packaging, 174–175
Stone fruit juice, processing of, 93–94
Stone fruits, 5–6
controlled storage, 6
types of, 5
Storage of fruit
and atmospheric changes, 66–67

chilling-sensitive/chilling-insensitive fruit, 59
controlled atmosphere technology, 68–71
cool storage, 63–65
coolroom requirements, 65–66
forced-air cooling, 61–63
hydrocooling, 63
mixed storage, 66
precooling for field-heat removal, 60–61
and temperature, 58–60
temperature zones for, 55
vacuum cooling, 63
Storage of fruit problems
chemical hazards, 80
disease, 73–76
disorders, 76–78
injury, 78–79
microbial hazards, 80
physical contamination, 80
Strawberries
canning, 169–170
new varieties, 15
Submerged aerated media (SAM), effluent treatment, 298–299
Sugars
and blood glucose level, 47
content in fruit, 38–40, 47
and flavor of fruit, 15–16
and glacé fruits, 193–195
Sulfur dioxide, 51, 89–90, 102
in apple cider, 133–134
in apple juice, 126–127
dried fruits, 51, 200
and glacé fruits, 193
Sultanas, 197
Sweetening agents, preserves, 180, 183, 184, 185
Syrup
in canned fruit, 152–153, 154
in preserves, 180

T

Tannins, in apple cider, 134
Tartaric acid, 10
Temperature
and disorders, 76–77
and storage. See Storage of fruit
Terpenes, 234

Tetra Pak system, 101
Thermal-accelerated-short-time-evaporators (TASTE), 106
Thermal treatment, fruit juice, 100
Tin-free steel, cans, 213
Tinplate, cans, 213
Tocopherols, 49
Tomato
  purée, processing of, 201–203
  types of, 201
Tropical fruit
  fungal disease, 74
  types of, 53
Tropical fruit juices, processing of, 94–99

V

Vacuum, in canning, 157
Vacuum cooling, 63
Vine fruit, drying, 196–199
Vintage cider, 124
Vitamin P activity, 46
Vitamins and minerals
  content in fruit, 42–45, 46–47
  lacking in fruits, 44, 46
Volatile Compounds in Food, 9

W

Water, content in fruit, 38
Water quality, importance of, 80
Water supply
  adjustment of pH value, 283
  carbon adsorption, 284
  color removal, 283
  effluent characterization, 291–292
  effluent planning, 288–291
  effluent treatment, 292–295
  filtration, 283–284
  primary disinfection, 285
  for process use, 287
  screening of, 282–283
  secondary treatment of, 286–287
  for special applications, 287
  storage of, 282
  water quality standards, 288
White cider, 113, 124
Wine. *See* Fruit wines

Y

Yeasts
  apple cider fermentation, 127–128
  desirable properties, 128
  strains for fermentation, 127